I0034291

Advances in Heat Transfer Augmentation Techniques in Single-Phase Flows

Augmentation of heat transfer is important in energy conservation and developing sustainable energy systems. This book provides the science necessary to understand the basics of heat transfer augmentation in single-phase engineering systems. It considers theory and practice including computational and experimental procedures, evaluation techniques for performance, and new trends. Several applications of augmentation methods like surface modification, introduction of vortex flow and impinging jets, opportunities of ultrasound and magnetic fields, pulsatile flows, heat exchangers, and nanofluids are provided. Details of basic phenomena and mechanisms are highlighted.

Key features:

- Provides the fundamental science needed to understand and further develop heat transfer augmentation for future energy systems.
- Give examples of how ultrasound and magnetic fields, vortex flow, impinging jets, surface modification, and nanofluids can augment heat transfer.
- Considers basic issues of computational and experimental methods for analysis, design, and evaluation of efficient and sustainable heat transfer.

It is an ideal reference text for graduate students and academic researchers working in the fields of mechanical, aerospace, industrial, manufacturing, and chemical engineering.

Advances in Heat Transfer Augmentation Techniques in Single-Phase Flows

Varun Goel

Wei Wang

Bengt Sunden

CRC Press
Taylor & Francis Group
Boca Raton London New York

CRC Press is an imprint of the
Taylor & Francis Group, an **informa** business

Front cover image: brumhildich/Shutterstock

First edition published 2024
by CRC Press
6000 Broken Sound Parkway NW, Suite 300, Boca Raton, FL 33487-2742

and by CRC Press
4 Park Square, Milton Park, Abingdon, Oxon, OX14 4RN

CRC Press is an imprint of Taylor & Francis Group, LLC

© 2024 Varun Goel, Wei Wang and Bengt Sunden

Reasonable efforts have been made to publish reliable data and information, but the author and publisher cannot assume responsibility for the validity of all materials or the consequences of their use. The authors and publishers have attempted to trace the copyright holders of all material reproduced in this publication and apologize to copyright holders if permission to publish in this form has not been obtained. If any copyright material has not been acknowledged please write and let us know so we may rectify in any future reprint.

Except as permitted under U.S. Copyright Law, no part of this book may be reprinted, reproduced, transmitted, or utilized in any form by any electronic, mechanical, or other means, now known or hereafter invented, including photocopying, microfilming, and recording, or in any information storage or retrieval system, without written permission from the publishers.

For permission to photocopy or use material electronically from this work, access www.copyright. com or contact the Copyright Clearance Center, Inc. (CCC), 222 Rosewood Drive, Danvers, MA 01923, 978-750-8400. For works that are not available on CCC please contact mpkbookspermissions@tandf.co.uk

Trademark notice: Product or corporate names may be trademarks or registered trademarks and are used only for identification and explanation without intent to infringe.

ISBN: 978-1-032-13561-8 (hbk)
ISBN: 978-1-032-13562-5 (pbk)
ISBN: 978-1-003-22986-5 (ebk)

DOI: 10.1201/9781003229865

Typeset in Sabon
by SPi Technologies India Pvt Ltd (Straive)

Contents

Preface

This book aims as an introduction and state-of-the-art of heat transfer augmentation in single-phase flows. The purpose of the book is to provide basic and applied concepts at undergraduate and graduate levels. As well, the book can be used in engineering design, R&D in industries to find improved heat transfer equipment by applying various augmentation techniques for achieving energy efficient and sustainable technical solutions but limited to single-phase convective flow and heat transfer.

The first chapter introduces convective and conductive heat transfer, extended surfaces, and heat exchangers. Then, the second chapter describes the general perspectives of heat transfer augmentation and briefly introduces performance evaluation criteria. The significance of augmented heat transfer is illustrated by showing the number of scientific publications in this topic area. In addition, the development of the number of filed patents is shown graphically. The third chapter summarizes various surface modification types to achieve improved heat transfer. Advantages and disadvantages are discussed. Flow fields with high vorticity are important in creating augmentation of convective single phase heat transfer and accordingly Chapter 4 deals with the mechanisms of vortex flows. Engineering applications are shown to depict the phenomena. In Chapter 5, the impact of pulsative flows is described, and typical examples are given. Ultrasound and magnetic forces appear in some relevant applications, and then Chapter 6 delivers a description of the involved physics as well the specific volume forces present. The effects on convective heat transfer are illustrated. In Chapter 7, the focus is on impinging jet flows and how the heat transfer can be augmented. The governing parameters are considered, and the results are showing the impact of various configurations of impinging jets. Base fluids with embedded nanoparticles are called nanofluids. Such fluids have emerged significantly recently and become of interest in heat transfer augmentation. Accordingly, Chapter 8 is devoted to convective heat transfer with nanofluids, and besides typical results, mechanisms, preparation, stability, etc. are discussed in detail. In Chapter 9, additional concepts for the evaluation of augmentation techniques for single-phase convective heat transfer are provided. It continues the introduction already provided in Chapter 2. In research and engineering work for new design as well as evaluation and performance tests, performing accurate experiments are important. Then, in Chapter 10, the most common experimental methods are described. Their uncertainties and limitations are discussed. In Chapter 11, the most common computational approach, that is, the so-called CFD, is described in detail, and examples illustrate the usefulness of this approach. All chapters provide citations of the most significant literature, both historically and as state of the art.

This comprehensive textbook on augmentation of single-phase convective heat transfer is unique and differs significantly from related chapters available in handbooks and older monographs. It provides a deeper analysis of convective heat transfer, and the relation to fluid mechanics is much more highlighted. Engineering problems of industrial and societal significance are illustrated in most of the chapters to highlight the relevance and importance of augmented heat transfer. In addition, the benefits of experimental and computational research efforts are highlighted in development of innovative solutions to achieve improved energy efficiency and sustainable future.

Dr. Varun Goel
Dr. Wei Wang
Dr. Bengt Sundén
March 2023

Acknowledgments

Professor Sunden acknowledges his wide cooperation with PhD students, postdocs, and senior research fellows from several countries. The financial supports of research projects over many years received from the Swedish Energy Agency, Swedish Scientific Council, companies, and the European Union are kindly acknowledged.

Dr. Varun acknowledges the support of students in the preparation of the manuscript. He especially thanks Mr. K.S. Mehra and Mr. Ankur Dwivedi for their help in typesetting and preparation of the draft. The support received from NIT Hamirpur is duly acknowledged.

Dr. Wang acknowledges the support by Dr. Wenke Zhao, PhD students Liang Ding, Bingrui Li, and Master's student Yameng Man. The financial support received from National Natural Science Foundation of China, China Postdoctoral Science Foundation, and Heilongjiang Provincial Science Foundation are kindly acknowledged.

We would like to express our deep and sincere gratitude and thanks to our present and former colleagues at the National Institute of Technology Hamirpur, Harbin Institute of Technology, Harbin, Chalmers and Lund Universities, Sweden. We also acknowledge the continuous support and encouragement of our families who made many sacrifices during the preparation of this book. We are also thankful to the publishing team (CRC) for its hard work and patience in waiting for the completion of the final manuscript.

Varun Goel

Wei Wang

Bengt Sunden

About the Authors

Varun Goel received his PhD in the year 2010, and works as an associate professor in the Department of Mechanical Engineering at the National Institute of Technology Hamirpur, India. His area of research includes heat transfer, CFD, renewable Energy, etc. He has published about 200 papers in journals and conferences. The h-index of the published work is 50, and the number of citations is more than 10,000.

Wei Wang received his PhD in Engineering Thermophysics in 2019, and became an associate professor in the School of Energy Engineering and Science in 2022, all from Harbin Institute of Technology, Harbin, China. He was visiting study at Lund University and learnt from Professor Bengt Sunden, at Lund, Sweden, in 2017–2018. His research activities include enhancement of heat transfer, heat dissipation in aerospace and electronics, heat and mass transfer in evaporation and condensation, and compact heat exchangers design.

Bengt Sunden received his MSc in 1973 and PhD in 1979, and became Docent in 1980, all from Chalmers University of Technology, Gothenburg, Sweden. He was appointed professor of Heat Transfer at Lund University, Lund, Sweden, in 1992, and served as head of the Department of Energy Sciences, Lund University, for 21 years (1995–2016). He is an active professor emeritus and was senior professor during 2016–2022. The research activities include heat transfer enhancement techniques, gas turbine heat transfer, computational modeling, analysis of multi-physics and multiscale transport phenomena for fuel cells, nano-and microscale heat transfer, boiling and condensation, aerospace heat transfer, and thermal management of batteries. He has supervised 51 PhD students and many postdocs as well as hosted/supervised many visiting scholars and PhD students. He serves as guest or honorary professor at several prestigious universities. He is a fellow of ASME, a regional editor *for Journal of Enhanced Heat Transfer* since 2007, an associate editor of *Heat Transfer Research* since 2011, *ASME J. Thermal Science, Engineering and Applications* in 2010–2016, *ASME J. Heat Transfer* in 2005–2008, *ASME J. Electrochemical Energy Conversion and Storage* in 2017–2023, and *ASME Open Journal of Engineering* in 2021–2024. He was a recipient of the ASME Heat Transfer Memorial Award 2011 and Donald Q. Kern Award 2016. He received the ASME HTD 75th Anniversary Medal 2013. He was awarded the 2023 AIAA Energy Systems Award. He has edited 35 books and authored 3 major textbooks. He has published about 650 papers in well-established and highly ranked scientific journals. The h-index is 61 and the number of citations is more than 17,000.

Nomenclature

A	area [m²]
AM	additive manufacturing
a	thermal diffusivity [m²/s], constant
a_{NB}	coefficient in algebraic equation
b	length, thickness [m], constant
C	heat capacity flow rate [W/K], coefficient of resistivity
C_D	drag coefficient [-]
C_f	convective flux [kg/s]
C_F	shear stress coefficient [-]
C_p	pressure coefficient [-]
c	specific heat [J/(kg·K)], mass concentration [-]
c_p	specific heat at constant pressure [J/(kg·K)]
D	diameter [m]
D_f	diffusive flux [kg/s]
D_h	hydraulic diameter [m]
d_{ij}	deviatoric stress tensor [N/m²]
E	energy [J]
$\dot{E}$	energy per time unit [W]
E	voltage [V]
E	augmentation or enhancement ratio
EMF	electromotive force [V]
F	force [N]
F	correction factor [-]
Fo	Fourier number [-]
f	Darcy friction factor [-]
Gr	Grashof number [-]
Gr*	modified Grashof number [-]
g	gravity constant [m/s²]
H	enthalpy [J]
$\dot{H}$	enthalpy per time unit [W]
Hue	color value
h	specific enthalpy [J/kg]
h	heat transfer coefficient [W/m²K]

I	luminous intensity
I	current [A]
IR	infrared
i, j, k	unit vectors
j	Colburn factor [-], $St \cdot Pr^{2/3}$
K_c	contraction coefficient [-]
K_e	expansion coefficient [-]
k	thermal conductivity [W/mK]
L	length, thickness [m]
L_i	entrance length [m]
LCT	liquid crystal thermography
LMTD	logarithmic mean temperature difference [K], [°C]
l_m	mixing length [m]
Ma	Mach number
m	mass [kg]
$\dot{m}$	mass flow rate [kg/s]
N	number of tube rows
N_s	entropy generation number
NTU	number of transfer units [-]
Nu	Nusselt number [-]
n	normal vector
P	efficiency parameter [-]
PEC	performance evaluation criterion
PIV	particle image velocimetry
Pr	Prandtl number [-]
Pr_t	turbulent Prandtl number, [-]
PSP	pressure sensitive paint
p	pressure [Pa]
p'	fluctuating pressure [Pa]
$\bar{p}$	time averaged pressure [Pa]
Q	heat [J]
$\dot{Q}$	heat transfer rate [W]
q	heat flux [W/m²]
R	radius [m]
R	gas constant [J/(kgK)]
R	heat capacity flow rate ratio [-]
R	residual
R_p	surface roughness [μm]
R_w	wire resistance
Ra	Rayleigh number [-]
Re	Reynolds number [-]
r	radius [m]
S	source term [/m³]
S_{gen}	entropy generation
S_L	lateral tube pitch [m]

Sr	Strouhal number [-]
St	Stanton number [-]
S_T	longitudinal tube pitch [m]
T	absolute temperature [K]
T^+	dimensionless temperature [-]
TC	thermocouple
TEF	thermal enhancement factor
TLC	thermocromic liquid crystal
TPF	thermal performance factor
TR	thermal resistance [K/W]
t	temperature [°C]
t'	fluctuating temperature [°C]
$\bar{t}$	time averaged temperature [°C]
U	internal energy [J]
$\dot{U}$	internal energy per time unit [W]
U	(mean) velocity [m/s]
U_∞	freestream velocity [m/s]
U	overall heat transfer coefficient [W/(m²K)]
u, v, w	local velocity [m/s]
u', v', w'	fluctuating velocity [m/s]
$\bar{u}, \bar{v}, \bar{w}$	time averaged velocity [m/s]
u^+	dimensionless velocity [-]
u_m	mean velocity [m/s]
$u\tau, u^*$	friction velocity [m/s]
V	volume [m³]
$\dot{V}$	volume flow rate [m³/s]
VG	variable geometry
W	work [J]
$\dot{W}$	work per time unit [W]
w	mean flow velocity [m/s]
X	coordinate [m]
x, y, z	coordinates [m]
y^+	dimensionless coordinate perpendicular to a solid surface [-]
Z	width, length [m]

GREEK

α	heat transfer coefficient [W/(m²K)]
β	thermal expansion coefficient [1/K]
γ	ratio of specific heats, c_p/c_v [-]
δ	boundary layer thickness [m]
δ_{ij}	Kronecker's delta [-]
Δp	pressure drop [Pa]
$\Delta\theta, \Delta T$	temperature difference

ε	efficiency [-]
ε_m	turbulent kinematic viscosity [m²/s]
ε_q	turbulent diffusivity [m²/s]
Φ	arbitrary variable
Γ	diffusivity
η	dimensionless coordinate [-]
η	fin effectiveness [-]
η	film cooling effectiveness [-]
θ	dimensionless temperature [-]
θ	thermal length [-]
ϑ	temperature [°C]
κ	von Karman constant [-]
λ	parameter [-]
λ	thermal conductivity [W/(mK)]
μ	dynamic viscosity [kg/(ms)]
ν	kinematic viscosity [m²/s]
ρ	density [kg/m³]
σ	shear stress [N/m²]
σ	area ratio [-]
τ	time [s]
φ	fin efficiency [-], (3-49)
ψ	stream function [s⁻¹]
f	Fanning friction factor
f	Darcy friction factor

INDICES

aw	adiabatic wall
B	background
f	at film temperature, fluid
g	gas
i	inner
m	mean
o	outer
R	reference
w	wall
∞	freestream

Abbreviations

AP	Beginning point of adverse pressure gradient region
AP'	Ending point of adverse pressure gradient region
Be	Bejan number
BEM	Boundary element method
BF	Base fluid
CAD	Computer-aided design
CCD	Central composite design
CCD	Charge coupled device
CFD	Computational fluid dynamics
CMC	Ceramic matrix composites
CVFEM	Control volume finite element method
DC	Direct current
DNS	Direct numerical simulation
DWNT	Double-walled carbon nanotube
EDE	Entransy dissipation extremum principle
EG	Ethylene glycol
EGM	Entropy generation minimization
Fc	Field synergy number
FEM	Finite element method
FG	Fixed geometry
FLIR	Forward looking infrared
FN	Fixed flow area
FSP	Field synergy principle
FVM	Finite volume method
HCT	Helically corrugated tube
µHEX	Micro heat exchanger
HX	Heat exchanger
IEP	Isoelectric point
ILBNF	Ionic liquid base nanofluid
MMC	Metallic matrix composites
MSBNF	Molten salt base nanofluid
MTHX	Multi-tube heat exchanger
MWCNT	Multiwalled carbon nanotube

NB	Neighbor
NF	Nano fluid
NP	Nano particle
PCHE	Printed circuit heat exchanger
PFHE	Plate-fin heat exchanger
PHE	Plate heat exchanger
PISO	Pressure implicit splitting operators
QUICK	Quadratic upstream interpolation for convective kinetics
RANS	Reynolds-averaged Navier–Stokes
RP	Reattachment point of secondary flow
RSM	Reynolds stress method
RSM	Response surface method
Se	Intensity of secondary flow
SIMPLE	Semi-implicit method for pressure-linked equations
SIMPLER	SIMPLE revised
SIMPLEX	SIMPLE extended
SP	Separation point of secondary flow
SST	Shear stress transport
SWNT	Single-walled carbon nanotube
TCT	Transverse corrugated tube
TDMA	Tridiagonal matrix algorithm
TKE	Turbulent kinetic energy

Chapter 1

Introduction to heat transfer

1.1 INTRODUCTION

It is of great importance to be able to determine temperature distributions and heat fluxes in most branches of engineering and technology. In design, sizing, and rating of heat exchangers, for example condensers, evaporators, radiators, and others, analysis of the heat transfer process is needed. Heat exchanger equipment appears frequently in heat and power generation, process industries of various kind, automotive engineering, etc. Design and sizing of air conditioning equipment, electronics cooling, and insulation of buildings require understanding and knowledge of heat transfer. For vehicles, many heat transfer problems are present.

Successful stress and strain analysis in equipment exposed to high temperature requires accompanied analysis of the temperature distribution and heat loads.

In manufacturing, production, and thermal or mechanical treatment of materials, heat transfer is an important issue.

Equipment carrying electrical currents (electronics, electric motors, and transformers) commonly need cooling. In energy conversion devices like electrochemical apparatus (fuel cells, batteries, and electrolyzers) and combustion units, the significance of heat transfer is vital.

Food processing and its treatment is another area where analysis of heat and mass transfer is required.

1.2 MECHANISMS OF HEAT TRANSFER

Energy transferred from the hot to the cold part of a substance or from a high temperature body to another body kept at a lower temperature is generally labeled as heat.

Application of basic relations of thermodynamics and fluid mechanics can in some cases easily determine the amount of transferred heat. When the mechanisms are not completely known, analogical or empirical methods based on experiments might be applicable.

DOI: 10.1201/9781003229865-1

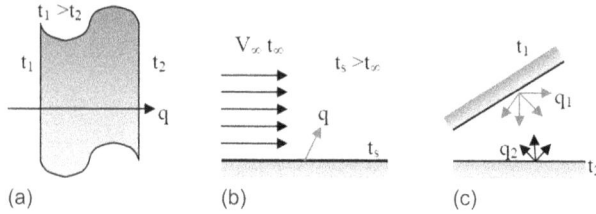

Figure 1.1 Mechanisms of heat transfer: (a) heat conduction; (b) convection; and (c) thermal radiation.

Three different mechanisms of the transfer of heat have been identified. These are heat conduction, convection, and thermal radiation (see Figure 1.1).

The mechanism of heat conduction is a process where energy is transported by the molecular motion as inside solid bodies and fluids (liquids and gases) at rest and by the electron movement in metals. The rate of heat transfer $\dot{Q}$ for the plane wall in Figure 1.1 is written as follows:

$$\dot{Q} = \lambda A \frac{t_1 - t_2}{b} \tag{1.1}$$

In Eq. (1.1), λ (or k) is the thermal conductivity (W/mK) of the wall material, A the heat transferring area, and b the thickness of the wall, and t represents temperature.

Values of thermal conductivity are available in the literature, for example [1, 2]. Some values are provided in Table 1.1.

The macroscopic motion of a fluid along an exterior surface or inside a duct will affect the heat exchange between the fluid and the solid surface. This mechanism of heat transfer is called convection. The macroscopic fluid movement can be created differently and accordingly forced convection or free (natural) convection prevails. However, the forced convection and free convection may occur simultaneously. The process is then called combined forced and free convection or mixed convection. In determining the rate of

Table 1.1 Values of thermal conductivity (λ) in W/mK and specific heat in J/kgK for some materials

Material	λ at 20°C (W/mK)	λ at 100°C (W/mK)	λ at 300°C (W/mK)	Specific heat at 20°C (J/kgK)
Aluminum	204	206	228	896
Carbon steel, carbon 1.0 %	43	43	40	473
Copper	386	379	369	381
Brass, Cu 85%, Zn 30%	111	128	147	385
AISI304, stainless steel	14	17	19	477
Water	0.613	0.683	—	4,239
Air	0.0261	0.0331	0.0456	1,014

Table 1.2 Order of magnitude of the convective heat transfer coefficient for air and water

Fluid	Convection process	Heat transfer coefficient (W/m²K)
Air	Free, natural	2–25
Air	Forced	25–250
Water	Free, natural	50–1,000
Water	Forced	50–20,000
Water	Boiling	2,500–100,000
Water, vapor	Condensation	5,000–100,000

heat transfer $\dot{Q}$, a convective heat transfer coefficient h (or α) (W/m²K) is introduced via Newton's law of cooling, i.e.,

$$\frac{\dot{Q}}{A} = \alpha\left(t_f - t_w\right) = h\left(t_f - t_w\right) \tag{1.2}$$

In Eq. (1.2), the fluid temperature t_f and the wall or surface temperature t_w appear.

In the literature, for example [1–2], values (orders of magnitude) of h or α are available. In Table 1.2, representative values for air and water are provided.

When forced convection prevails, the α or h, depends on the fluid velocity, fluid type as well as geometry. The convective flow field is commonly characterized as laminar or turbulent. In laminar cases, the flow velocities are low, while in turbulent flow high velocities prevail. Conventionally, a Reynolds number, Re, is used to determine whether a flow field is to be regarded as laminar or turbulent.

1.2.1 Dimensionless numbers in analysis of convection heat transfer

1.2.1.1 Reynolds number

The Reynolds number (Re) is a common dimensionless number used to determine whether a flow is laminar or turbulent. For tube or pipe flow, Re is defined,

$$Re = \rho V D_i / \mu \tag{1.3}$$

where ρ is density of the fluid, V average flow velocity, D_i tube or pipe inside diameter, and μ the dynamic viscosity of the fluid.

In a variety of compact heat exchangers, noncircular cross section of the tubes appears. Then a so-called hydraulic diameter, D_h, is introduced. It is defined as,

$$D_h = \frac{4A}{P_m} \tag{1.4}$$

where A and P_m are the cross-sectional area and the wetted perimeter of the tube, respectively. Because of the specific geometry of noncircular tubes, a greater surface area per unit volume can be achieved in compact heat exchangers.

1.2.1.2 Nusselt number

For a particular tube diameter and flow conditions, empirical correlations have been established. To generalize such and enable application of these for equipment of various sizes and different flow conditions, a nondimensionalized heat transfer coefficient is introduced by the Nusselt number, Nu. This is defined as,

$$\text{Nu} = \frac{hL}{k_f} = \frac{\alpha L}{k_f} \tag{1.5}$$

where L is a representative length of the considered flow geometry, that is, the tube internal diameter (D_i) for flow inside a tube, and k_f is the fluid thermal conductivity.

1.2.1.3 Prandtl number

A fluid property appearing in correlations is the Prandtl number, Pr. Physically, this is defined as the ratio of the kinematic viscosity, ν, to the thermal diffusivity, a. It can be expressed as,

$$\text{Pr} = \frac{\nu}{a} = \frac{\mu c_p}{k_f} \tag{1.6}$$

The Prandtl number is a thermophysical property of a fluid independent of the flow conditions. For air and water at room temperature, the Prandtl number is approximately 0.7 and 7.0, respectively. The air Prandtl number remains rather constant with increasing temperature, whereas for water it decreases significantly. Other heat transfer fluids like glycerine, ethylene glycol, and engine oil have relatively high Prandtl numbers; see [1, 2].

1.2.2 Convective heat transfer correlations

Some correlations for convective heat transfer coefficients, in terms of the Nusselt number, for external and internal flows are presented in Tables 1.3 and 1.4, respectively. The friction factor is important to determine the pumping power requirement. Corresponding values of the friction factor can be found in [1–3].

For laminar flow in noncircular ducts, convective heat transfer coefficients can be calculated from empirical or analytical expressions of the Nusselt numbers, as exemplified in Table 1.5. In turbulent flow, the concept of hydraulic diameter, as defined in Eq. (1.4), may be used, followed by the application of correlations for circular ducts.

Table 1.3 External flow convective heat transfer correlations, [1, 2]

Flow condition	Nusselt number	Reynolds number	C	m
Laminar flow over a flat surface[a]	$0.664 \, Re_L^{0.5} \, Pr^{1/3}$	$\leq 5 \times 10^5$	—	—
Circular cylinder[b]	$C \, Re_d^m \, Pr^{1/3}$	1–40	0.75	0.4
Circular cylinder[b]	$C \, Re_d^m \, Pr^{1/3}$	40–1,000	0.51	0.5
Circular cylinder[b]	$C \, Re_d^m \, Pr^{1/3}$	1×10^3–2×10^5	0.26	0.6
Circular cylinder[b]	$C \, Re_d^m \, Pr^{1/3}$	2×10^5–10^6	0.076	0.7

Note:
[a] Average value.
[b] Nusselt number based on cylinder diameter d.

Table 1.4 Internal tube flow convective heat transfer correlations, [1, 2]

Fully developed condition	Nusselt number	Prandtl number
Laminar	3.656^a, 4.364^b	No range
Turbulent, smooth tube: $Re_d \geq 10,000^c$	$0.027 \, Re_d^{0.8} \, Pr^{0.33} \left(\mu / \mu_w \right)^{0.14}$	$0.7 \leq Pr \leq 16,700$
Turbulent, smooth tube: $Re_d \geq 10,000^d$	$0.023 \, Re_d^{0.8} \, Pr^n$	$0.7 \leq Pr \leq 160$

Note:
[a] Uniform wall temperature.
[b] Uniform wall heat flux.
[c] Viscosity temperature dependent.
[d] $n = 0.3$ for cooling, $n = 0.4$ for heating.

Table 1.5 Noncircular laminar duct flow correlations of heat transfer coefficients, [1, 2]

Duct cross section	Nu_{H2}	Nu_T
Square	3.091	2.976
Rectangular, aspect ratio 0.5	3.017	3.391
Rectangular, aspect ratio 0.25	4.35	3.66
Triangular, Isosceles	1.892	2.47

Note: Nu_{H2} concerns uniform wall heat flux, Nu_T concerns uniform wall temperature.

1.2.2.1 Composite walls and tubes

In engineering applications, commonly plane walls and circular tubes are composed of layers of different material as shown in Figures 1.2 and 1.3, respectively. Convective heat transfer occurs on the hot and cold sides. All thermal resistances are in series; thus, for a composed plane wall, the total rate of heat transfer is given in Eq. (1.7).

$$\dot{Q} = \frac{t_{f_1} - t_{f_2}}{\dfrac{1}{\alpha_1 A} + \dfrac{b_1}{\lambda_1 A} + \dfrac{b_2}{\lambda_2 A} + \dfrac{b_3}{\lambda_3 A} + \dfrac{1}{\alpha_2 A}} \tag{1.7}$$

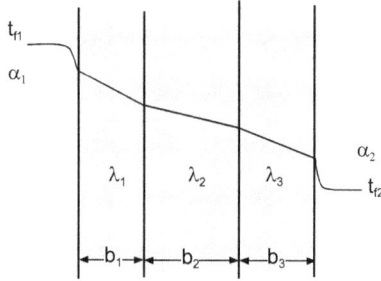

Figure 1.2 The flow of heat across a composite plane wall with heating and cooling by convection.

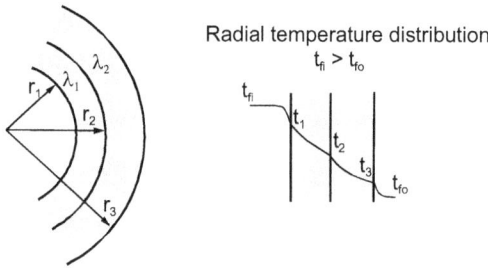

Figure 1.3 Schematics of a composite tube wall with convection internally and externally.

For the composite tube wall in Figure 1.3, the corresponding expression for the rate of heat transfer is given in Eq. (1.8).

$$\dot{Q} = \frac{t_{f_i} - t_{f_o}}{\dfrac{1}{2\pi r_1 L \alpha_i} + \dfrac{1}{2\pi \lambda_1 L}\ln\dfrac{r_2}{r_1} + \dfrac{1}{2\pi \lambda_2 L}\ln\dfrac{r_3}{r_2} + \dfrac{1}{2\pi r_3 L \alpha_o}} \tag{1.8}$$

1.2.2.2 Extended surfaces: fins

In many applications, extended surfaces are used to augment the rate of heat transfer between a solid and an adjacent fluid. Extended surfaces are also referred to as fins, and various fin configurations exist. Figure 1.4 shows annular fins (a) on a circular tube and rectangular fins (b) on a plane wall.

Evaluation of the performance of fins relies on the concepts of fin effectiveness and efficiency. The effectiveness expresses the ratio of the fin heat transfer rate to the heat transfer rate of the surface without any fin. Its value should be greater than unity. The fin efficiency φ is expressing the ratio of the fin heat transfer rate to the corresponding rate of an identical fin but with infinite thermal conductivity. Accordingly, the value of φ occupies the interval $(0, 1)$. The heat transfer rate for the configurations in Figure 1.4 is written as,

$$\dot{Q} = \alpha(t_b - t_f)(A_b + \varphi A_{\text{fins}}) \tag{1.9}$$

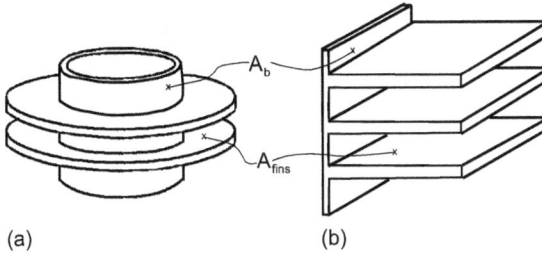

Figure 1.4 Typical fin configurations: (a) annular fins and (b) rectangular fins.

The fin efficiency of a rectangular fin can be shown as follows:

$$\phi = \frac{\tanh mL}{mL} \tag{1.10}$$

where parameter m is $m = \sqrt{\dfrac{2\alpha}{\lambda b}}$ (where α is the heat transfer coefficient, λ the fin thermal conductivity, and b the thickness of the fin).

The fin efficiency for annular fins depicted in Figure 1.4; φ can be determined from Figure 1.5 (or available correlations).

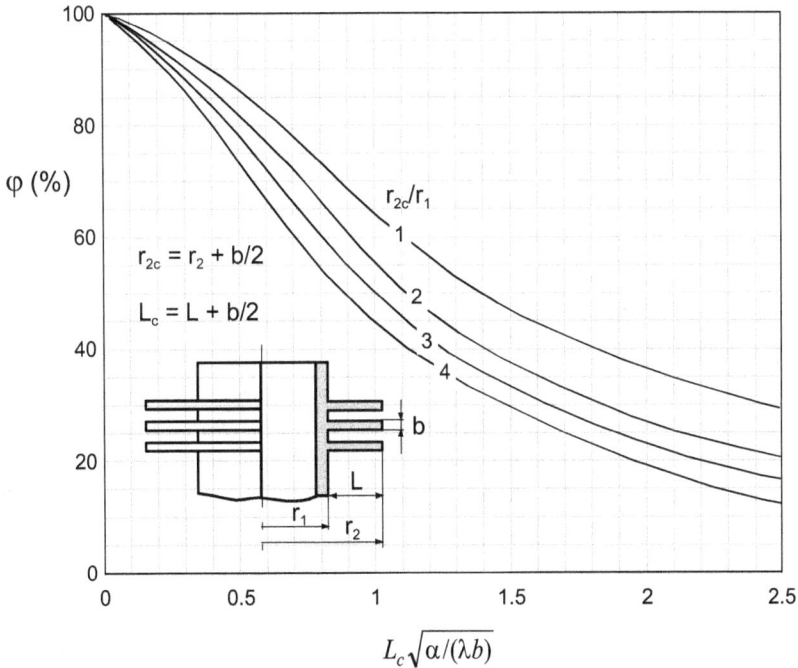

Figure 1.5 Fin efficiency of annular fins.

1.3 THERMAL RADIATION

Thermal radiation may occur between surfaces, between a surface and a participating medium like a gas. The radiative heat exchange is governed by electromagnetic waves according to the theory of Maxwell or by discrete photons according to the hypothesis of Planck. For enclosures, the radiative heat exchange for any of the involved surfaces can be calculated by Eqns. (1.11) and (1.12).

$$\dot{Q}_i = A_i \frac{\varepsilon_i}{1 - \varepsilon_i} \left(E_{B,i} - J_i \right) \tag{1.11}$$

$$\dot{Q}_i = A_i \sum_k F_{ik} \left(J_i - J_k \right) \tag{1.12}$$

In these equations, $E_{B,i}$ represents the blackbody radiation of surface i (by the Stefan–Boltzmann law), J_i denotes the radiosity of surface i, F_{ik} is the view factor between the surfaces i and k, A_i is the area of surface i, while ε_i stands for the emissivity of surface i. In most conventional heat exchangers, radiative heat exchange is not significant but in high-temperature applications it might be important, and, in some cases, gaseous and soot radiation (in, that is, combustion) must be considered. Thermal radiation can be studied in Modest [4] and Howell et al. [5].

1.4 INTRODUCTION TO HEAT EXCHANGERS

1.4.1 General

Heat exchangers of various kinds have been developed and found applications in power plants, chemical process industries, refrigerators, vehicle heat exchangers, heat pumps, air conditioning, etc. In common shell-and-tube heat exchangers and for vehicle radiators, the mechanism of convection and conduction from a hot fluid to a cold one is the primary mode of heat transfer. A metallic wall usually separates the hot and cold fluids. The processes of evaporation and condensation are the primary mechanisms in so-called evaporators and condensers. In cooling towers, direct mixing of the hot fluid (water) and the cold fluid (air) constitutes the cooling process. Methods for design, sizing, and rating of heat exchangers are complicated but of significance for engineering analysis. Accordingly, knowledge of convection, pressure drop, overall thermal performance, as well as cost issues are important at the design stage. In space and aircraft applications, weight and size (compactness) might be very important. In contrast, for huge units in power plants or chemical process industries, cost (investment and operation) might be the most important issue. This section introduces classification and functioning of heat exchangers and describes engineering methods for analysis, sizing, and rating.

1.4.2 Heat exchanger classifications

In this chapter, heat exchangers are classified according to the following:

a. The process of heat transfer
b. Compactness (area per unit volume)
c. Design type
d. Flow process
e. Heat exchange mechanisms

1.4.2.1 Process of heat transfer

In direct contact exchangers, the heat transfer occurs between two immiscible fluids, for example, a liquid and a gas, which are forced into direct contact. Examples are provided by cooling towers.

Commonly both natural convection and forced convection occur in cooling towers. A typical application is to recover waste heat from industrial processes.

For heat exchangers operating with indirect contact (the most common heat exchanger type), an impermeable surface separates the hot and cold fluids. These fluids are not mixed. Such heat exchangers are called surface heat exchangers.

1.4.2.2 Compactness

A measure of the compactness of a heat exchanger is introduced as the ratio between the heat transfer area (on one side of the heat exchanger) and the volume. If this ratio, A/V, is greater than 700 m^2/m^3, the heat exchanger is regarded as compact. A radiator in a private car typically has A/V = 1,100 m^2/m^3, while in some glass–ceramic heat exchangers A/V might reach a value as high as 6,500 m^2/m^3. The very common shell-and-tube heat exchangers have $A/V \approx$ 70–500 m^2/m^3, and are usually not said to be compact.

Compact heat exchangers with high A/V – value diminish the volume of the exchangers. Heat exchangers where the weight and size (volume) and thus compactness is a key issue find applications in private cars, trucks and buses, marine vehicles, aircraft and spaceships, cryogenic systems as well as in air conditioning systems. In heat exchangers for gas-liquid heat exchange, the convective heat transfer coefficient on the gas side is normally much less than that on the liquid side. Then, to guarantee transfer of a certain amount of heat between the fluids, the surface on the gas side must be larger than that on the liquid side. Extended surfaces of various geometries are then used. The amount of heat transferred per unit volume (W/m^3) can also be applied as a measure of compactness.

1.4.2.3 Design types

In terms of design types, one distinguishes between shell-and-tube heat exchangers, finned tubular heat exchangers, plate heat exchangers (PHEs), finned plate heat exchangers, regenerative heat exchangers, and others.

The most common heat exchangers are the so-called *Shell-and-tube heat exchangers*, and these are manufactured in wide ranges of sizes, arrangement of flow, etc. The manufacturing process is relatively simple, and if conventional carbon steels can be used, the heat exchanger creates a cheap alternative. Figure 1.6 depicts a conjectured sketch of a simple shell-and-tube heat exchanger where one of the media is flowing outside of the tubes (shell side), while the other is flowing in the tubes (tube side). The components of such heat exchangers are called tube bundle, shell, distribution, and collection headers and baffles.

Baffles are employed to mechanically support the tubes. They also provide guidance of the fluid flow, and then basically the outer tube surfaces are approached in cross flow. In addition, the turbulence in the flow field is increased by the baffles. Various types and design of baffles exist, and the proper baffle distance as well as the geometry of the baffles depend on the flow regime, permitted pressure drop on the shell side, required tube support, and flow-induced vibration risk. The media involved in the heat exchange process are liquid-to-liquid, liquid-to-gas, or gas-to-gas. Two-phase flows on either side occur frequently. Liquid-liquid is the most common case but liquid-gas is also very common. For the liquid-gas case, it is common to install extended surfaces or fins on the gas side as the convective heat transfer coefficient is relatively low on that side.

PHEs consist of several thin plates assembled into a package. The plates might be smooth, but, more commonly, they are corrugated in a unique manner. PHEs operate generally at lower pressure and temperature than shell-and-tube heat exchangers. This limitation is caused by the gasket being

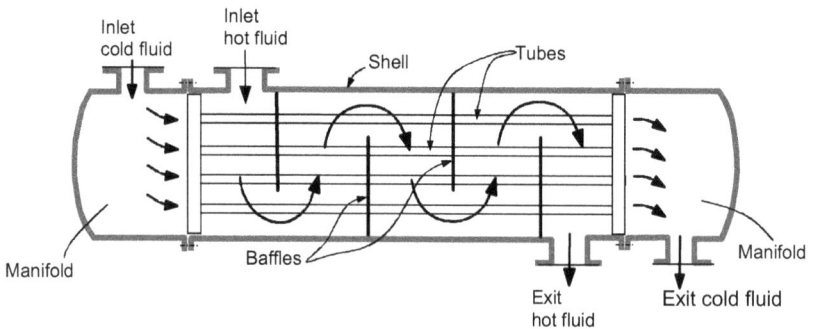

Figure 1.6 Conjectured sketch of a shell-and-tube heat exchanger with one shell pass and one tube pass.

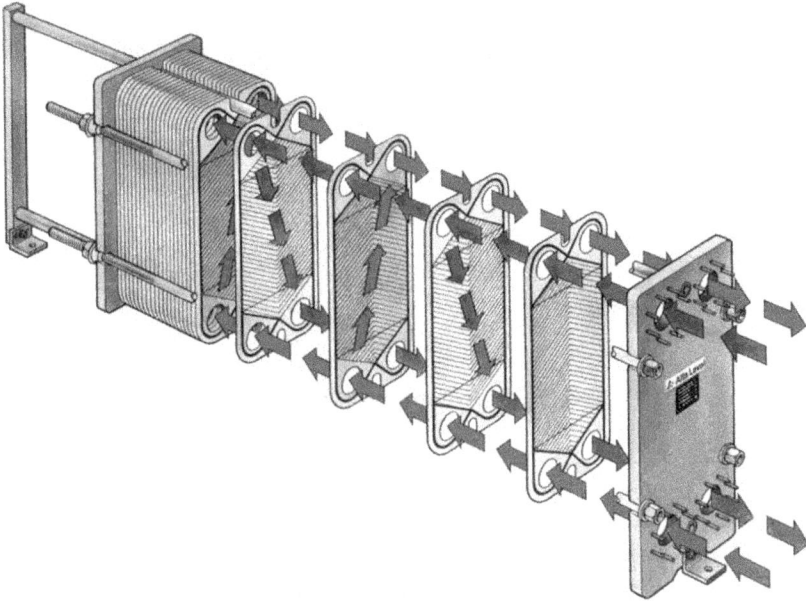

Figure 1.7 Plate heat exchanger. (From Alfa Laval AB).

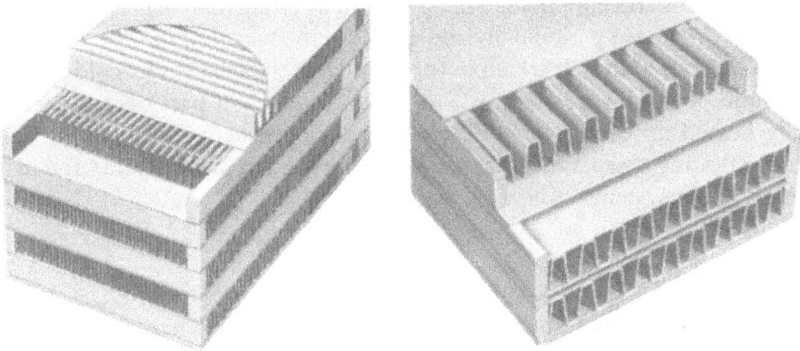

Figure 1.8 Finned plate heat exchangers. (From Harrison Radiator Division, General Motors.)

used to seal between adjacent plates. The compactness value, A/V, is about 120–250 m²/m³. Figure 1.7 illustrates a typical PHE.

Figure 1.8 illustrates *Finned plate heat exchangers*, which may have a compactness value up to 6,000 m²/m³. Such heat exchangers are common for gas-to-gas heat exchange. Fin designs like louvered, perforated, or offset-strip fins are employed to create flow channels and separate the plates, and counterflow, parallel flow, or crossflow arrangements occur.

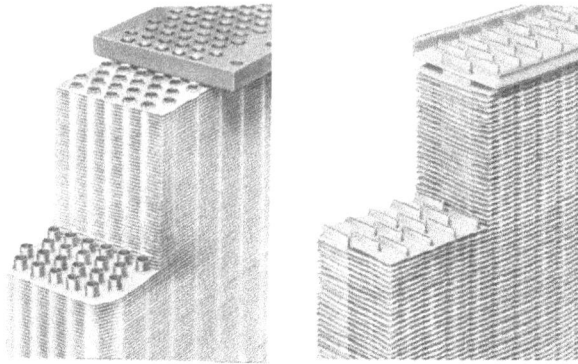

Figure 1.9 Finned tubular heat exchangers. (From Harrison Radiator Division, General Motors.)

Tubular heat exchangers with fins find applications where high operating pressure prevails for one of the media or in cases where finned surfaces are needed to enable the heat exchange process, for. between a liquid and a gas. Figure 1.9 depicts two common configurations, one with circular tubes and another one with flat tubes. The compactness value is less than 350 m²/m³.

Heat exchangers of regenerative type can be regarded as static or dynamic. In the static type, a core of a porous material provides channels through which the hot and cold fluids are flowing in alternate manner. The exchange process is repeated periodically and works in the way that the hot fluid heats up the porous material, which in turn transfers the heat to the cold fluid. In dynamic operating regenerators, the core rotates and part of it is periodically heated up by the hot fluid and then successively cooled by the cold fluid. Rotating regenerators are used as preheaters in power plants and in air conditioning systems as heat recovery units.

1.4.2.4 Flow process

The flow processes can be considered to classify heat exchangers.

Parallel flow or co-current flow implies that the fluids enter the heat exchanger at the same place and are flowing unidirectionally along the exchanger and finally leave the heat exchanger at a common exit location. Figure 1.10 provides a conjectured picture of a parallel flow heat exchanger.

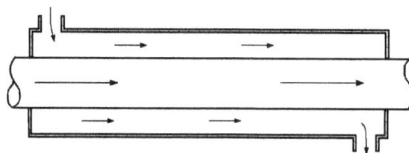

Figure 1.10 Conjecture of a co-current or parallel flow heat exchanger.

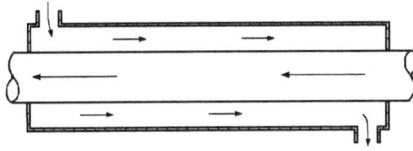

Figure 1.11 Counterflow arrangement.

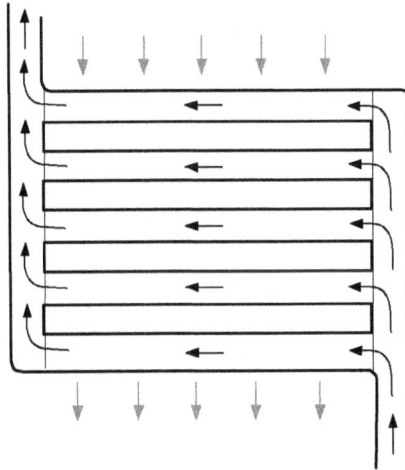

Figure 1.12 Crossflow arrangement of a heat exchanger.

Counter flow means that the hot and cold fluids are entering the heat exchanger at different locations and accordingly flow in opposite directions as illustrated in Figure 1.11.

In Figure 1.12, a crossflow arrangement of a heat exchanger is shown. The fluids flow perpendicular to one another.

The media can be said to be mixed or unmixed. If both the hot and cold fluids flow through individual channels, that is, the fluid streams cannot pass in the transversal direction, the fluids are said to be unmixed. It may happen in other cases, that one medium flows in tubes and cannot move in the transversal direction and obviously that medium is unmixed. Another fluid may flow across the outer surfaces of tubes and then it is free to move in the transversal direction. This medium is accordingly regarded as mixed.

Multi-pass flow fields appear in some heat exchangers, for example, the shell-and-tube heat exchangers. Figure 1.13 provides the conjectured sketches of a few arrangements.

In Figure 1.13a, a case with one shell pass and two tube passes is shown, while in Figure 1.13b an arrangement with two shell passes and four tube passes is illustrated.

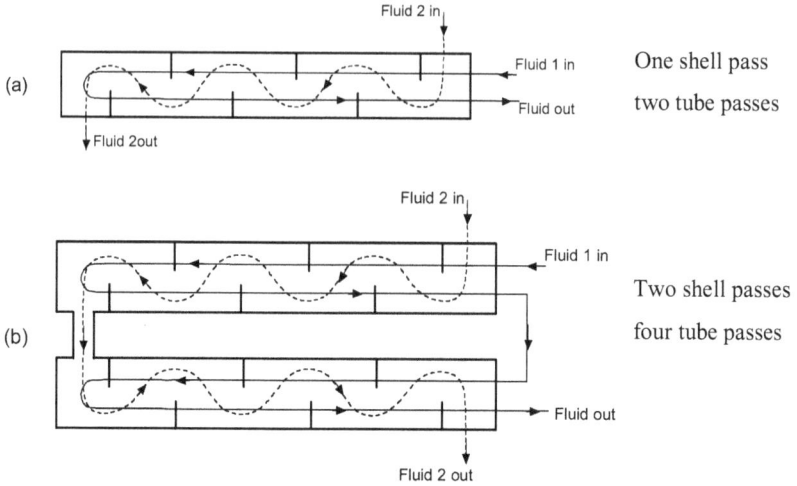

Fluid 2 in

Fluid 1 in

(a)

Fluid out

Fluid 2out

One shell pass
two tube passes

Fluid 2 in

Fluid 1 in

(b)

Fluid out

Fluid 2 out

Two shell passes
four tube passes

Figure 1.13 Arrangements of multi-pass flow: (a) one shell pass two tube passes and (b) two shell passes four tube passes.

1.4.2.5 Mechanisms of the heat exchange

The heat transfer mechanisms can also be described as follows:

  a. Forced or free (natural) convection in single-phase flow.
  b. Phase change process, that is, boiling or condensation.
  c. Radiative hear transfer or combined modes.

Further general information about heat exchangers can be found in [6, 7].

1.4.3 The overall heat transfer coefficient

In Figure 1.14, the heat transfer process from a hot fluid to a cold fluid is illustrated.

The heat transfer rate between the fluids is written as,

$$\dot{Q} = UA \cdot \Delta t_m = \frac{1}{TR} \cdot \Delta t_m \qquad (1.13)$$

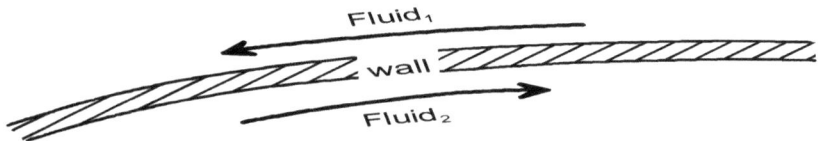

Fluid₁

wall

Fluid₂

Figure 1.14 Illustration of the heat transfer process between two fluids with a separating wall.

where the overall heat transfer coefficient U is introduced. The area through which the heat is passing is denoted by A, and Δt_{m} represents the mean temperature difference between the fluids. Eq. (1.13) is comparable with Eqns. (1.7) and (1.8). TR = 1/UA is called the total thermal resistance. Several resistances appear between the hot and cold fluids (see [2]), and these are in series and then TR is expressed as (see also Figures 1.2 and 1.3)

$$\frac{1}{UA} = TR = \frac{1}{\alpha_{i} A_{i}} + \frac{1}{\alpha_{Fi} A_{i}} + \frac{b_{w}}{\lambda_{w} A_{vl}} + \frac{1}{\alpha_{Fo} A_{o}} + \frac{1}{\alpha_{o} \cdot A_{o}} \qquad (1.14)$$

where α_{i} is the inside convective heat transfer coefficient, A_{i} the inside convective heat transfer area, α_{Fi} the inside fouling factor, b_{w} the intermediate solid wall thickness, λ_{w} the thermal conductivity of the material of the solid wall, A_{vl} the heat conducting area, α_{Fo} the outer surface fouling factor, α_{o} the outer surface convective heat transfer coefficient, and A_{o} the outer side convective heat transfer area. The expression of the heat conduction resistance in Eqns. (1.14) is valid only for a plane wall or, in other cases, if the thickness of the material is very small.

The determination of the heat transfer coefficients α_{i} and α_{o} follows the methods presented in, that is, [1–3].

In real operation, fouling of the heat transfer surfaces appears by the fluids due to various mechanisms. The type of fluid also plays a major role. The fouling acts as a thermal resistance as indicated in Eqn. (1.14). The fouling factors are introduced as $1/\alpha_{Fi}$ and $1/\alpha_{Fo}$, respectively, to account for this; see, that is, [8, 9]. Table 1.6 (data from Ref. [8]) provides a few values of fouling factors.

1.4.4 Analysis of heat exchangers by the LMTD method

The total heat transfer rate $\dot{Q}$ [W] is of primary interest. If the overall heat transfer coefficient U is assumed to be constant in the whole heat exchanger

Table 1.6 Fouling factors available in the literature

Fluid	$1/\alpha_{F}$ [m²K/W]
Distilled water	1×10^{-4}
Seawater ($T < 325$ K)	1×10^{-4}
Seawater ($T > 325$ K)	2×10^{-4}
Feed water to furnaces	2×10^{-4}
Oil	9×10^{-4}
Dirty air	3.5×10^{-4}

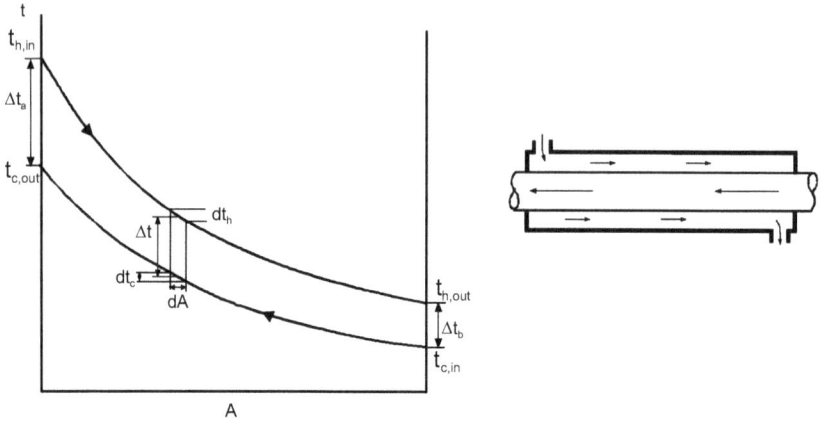

Figure 1.15 Conjectured distributions of temperature in a counterflow arrangement.

(average value), then the total heat transfer rate is written as follows (see also Eqns. (1–13)):

$$\dot{Q} = UA \cdot \Delta t_m \tag{1.15}$$

However, one must find an expression for Δt_m so that Eq. (1.15) becomes meaningful. Here, counter flow and parallel flow are considered.

1.4.4.1 Counterflow heat exchangers

Figure 1.15 depicts the conjectured temperature distributions of the fluids in a single-pass counterflow arrangement.

The heat balance of a small element dA reads,

$$d\dot{Q} = U dA \cdot \Delta t = -\left(\dot{m}c_p\right)_h dt_h = -\left(\dot{m}c_p\right)_c dt_c \tag{1.16}$$

where index h indicates hot fluid, index c indicates cold fluid.

For heat capacity flow rates, i.e., $\dot{m}c_p$, notations are introduced as,

$$C_h = \left(\dot{m}c_p\right)_h, \ C_c = \left(\dot{m}c_p\right)_c \tag{1.17}$$

The total heat transfer rate $\dot{Q}$ is expressed in the following ways.

$$\dot{Q} = C_h \left(t_{h_{in}} - t_{h_{out}}\right) \tag{1.18}$$

$$\dot{Q} = C_c \left(t_{c_{out}} - t_{c_{in}}\right) \tag{1.19}$$

Δt in Eqn. (1.16) is given by,

$$\Delta t = t_h - t_c \tag{1.20}$$

After some algebra and integration along the whole heat exchanger, one finds for Δt_m,

$$\Delta t_m = \text{LMTD} = \frac{\Delta t_b - \Delta t_a}{\ln \dfrac{\Delta t_b}{\Delta t_a}}$$

or,

$$\Delta t_m = LMTD = \frac{\left(t_{h_{out}} - t_{c_{in}}\right) - \left(t_{h_{in}} - t_{c_{out}}\right)}{\ln \dfrac{\left(t_{h_{out}} - t_{c_{in}}\right)}{\left(t_{h_{in}} - t_{c_{out}}\right)}} \tag{1.21}$$

Equation (1.21) provides the formula to calculate Δt_m for a heat exchanger in counter flow. This temperature difference is denoted as LMTD, i.e., the logarithmic mean temperature difference.

1.4.4.2 Heat exchangers operating in parallel flow

In Figure 1.16, the conjectured temperature distributions of parallel flow in a single-pass heat exchanger are shown.

Using the notations in Figure 1.16, one may derive the expression for Δt_m of a heat exchanger with parallel flow of the streams. The result is,

$$\Delta t_m = \frac{\Delta t_b - \Delta t_a}{\ln \dfrac{\Delta t_b}{\Delta t_a}}$$

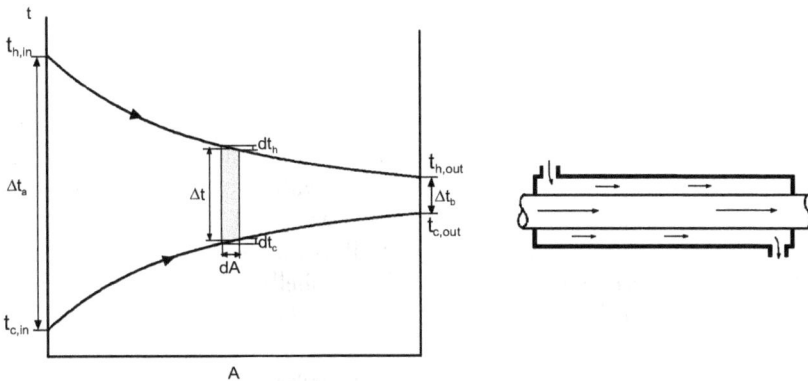

Figure 1.16 Conjectured temperature distributions in a heat exchanger with parallel flow.

or in details,

$$\Delta t_{\mathrm{m}} = \frac{\left(t_{h_{in}} - t_{c_{in}}\right) - \left(t_{h_{out}} - t_{c_{out}}\right)}{\ln\dfrac{\left(t_{h_{in}} - t_{c_{in}}\right)}{\left(t_{h_{out}} - t_{c_{out}}\right)}} \tag{1.22}$$

1.4.4.3 Correction factors of LMTD for heat exchangers not operating in counter flow

Commonly the temperature difference LMTD is used in engineering analysis and design independent of the heat exchanger type and the flow arrangement. However, the heat transfer rate is then written as,

$$\dot{Q} = UA \cdot F \cdot LMTD \tag{1.23}$$

where the correction $F(0 < F \leq 1)$ considers the deviation from the corresponding counterflow operation.

Two additional parameters, P and R, are introduced. P represents efficiency and R is the ratio between the heat capacity flow rates of the streams.

For P one has,

$$P = \frac{t_{c_{out}} - t_{c_{in}}}{t_{h_{in}} - t_{c_{in}}} \tag{1.24}$$

and for R,

$$R = \frac{\left(\dot{m}c_{\mathrm{p}}\right)_{\mathrm{c}}}{\left(\dot{m}c_{\mathrm{p}}\right)_{\mathrm{h}}} \tag{1.25}$$

Using Eqns. (1.18) and (1.19), the ratio R can be alternatively written as,

$$R = \frac{t_{h_{in}} - t_{h_{out}}}{t_{c_{out}} - t_{c_{in}}} \tag{1.26}$$

The correction factor F is a function of P and R but also the heat exchanger type influences the value. In the international literature, analytical expressions are available for several cases. The derivation of such expressions is quite extensive and based on mathematics and algebra. In this chapter, only a single figure, Figure (1.17), is presented to illustrate how F can be estimated.

Figure 1.17 presents F versus P and R for a shell-and-tube heat exchanger with only one shell pass but multiples of two tube passes (i.e., 2, 4, 6, 8, ..., $2n$, tube passes).

For engineering applications, it is not recommended to use a heat exchanger with $F \leq 0.75$. If $F > 0.75$ is impossible to achieve for a certain design, one must try another type of heat exchanger. When $F < 0.75$, it is obvious that the curves tend to become vertical, which implies that small

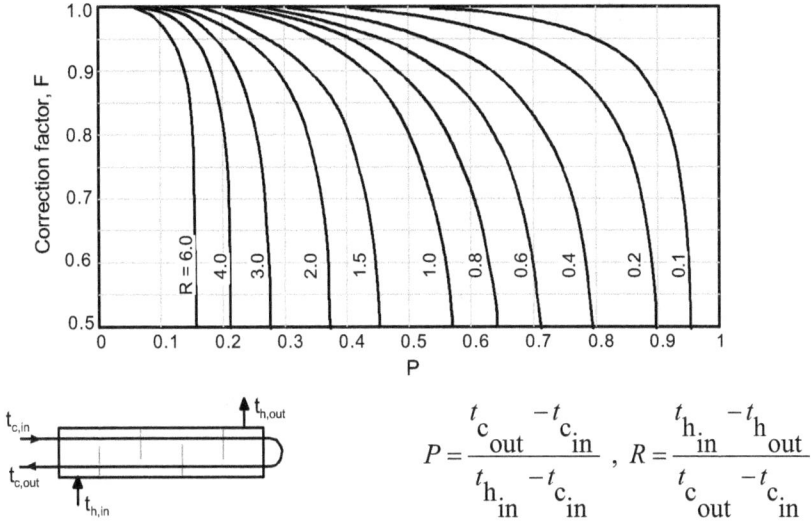

$$P = \frac{t_{c_{out}} - t_{c_{in}}}{t_{h_{in}} - t_{c_{in}}} \ , \ R = \frac{t_{h_{in}} - t_{h_{out}}}{t_{c_{out}} - t_{c_{in}}}$$

Figure 1.17 Correction factor F for a shell-and-tube heat exchanger with one shell pass and two tube passes. (Based on [2].)

variations in temperature or flow rate will result in degradation of the heat exchanger performance.

Further, information about the correction factor F is available in Refs. [7–13].

1.4.5 Analysis of heat exchangers by the ε-NTU method

The LMTD method is quite suitable if the inlet and outlet temperatures of the hot and cold fluids are given. For other cases, the so-called ε-NTU method is more appropriate. This method was developed by Kays and London already in 1955; see Ref. [14].

An efficiency or effectiveness ε is introduced as,

$$\varepsilon = \frac{\text{real heat transfer rate}}{\text{maximum possible heat transfer rate}} = \frac{\dot{Q}}{\dot{Q}_{max}} \tag{1.27}$$

The maximum possible heat transfer rate is received or released by the fluid with the lowest heat capacity flow rate if its outlet temperature becomes equal to the inlet temperature of the other fluid. This implies that $\dot{Q}_{max} = C_{min}\left(t_{h_{in}} - t_{c_{in}}\right)$. Using the notations introduced already, one finds,

$$\varepsilon = \frac{C_h\left(t_{h_{in}} - t_{h_{out}}\right)}{C_{min}\left(t_{h_{in}} - t_{c_{in}}\right)} = \frac{C_c\left(t_{c_{out}} - t_{c_{in}}\right)}{C_{min}\left(t_{h_{in}} - t_{c_{in}}\right)} \tag{1.28}$$

where C_{min} is the smallest heat capacity flow rate.

From Eqns. (1.27) and (1.28), one finds that

$$\dot{Q} = \varepsilon C_{min} \left(t_{h_{in}} - t_{c_{in}} \right) \tag{1.29}$$

The number of transfer units, NTU, is defined as,

$$NTU = \frac{UA}{C_{min}} \tag{1.30}$$

The NTU (originally introduced by W. Nusselt) can be interpreted as the ratio between the heat capacity of the heat exchanger [W/K] and the smallest heat capacity flow rate $C_{min} = \left(\dot{m} c_p \right)_{min}$.

For a counterflow heat exchanger, and if one assumes that $C_{min} = C_c$, which means that $C_{max} = C_h$, one achieves after a few algebraic steps,

$$\varepsilon = \frac{1 - \exp\left[-\left(1 - C_{min}/C_{max} \right) NTU \right]}{1 - C_{min}/C_{max} \exp\left[-\left(1 - C_{min}/C_{max} \right) NTU \right]} \tag{1.31}$$

If C_{min} is set to C_h, the same result will be obtained. Thus, the ratio C_{min}/C_{max} is the proper parameter.

For other heat exchanger configurations, similar calculations can be carried out and corresponding results can be found in the literature. Expressions of the forms $\varepsilon = function\left(NTU, \dfrac{C_{min}}{C_{max}} \right)$ and $NTU = function\left(\varepsilon, \dfrac{C_{min}}{C_{max}} \right)$, respectively, are provided for some typical cases in Tables 1.7 and 1.8. Nevertheless, one should note that some of the formulas are exact while others are only approximate. Solutions can also be presented graphically, so a few are shown in Figures 1.18 and 1.19.

It should be noted that in Tables 1.7 and 1.8 as well as in Figures 1.18 and 1.19 the parameter C is the ratio of heat capacity flow rates, i.e., $C = C_{min}/C_{max}$.

1.4.6 Compact heat exchangers

1.4.6.1 Heat transfer and friction factor

As stated earlier, a compact heat exchanger has a ratio between the heat transfer area and the volume, (A/V), larger than 700 m²/m³. In such heat exchangers at least one of the fluids is a gas. The variety in configuration is huge and data of heat transfer and pressure drop can be found in, that is, Ref. [15]. Here a few configurations are considered, and the associated heat transfer and pressure drop data are provided. In Figure 1.20, the heat transfer coefficient and friction factor on the gas side are shown for a so-called tubular heat exchanger with plane lamellas while Figure 1.21 presents corresponding data for a heat exchanger equipped with circular or annular fins.

Table 1.7 Relations of ε-NTU for common heat exchangers (HEX)

HEX-type	ε	Figure References
Parallel flow	$\varepsilon = \dfrac{1 - \exp\left[-NTU\left(1+C\right)\right]}{1+C}$	Figure 1.18b
Counter flow	$\varepsilon = \dfrac{1 - \exp\left[-NTU\left(1-C\right)\right]}{1 - C\exp\left[-NTU\left(1-C\right)\right]}\ C < 1$	Figure 1.18a
	$\varepsilon = \dfrac{NTU}{1+NTU}\ C = 1$	
(Shell-and-tube HEX)		
I Shell pass 2, 4, 6, ... tube passes	$\varepsilon_1 = 2\left\{1+C+\left(1+C^2\right)^{1/2} \times \dfrac{1 + \exp\left[-NTU\left(1+C^2\right)^{1/2}\right]}{1 - \exp\left[-NTU\left(1+C^2\right)^{1/2}\right]}\right\}^{-1}$	Figure 1.19a
n Shell passes 2n, 4n, ...tube passes	$\varepsilon_n = \left[\left(\dfrac{1-\varepsilon_1 C}{1-\varepsilon_1}\right)^n - 1\right]\left[\left(\dfrac{1-\varepsilon_1 C}{1-\varepsilon_1}\right)^n - C\right]^{-1}$	Figure 1.19b
Cross flow (Single pass)		
Both fluids unmixed	$\varepsilon \approx 1 - \exp\left[C^{-1}\left(NTU\right)^{0.22}\left\{\exp\left[-C\left(NTU\right)^{0.78}\right]-1\right\}\right]$	
Both fluids mixed	$\varepsilon = NTU\left[\dfrac{NTU}{1-\exp(-NTU)} + \dfrac{C(NTU)}{1-\exp\left[-C(NTU)\right]} - 1\right]^{-1}$	
C_{min} unmixed C_{max} mixed	$\varepsilon = C^{-1}(1 - \exp[-C\{1-\exp(-NTU)\}])$	
C_{min} mixed C_{max} unmixed	$\varepsilon = 1 - \exp(-C^{-1}\{1-\exp[-C(NTU)]\})$	
All heat exchangers $C = 0$	$\varepsilon = 1 - \exp(-NTU)$	

Note: $C = C_{min}/C_{max}$

It must be observed that the data are only valid for the dimensions given in these figures. The Stanton number St and Reynolds number Re used in these figures are defined as,

$$St = \frac{\alpha}{Gc_p}, Re = \frac{GD_h}{\mu} \tag{1.32}$$

Table 1.8 Relations of NTU-ε for common heat exchangers (HEX)

HEX-type	NTU	Figure references
Parallel flow	$NTU = -\dfrac{\ln\left[1-\varepsilon(1+C)\right]}{1+C}$	Figure 1.18b
Counter flow	$NTU = \dfrac{1}{C-1}\ln\left(\dfrac{\varepsilon-1}{\varepsilon C-1}\right)\ C<1$	Figure 1.18a
	$NTU = \dfrac{\varepsilon}{1-\varepsilon}\qquad C=1$	
(Shell-and-tube HEX)		
1 Shell pass 2, 4, 6, …tube passes	$NTU = -\left(1+C^2\right)^{-1/2}\ln\left(\dfrac{E-1}{E+1}\right)$ $E = \dfrac{2/\varepsilon_1 - (1+C)}{\left(1+C^2\right)^{1/2}}$	Figure 1.19a
n Shell passes 2n, 4n, …tube passes	Use the expression for one shell pass but with $\varepsilon_1 = \dfrac{F-1}{F-C}$ where $F = \left(\dfrac{\varepsilon C-1}{\varepsilon-1}\right)^{1/n}$	Figure 1.19b
Cross flow (Single pass)		
C_{min} unmixed C_{max} mixed	$NTU = -\ln\left[1+C^{-1}\ln(1-\varepsilon C)\right]$	
C_{min} mixed C_{max} unmixed	$NTU = -\,C^{-1}\ln\left[C\ln(1-\varepsilon)+1\right]$	
All heat exchangers $C = 0$	$NTU = -\ln(1-\varepsilon)$	

Note: $C = C_{min}/C_{max}$.

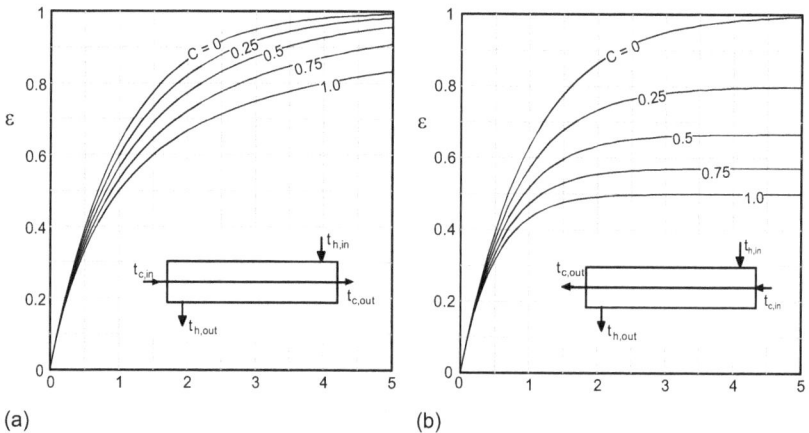

(a)   (b)

Figure 1.18 (a) ε-NTU for a heat exchanger in counter flow; (b) ε-NTU for a heat exchanger in parallel flow. (From Ref. [2].)

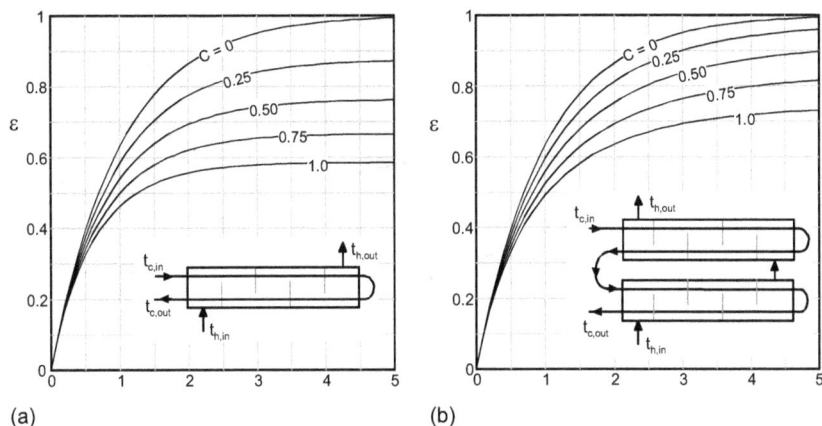

Figure 1.19 (a) ε-NTU for a shell-and-tube heat exchanger with one shell pass and two tube passes; (b) ε-NTU for a shell-and-tube heat exchanger with two shell passes and four tube passes. (From Ref. [2].)

where G is the mass velocity, which is calculated by,

$$G = \frac{\dot{m}}{A_{\min}} \qquad (1.33)$$

where $\dot{m}$ is the rate of mass flow, $A_{\min}$ the minimum area of the cross flow. The definition of the hydraulic diameter reads,

$$D_h = 4 \frac{A_{\min} L}{A} \qquad (1.34)$$

where A is the total heat transferring area and L the depth of the heat exchanger in the main flow direction.

Several correlations for heat transfer and friction factor have been presented in the literature, see, that is, Eriksson and Sundén [16]. The following correlation for the Stanton number has been suggested by Gray and Webb [15].

$$St = 0.14 \, Re_D^{-0.328} \left(S_t / S_l\right)^{-0.502} \left(s/D\right)^{0.0312} Pr^{-2/3} \qquad (1.35)$$

This equation is valid if four or more tube rows are placed in the main flow direction. For a lower number (N) of tube rows, a modified equation should be used,

$$St_N / St = 0.991 \left(2.24 \, Re_D^{-0.092} \left(N/4\right)^{-0.031}\right)^{0.607(4-N)} \qquad (1.36)$$

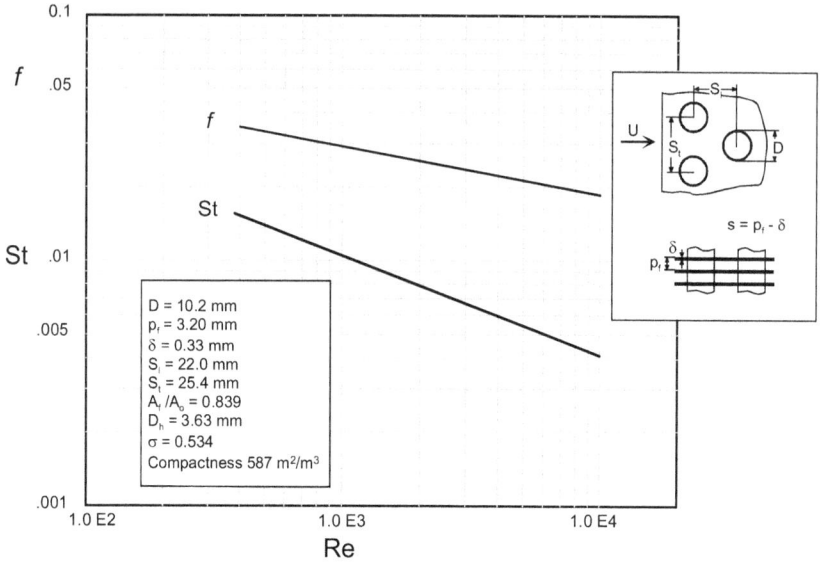

Figure 1.20 Stanton number, St, and friction factor, f, versus Reynolds number Re for a finned tubular heat exchanger. (From Ref. [18].)

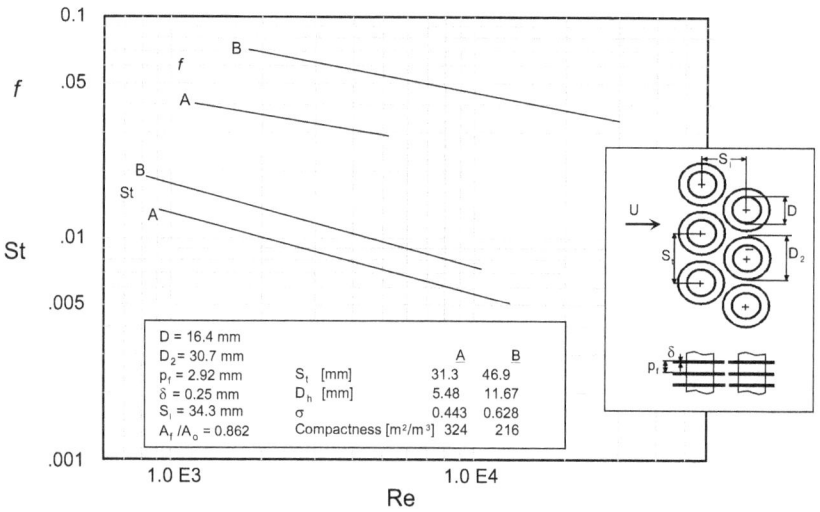

Figure 1.21 Stanton number, St, and friction factor, f, versus Reynolds number Re for a tubular heat exchanger with circular or annular fins. (From Ref. [18].)

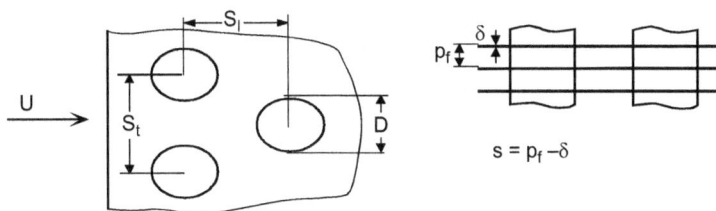

Figure 1.22 Notations for a tubular heat exchanger with plane fins.

The Reynolds number can be calculated by,

$$\mathrm{Re_D} = \frac{GD}{\mu} \qquad (1.37)$$

where D is the tube outer diameter. Additional notations are shown in Figure 1.22.

The pressure drop has two parts. These are related to the pressure drop over the fins represented by the friction factor, f_f, and the pressure drop over the tubes, represented by the friction factor, f_t, respectively. Calculation of the friction factor f_f is as follows:

$$f_f = 0.508\,\mathrm{Re_D}^{-0.521}\left(S_t/D\right)^{1.318} \qquad (1.38)$$

While f_t can be calculated by a correlation for tube bundles given by Zukauskas and Ulinskas (see [13]), a simpler correlation was suggested by Jakob and is reported in [15]. This correlation is as follows:

$$f_t = \frac{4}{\pi}\left(0.25 + \frac{0.118}{\left(S_t/D - 1\right)^{1.08}}\,\mathrm{Re_D}^{-0.16}\right)\left(S_t/D - 1\right) \qquad (1.39)$$

A total friction factor can be calculated according to the expression,

$$f = f_f\,\frac{A_f}{A_o} + f_t\left(1 - \frac{A_f}{A_o}\right)\left(1 - \frac{\delta}{p_f}\right) \qquad (1.40)$$

where A_f is the area of the fins and A_o means the total area on the gas side, i.e., the tube surface area plus the fin area.

The validation ranges of the Gray and Webb correlation are $400 \le \mathrm{Re} \le 24700$, $1.97 \le S_t/D \le 2.55$, $1.70 \le S_l/D \le 2.58$, and $0.08 \le s/D \le 0.64$.

Correlations for friction factors and heat transfer on the gas side for tubular heat exchangers with plane fins, Figure 1.21, can be found in Refs. [17, 18]. In Refs. [18–20], data for other fin geometries are available.

With the heat transfer coefficient (α from St) for the gas side from above and the heat transfer coefficient for tube inside flow (from, that is, Ref. [2]), the overall thermal resistance ($1/UA$) can be calculated. Then, the ε-NTU method or the LMTD method can be used to design and size a new heat exchanger or rate an existing one.

The fin efficiency must be considered as the thermal resistance on the gas side is determined; see Ref. [2]. The total resistance (not considering fouling) is then

$$\frac{1}{UA} = \frac{1}{\phi_o A_o \alpha_o} + \frac{b_w}{\lambda_w A_{vl}} + \frac{1}{\alpha_i A_i} \tag{1.41}$$

In Eqn. (1.46), the gas side (outer surface) has fins while the inside is smooth.

The overall efficiency ϕ_o on the gas side is related to the fin efficiency, as shown in the following equation:

$$\phi_o = 1 - \frac{A_f}{A_o}(1 - \phi) \tag{1.42}$$

A_o is the total heat transfer area on the gas side.

1.4.6.2 On the pressure drop in compact heat exchangers

On the gas side in compact heat exchangers, the pressure drop is commonly split up into three components, i.e., the frictional loss, acceleration of the fluid, and losses at the inlet and outlet. For the exchanger in Figure 1.21 and if the gas is in cross flow, the pressure drop is calculated as,

$$\Delta p = \frac{G^2}{2\rho_{in}}\left[\left(1 + \sigma^2\right)\left(\rho_{in}/\rho_{out} - 1\right) + f\frac{A}{A_{min}}\frac{\rho_{in}}{\rho_m}\right] \tag{1.43}$$

In Eqn. (1.48), ρ_{in} stands for the density at the inlet and ρ_{out} as the outlet density, and G is calculated by Eqn. (1.37). σ is the area ratio, which is determined according to the following:

$$\sigma = \frac{A_{min}}{A_{front}} \tag{1.44}$$

The average density ρ_m is calculated by,

$$\frac{1}{\rho_m} = \frac{1}{2}\left(\frac{1}{\rho_{in}} + \frac{1}{\rho_{out}}\right) \tag{1.45}$$

In Eqn. (1.43), the friction factor f includes the outlet and inlet losses.

The pressure drop for finned PHEs, depicted in Figure 1.8, is calculated by,

$$\Delta p = \frac{G^2}{2\rho_{in}} \left[\underbrace{\left(K_c + 1 - \sigma^2\right)}_{\text{inlet}} + \underbrace{2\left(\rho_{in}/\rho_{out} - 1\right)}_{\text{acceleration}} + \underbrace{f \frac{A}{A_{min}} \frac{\rho_{in}}{\rho_m}}_{\text{friction}} - \underbrace{\left(1 - K_e - \sigma^2\right)}_{\text{exit}} \frac{\rho_{in}}{\rho_{out}} \right]$$

(1.46)

In Eqn. (1.46), K_c is the contraction coefficient at the inlet and K_e is the expansion coefficient at the outlet. In Ref. [14], typical values of K_c and K_e are available.

1.4.7 Shell-and-tube heat exchangers

The most common heat exchanger type in process industries is the shell-and-tube heat exchanger. Typically, the heat power is higher than 1 MW, and the active area can be as high as 5,000 m². Figure 1.23 a, b, c, and d shows some common design layouts.

Advantages claimed for shell-and-tube heat exchangers are as follows:

- Flexible operating conditions, that is, phase change, evaporation, and condensation
- Huge operating pressure range
- Robust equipment
- Fins can be used on the tube surfaces and then increased heat transfer area
- Thermal stresses handled by proper selection of material

Disadvantages associated with shell-and-tube heat exchangers are stated as:

- An accurate design is limited as the correlations on the shell side suffice of uncertainties
- Risk for flow-induced vibrations

The tube length is in the range of 1–20 m, and the shell diameter may occupy a range of 0.25–3.1 m. The tube outer diameter is in the range of 6–51 mm.

1.4.7.1 Common aspects in practical design

Temperature differences: $t_{hin} - t_{cout} > 20°C$, $t_{hout} - t_{cin} > 5°C$
Temperature level: The high temperature fluid should be on the tube side as this limits the number of components to be manufactured in high-temperature-resistant material.

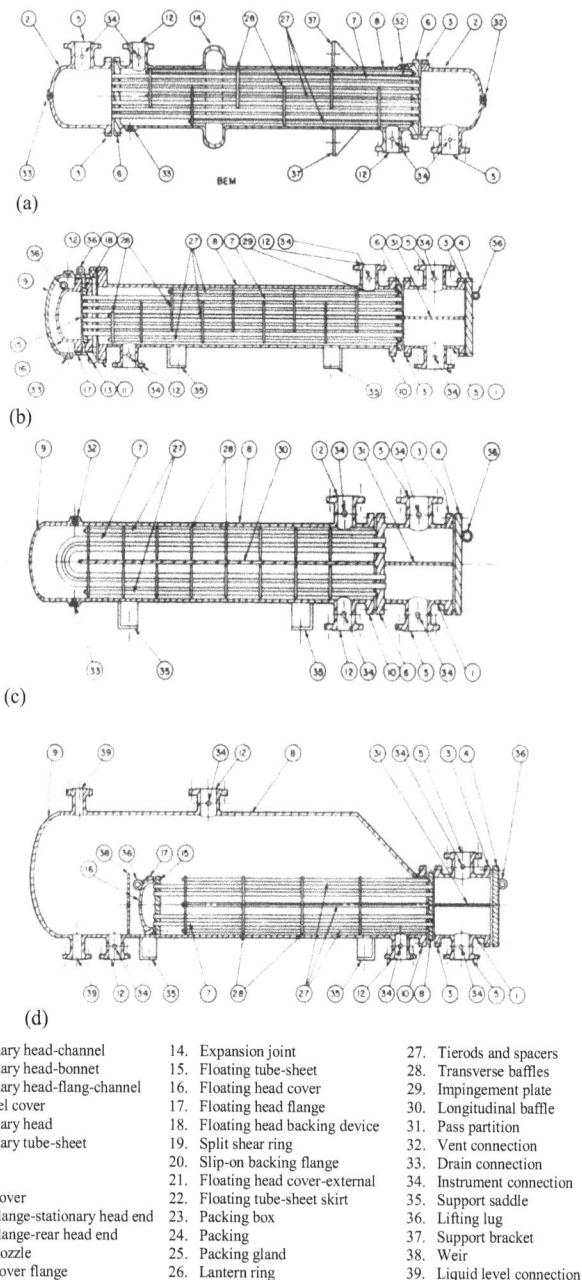

(a)

(b)

(c)

(d)

1. Stationary head-channel	14. Expansion joint	27. Tierods and spacers
2. Stationary head-bonnet	15. Floating tube-sheet	28. Transverse baffles
3. Stationary head-flang-channel	16. Floating head cover	29. Impingement plate
4. Channel cover	17. Floating head flange	30. Longitudinal baffle
5. Stationary head	18. Floating head backing device	31. Pass partition
6. Stationary tube-sheet	19. Split shear ring	32. Vent connection
7. Tubes	20. Slip-on backing flange	33. Drain connection
8. Shell	21. Floating head cover-external	34. Instrument connection
9. Shell cover	22. Floating tube-sheet skirt	35. Support saddle
10. Shell flange-stationary head end	23. Packing box	36. Lifting lug
11. Shell flange-rear head end	24. Packing	37. Support bracket
12. Shell nozzle	25. Packing gland	38. Weir
13. Shell cover flange	26. Lantern ring	39. Liquid level connection

Figure 1.23 Examples of shell-and-tube heat exchangers: (a) one shell pass and one tube pass; (b) one shell pass and two tube passes, mixing occurs between the tube passes; (c) one shell pass and two tube passes, no mixing between the tube passes; (d) a Kettle reboiler. From TEMA-Tubular Exchanger Manufacturer Association [8].

Pressure drops: Pressure drop Δp ranges typically from 10 to 500 kPa on both tube and shell sides but commonly somewhat less on the shell side.

Pressure level: On the tube side, the highest pressure can be accommodated.

Viscosity: On the shell side, the most viscous fluid should be flowing.

Mass flow rate: On the shell side, the fluid with the smallest mass flow rate should be flowing.

Corrosion: To minimize the damage effect of corrosion, the most corrosive fluid should be kept on the tube side

Fouling: It is recommended that the fluid suspected to foul the surfaces should flow on the tube side.

1.4.7.2 Heat transfer and pressure drop on the tube side

The tube-side pressure drop consists of friction losses in straight tubes, expansion and contraction losses at inlets and outlets, as well as losses in U-bends or mixing chambers. The friction loss dominates long tubes. This loss can be calculated by conventional methods; see Ref. [2]. The other losses can be calculated using handbooks, that is, Ref. [11]. The convective heat transfer coefficient in smooth tubes can be determined by conventional methods available in many textbooks, that is, Ref. [2].

1.4.7.3 Shell-side heat transfer and pressure drop

The shell-side flow field is quite complex. A conjectured sketch is depicted in Figure 1.24. The fluid enters the shell side through an inlet pipe and passes across a tube bundle. Sometimes an impingement plate is located just below the inlet pipe to retard the fluid flow and limit the mechanical force on the tubes. The baffles support the tubes but most importantly direct the fluid flow. Gaps are present between the baffles and the inner shell surface as well as between the baffle holes and the tubes. Through these gaps, leakage occurs. As an effect, part of the mass flow does not pass the tubes in cross flow.

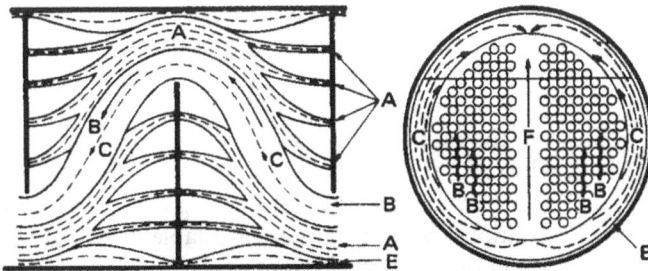

Figure 1.24 Flow field sketch according to Tinker on shell side of a shell-and-tube heat exchanger. (See [21–23]).

The nonuniformity of the flow field and flow direction as well as the leakage flows make accurate calculations of the convective heat transfer coefficient and pressure drop hard. In Figure 1.24, the notations A–F represent the following:

A: leakage flow created by the gaps between baffle holes and tubes
B: main flow path, ideally cross flow
C: bypass-flow in the gap between the tube bundle and inner shell surface
E: leakage flow created by gaps between baffles and inner shell surface
F: (not marked in Figure 1.24): bypass-flow in streaks due to missing tubes in some regions

Various baffle designs appear but typically the baffle distance in the axial direction is taken as 0.25 × shell diameter.

The convective heat transfer coefficient α_s is calculated by,

$$\alpha_s = c\alpha_{\text{tube bundle}} \tag{1.47}$$

where $\alpha_{\text{tube bundle}}$ represents the coefficient for a tube bundle in cross flow. Such coefficients can be found in Refs. [2, 3, 11]. The correction factor c is based on several phenomena like leakage, bypass flow, etc., see, that is, Ref. [11]. The order of magnitude of the correction factor c is $c \approx 0.6$.

The shell-side pressure drop is split into the pressure drops at the inlet and outlet, tube bundle cross flow, and flow in the window sections. A window section is the region created between the inner shell surface and the position where a baffle ends; see also Figure 1.24. For the actual tube bundle cross flow, the pressure drop is calculated by,

$$\Delta p_c = \Delta p_{\text{tube bundle}} (N_b - 1) R_1 \tag{1.48}$$

where $\Delta p_{\text{tube bundle}}$ is the ideal pressure drop across the tube bundle (see Refs. [3] and [11]); N_b is the number of baffles and the correction factor R_1 takes into account the leakage and bypass flow (see Ref. [11]).

The pressure drop for the inlet and outlet is written as,

$$\Delta p_e = \Delta p_{\text{tube bundle}} \frac{N_c + N_{cw}}{N_c} R_2 \tag{1.49}$$

where N_c represents the number of tube rows in cross flow between two adjacent baffles, while N_{cw} represents the number of tube rows in the window section, and R_2 a correction factor considering the bypass flow and the inlet and outlet conditions. More details are available in Ref. [11].

The pressure drop in the window section is calculated by,

$$\Delta p_w = \Delta p_{tkw} N_b R_3 \tag{1.50}$$

where Δp_{tkw} is the pressure drop across the tube bundle in the window section and R_3 is another correction factor for leakage flow.

The total pressure drop Δp_{tot} is then found by summing up the various contribution as,

$$\Delta p_{tot} = \Delta p_c + \Delta p_e + \Delta p_w \tag{1.51}$$

1.4.7.4 Estimation of errors

In the design procedure, one needs to know the accuracy in the determination of the convective heat transfer coefficient α_s and the overall pressure drop Δp_{tot} on the shell side. The estimated uncertainty of α_s is within 25% while for Δp_{tot} the uncertainty is within 40–75%.

1.4.8 Plate heat exchangers

Figure 1.7 showed PHEs of the plate-and-frame type. This is the second most common type of heat exchangers. In general, PHEs have higher overall heat transfer coefficients and higher compactness value than shell-and-tube heat exchangers. In addition, PHEs can more easily be cleaned as the package can be disassembled (for gasketed PHEs) conveniently. Nevertheless, the operating temperatures and pressures are lower.

Many plates (see Figures 1.7 and 1.25) assembled in a package create the heat transfer surface. The plates are equipped with a gasket which is fixed in grooves on one side along a plate – border. The gasket seals between adjacent plates. As shown in the figures, holes are located at the corners of a plate. With the gasket layout and its placement, the flow is controlled in the way that a stream either enters or passes the channel between two adjacent plates. The package created by assembling the plates (see Figures 1.7 and 1.25) with the holes and the gaskets establish a channel system. The cold and hot fluids pass and enter every second channel. Counter flow prevails commonly. The distance between adjacent plates is very small, typically only a few millimeters. The heat transfer area per unit volume is high.

The surfaces of the plates are corrugated creating turbulence, a complex flow structure which enhances mixing and accordingly high heat transfer coefficient is achieved. The stiffness of the plates is improved by corrugation, which means that the plate material thickness can be kept small (typically, 0.5–0.6 mm). Accordingly, the thermal resistance will then be small in the plates. Two main patterns of the plate surface, namely the so-called herringbone pattern and the washboard pattern, are commonly used. Figures 1.25 and 1.26 visualize a few patterns. Other patterns exist and combinations of different patterns are possible.

The frame of a PHE arranged by one fixed and one adjustable end plate. These thick end plates are bolted together, and the plate package appears in between.

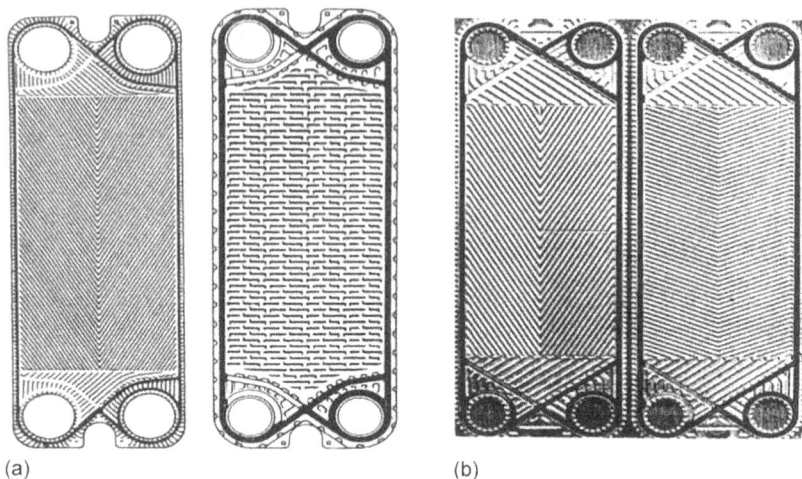

(a)   (b)

Figure 1.25 Plate pattern.

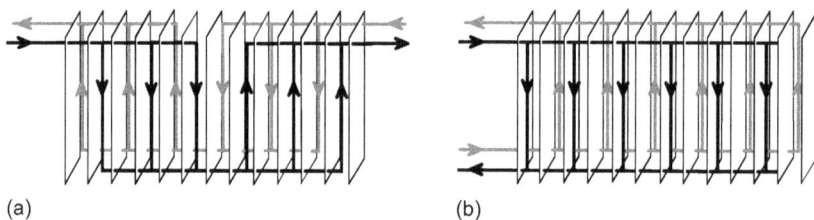

(a)   (b)

Figure 1.26 Plate pattern: (a) two streaks (2 × 3/2 × 3) and (b) one streak (1 × 6/1 × 6).

It is possible to couple the plates together in several ways. The packaging of plates can be divided into streaks for the fluids. In Figure 1.27, a few flow streaks and the couplings are shown. PHEs are nowadays also manufactured with gasket-free plates. The packages are then brazed or welded together, and accordingly higher operating pressures and temperatures can be permitted. The disadvantage is that it is not possible to disassemble the brazed PHEs. Figure 1.28 displays some brazed PHEs.

A thermal length θ is sometimes introduced in the analysis of PHEs. It is defined as,

$$\theta = \frac{\Delta t}{LMTD} \tag{1.52}$$

where LMTD is the logarithmic mean temperature difference for an ideal counterflow heat exchanger and Δt is the temperature difference of one of the fluids. If Δt is taken as the temperature difference for the fluid with the smallest heat capacity flow rate, then θ will be equal to NTU, i.e., the number of transfer units.

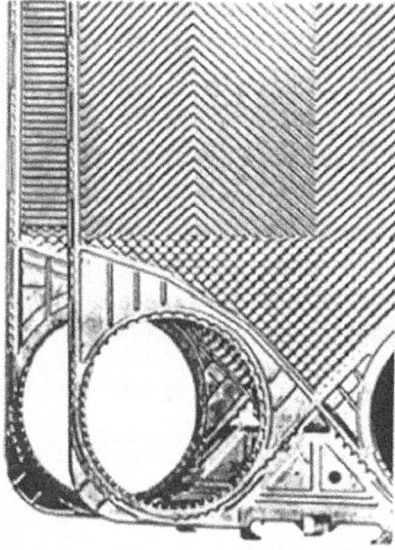

Figure 1.27 Flow streaks.

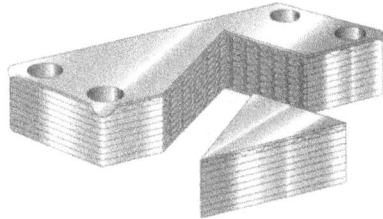

Figure 1.28 Examples of brazed PHEs.

A tight pattern of the plates results in a high pressure drop, but the heat transfer is very efficient. A thermally long channel (high θ) then exists. With a more open pattern, the pressure drop will be lower, but the heat transfer will then be worse. Then a thermally short channel (low θ) prevails. Plates with different patterns can be assembled. The result will be something in between a short and a long channel in terms of heat transfer and pressure drop performance.

Detailed descriptions of PHEs are available in Ref. [24].

1.4.9 Heat exchangers of regenerative type

In regenerative heat exchangers, a relatively large matrix of solid material is formed with flow channels inside. The cold and hot fluids are passing the matrix in an alternating manner. Heat is transferred to the matrix as the hot fluid passes the flow channels and the temperature of the matrix is increased. In the following step, heat is transferred from the matrix to the cold fluid as it passes through the matrix, and accordingly the matrix is cooled down.

The operation of regenerative heat exchangers can be static or dynamic. In the dynamic apparatus, moving parts are involved. Commonly, then, the matrix has the form of a drum or a planar disc. Sequentially, the matrix material receives and delivers heat, respectively, as it is exposed to the hot and cold fluids. In the most common applications, the matrix rotates. Such a regenerator is frequently used as preheater in heat and power plants (Ljungström preheater) as well as for heat recovery in air conditioning systems.

1.4.9.1 On rotating heat exchangers for air preheating and heat recovery

In Figure 1.29, the matrix (also called rotor or wheel) of a rotating regenerative heat exchanger is shown. The matrix consists of two foil systems arranged so that many small channels are created. The rotation speed of the wheel is relatively low, up to about 10 rpm. It is common that approximately 50% of the front surface is open for one of the fluids while the other

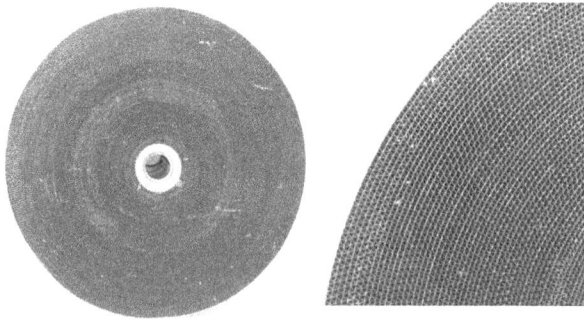

Figure 1.29 Pictures of the wheel or rotor of a specific rotating regenerative heat exchanger.

part is open for the other fluid. Most frequently, the counterflow operation prevails. The wheel is then periodically heated up and cooled down.
However, in continuous operation, the wheel will approach a certain mean
temperature with periodic fluctuations above and below this temperature.

The conventional theory for analysis was established in the 1930s by
Hausen. The overall heat transfer coefficient U is calculated as,

$$\frac{1}{U} = \left(\tau_1 + \tau_2 \right) \left(\frac{1}{\alpha_1 \tau_1} + \frac{\delta_m}{3\lambda_m} \left(\frac{1}{\tau_1} + \frac{1}{\tau_2} \right) + \frac{1}{\alpha_2 \tau_2} \right) \tag{1.53}$$

where τ_1 and τ_2 represent the time intervals the matrix is exchanging heat
with fluids 1 and 2, respectively. The material thickness is δ_m, while λ_m is
its thermal conductivity. The heat transfer coefficients in the channels are
denoted by α_1 and α_2 for fluids 1 and 2, respectively. The analysis carried
out by Hausen included some assumptions and later research has suggested
other and improved methods for the heat exchange analysis.

An alternative method for analysis of a rotating regenerator in counter
flow was presented in reference [23]. The procedure is outlined as,

$$\varepsilon = \varepsilon_{\text{counter flow}} \times \left[1 - \frac{1}{9 \left(C_r^* \right)^{1.93}} \right] \tag{1.54}$$

where $\varepsilon_{\text{counter flow}}$ is the effectiveness of a nonrotating counterflow heat
exchanger, while,

$$C_r^* = \frac{\left(mc_p \right)_{\text{heat exchanger}} \omega}{C_c} \tag{1.55}$$

$\left(mc_p \right)_{\text{heat exchanger}}$ is the matrix material heat capacity, ω the speed of rotation
[revolutions/s], while $C_c = \left(\dot{m}c_p \right)_{\text{cold fluid}}$.

For equal heat capacity flow rates of the hot and cold fluids, i.e., $C_c = C_h$,
one finds (see, that is, Table 1.7),

$$\varepsilon = \frac{NTU_0}{1 + NTU_0} \times \left[1 - \frac{1}{9 \left(C_r^* \right) 1.93} \right] \tag{1.56}$$

where,

$$NTU_0 = \frac{1}{C_{\min}} \left[\frac{1}{\left(1/\alpha A \right)_c + \left(1/\alpha A \right)_h} \right] \tag{1.57}$$

If the material of the heat exchanger has high thermal conductivity, a correction factor for longitudinal heat conduction is introduced. One then writes,

$$\varepsilon = \varepsilon \times correction_\lambda \tag{1.58}$$

Reference [23] gives if the heat capacity flow rates are equal, i.e., $C_c = C_h$,

$$correction_\lambda = 1 - \left[\cfrac{1}{1 + NTU_0 \cfrac{(1 + \lambda\Phi)}{(1 + \lambda NTU_0)}} - \frac{1}{1 + NTU_o} \right] \tag{1.59}$$

where $\lambda = \dfrac{\lambda_{heat\ exchanger} A_{vl}}{L C_{min}}$, $\lambda_{heat\ exchanger}$ is the heat exchanger material thermal conductivity, A_{vl} is the heat conducting area, and $\Phi \approx \sqrt{\dfrac{\lambda NTU_0}{1 + \lambda NTU_0}}$.

1.4.10 Frictional pressure drop

Inside a heat exchanger, the frictional pressure drop is created as the fluid elements move at different velocities because structural walls such as tubes, shell, channels, etc. are present. This pressure drop is calculated from the conventional expression, i.e.,

$$(\Delta p)_{friction} = f_D \frac{L}{D_h} \left(\frac{1}{2} \rho V^2 \right) \tag{1.60}$$

where f_D is the Darcy friction factor, D_h the hydraulic or equivalent diameter = 4 × (flow area)/(flow perimeter), and ½ ρV^2 represents the dynamic pressure. By a theoretical analysis of fluid flow in a closed conduit for laminar flow, the Darcy friction factor, f_D, can be found to be as follows:

$$f_D = \frac{64}{Re} \tag{1.61}$$

For turbulent flow the following formula can be used:

$$f_D = \frac{0.184}{Re^{0.2}} \tag{1.62}$$

The application of Eqn. (1.66) is valid for smooth-walled conduits in the range of 10,000 < Re < 120,000. The Darcy friction factor, f_D, is in general a function of the Reynolds number, Re, and the relative roughness of the conduit surface, ε_w/D_h, as shown in the Moody diagram, Figure 1.30.

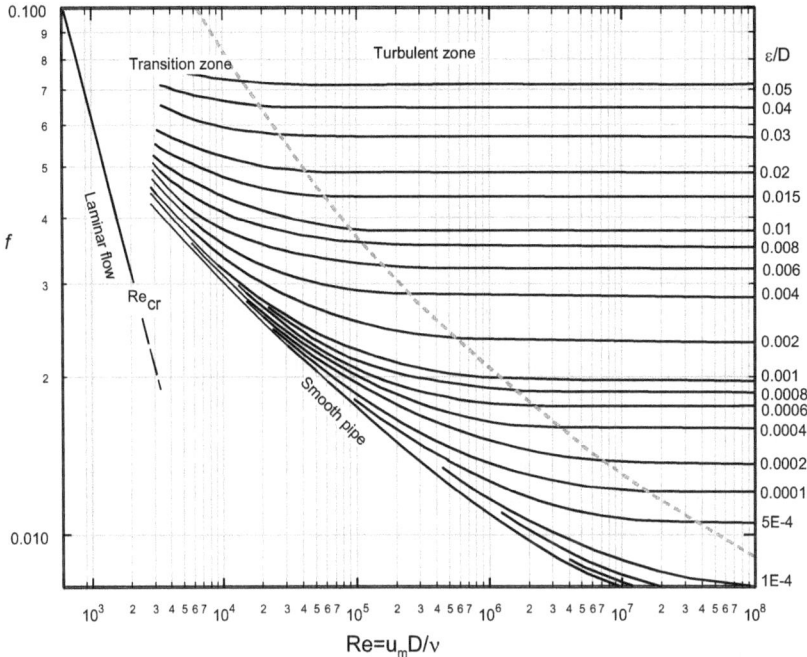

Figure 1.30 The Moody diagram for the friction factor in pipe flow, [25–27].

1.4.11 Applications of heat exchangers

In many engineering applications, heat exchangers are commonly appearing as already mentioned. In this book, other chapters will focus on augmented heat transfer in some application areas where heat exchangers are key elements. Here just some topics are listed, and the reader is referred to later chapters for more detailed information. Important applications appear in (a) high temperature and waste heat recovery exchangers, (b) low temperature difference heat exchangers, (c) direct contact heat exchangers, (d) heat exchangers using non-Newtonian fluids, (e) micro-heat exchangers, (f) electro-hydrodynamic-based heat exchangers, (g) nanofluids in heat exchangers, (h) heat exchangers for aerospace applications, (i) gas turbine cycles, (j) environmental control system, (k) and thermal management.

1.4.12 Selection of material

The material selection and the manufacturing are quite distinct for a heat exchanger. These depend on the specific application and, for instance, the heat duty, fluid types being involved, and heat transfer mode (i.e., single-phase flows, boiling, or condensation). Compatibility between the construction materials and working fluids is essential as corrosion aspects are considered. Criteria can also be formulated in terms of erosion and fouling. The properties of the required material, however, seldom affect the design that can be used.

Stainless steels, nickel alloys, titanium, and titanium alloys are commonly used for PHEs but graphite, copper and copper alloys, tantalum, and aluminum also appear to be used. In gasketed plate-and-frame heat exchangers, the gasket material is selected according to the operating conditions, such as fluid type, concentration, temperature, etc. The lifetime, reliability, and safety are influenced by the material choice. Commonly various elastic and formable materials such as rubber are used for the gaskets (see [24]).

A variety of materials, metallic and nonmetallic, can be used in the design of shell-and-tube-heat exchangers. In cases where carbon steel can be used, the manufacturing of the heat exchangers is cheap (see [13]).

Requirements of low-cost, lightweight, ease of manufacturing, and high-thermal conductivity need to be satisfied by compact heat exchangers, and where applicable aluminum is often selected.

1.4.12.1 New materials in heat exchangers

1.4.12.1.1 Foams

Lightweight materials are preferable in manufacturing of aerospace heat exchangers to save consumption of fuel. In the aerospace industry, PTHEs heat exchangers in aluminum alloys are common. Open-cell porous metallic foams and graphite foams have received increased attention for commercial heat exchangers as they can provide lightweight, improved thermal performance, high compactness, and attractive flexibility in forming complex shapes. In addition, good properties of stiffness and strength, low cost via metal sintering routes for mass production, and some materials for high temperature applications up to 1,200 K can be achieved. The interconnection of pores and tortuosity of open-cell foams improves the fluid mixing and augments convective heat transfer greatly. The porosity, pore density (pores per inch), mean pore diameter, surface area to volume ratio, effective thermal conductivity, and permeability are the essential parameters affecting the performance of the foams. The appropriate correlations for the flow and thermal transport processes in heat exchangers of metal foam are categorized in [28]. The foam material is attractive as its density ranges from 200 to 600 kg/m^3, which is much less than that of aluminum. On the other hand, the tensile strength of graphite foam is low and much less than that of metal foams. The development of the graphite foam heat exchangers has been limited due to the weak mechanical properties. Adding other material into the graphite foam and changing the fabrication process have been shown to improve the mechanical properties of the foam [29]. However, as pointed out by Muley et al. [30], the relatively high pressure drop penalty is a major disadvantage of open-cell foams. Furthermore, as the maximum pressure permitted for foam heat exchangers is relatively low, foam heat exchangers are not suitable for high pressure conditions.

1.4.12.1.2 Ceramic materials

Ceramic materials and ceramic matrix composites (CMCs) have advantages of temperature resistance and corrosion resistance, which are beneficial in heat exchanger construction for rocket nozzles and jet engines, heat shields of space vehicles, aircraft brakes, power plant gas turbines, fusion reactor walls, furnaces for heat treatment, heat recovery systems, and others. The widespread use of ceramics is blocked by ceramic–metallic mechanical sealing, manufacturing costs and procedures, and their brittleness in tension. The nonoxide CMCs, i.e., carbon/carbon, carbon/silicon carbide, and silicon carbide/silicon carbide, are the most common ones. A variety of heat exchangers like plate-fin heat exchangers (PFHEs), micro heat exchangers and primary surface heat exchangers can be manufactured in ceramic materials. Recuperators and intercoolers in gas turbine systems can be manufactured in ceramics and accordingly contribute to improve the cycle efficiency.

1.4.12.1.3 AM: Additive manufacturing or 3D printing

Over recent years, the so-called additive manufacturing (AM) technologies (also referred to as 3D printing) have progressed significantly and enabled the design and development of augmented heat transfer surfaces. Recent investigations in AM have revealed the potential of AM in the development and production of next-generation compact heat exchangers providing lightweight and low material volume. The AMs provide freedom in design of complex elements in heat exchangers. For instance, inserts or artificial roughness can be directly printed on a substrate and then excellent interface connections can be achieved in contrast to conventional brazing. The AM processes based on the powder bed fusion technologies like selective laser melting (SLM) and direct metal laser sintering (DMSL) are applicable. In these, the powder material is melted or sintered layer-by-layer to create the desired three-dimensional solid structure. Obviously, AM is crucial for developing the next generation of heat exchangers. It has been envisaged that enhanced micro-channels manufactured by AM may provide appreciable (20%–50%) volume/weight improvement over current state-of-the-art designs. Nevertheless, in-situ characterization of the AM surfaces is important. The significance of AM technologies as options for manufacturing efficient heat exchanging surfaces is evident from a recent review of the existing literature, [31].

REFERENCES

[1] T.L. Bergman, A.S. Levine, F.P. Incropera, D.P. DeWitt, *Fundamentals of Heat and Mass Transfer*, 8th Ed., J. Wiley & Sons, New York, 2017.

[2] J.P. Holman, *Heat Transfer*, 9th Ed., McGraw-Hill, New York, 2002.

[3] B. Sunden, *Introduction to heat transfer*, WIT Press, Southampton, UK, 2012.

[4] M.F. Modest, S. Mazumder, *Radiative heat transfer*, 4th Ed., Elsevier-Academic Press, 2021.

[5] J.R. Howell, R. Siegel, M.P. Menguc, *Radiation heat transfer*, 5th Ed., CRC Press Taylor & Francis, Boca Raton, Florida, USA, 2011.

[6] J.E. Hesselgreaves, *Compact heat exchangers – Selection, design and operation*, Pergamon Press, UK, 2001.

[7] R.K. Shah, D.P. Sekulic, *Fundamentals of heat exchanger design*, J. Wiley & Sons, New York, 2003.

[8] Tubular Exchanger Manufacturers Association, Standards, TEMA, New York, 1959.

[9] E.F.C. Somerscales, J.G. Knudsen, *Fouling of heat transfer equipment*, Hemisphere Publ. Corp., Washington, 1980.

[10] R.A. Bowman, A.C. Mueller, W.M. Nagle, Mean temperature difference in design, *Trans. ASME*, Vol. 62, 283–294, 1940.

[11] K.A. Gardner, Variable heat transfer rate correction in multipass exchangers, shell-side film controlling, *Trans. ASME*, Vol. 67, 31–38, 1945.

[12] R.A. Stevens, J. Fernandes, J.R. Woolf, Mean temperature difference in one, two and three-pass cross flow heat exchangers, *Trans. ASME*, Vol. 79, 287–297, 1957.

[13] *Heat Exchangers Design Handbook*, Hemisphere Publ. Corp., Washington, 1983.

[14] W.M. Kays, A.L. London, *Compact Heat Exchangers*, 3rd Ed., McGraw-Hill, Begell House, USA, 1984.

[15] L. Gray R.L. Webb, Heat transfer and friction factor for plate finned-tube heat exchangers, *Proc. 9i Int. Heat Transfer Conference*, Vol. 6, 2745–2750, 1986.

[16] D. Eriksson, B. Sundén, Plate fin-and-tube heat exchangers: A literature survey of heat transfer and friction correlations, in *Progress in engineering heat transfer* (Eds. B. Grochal, J. Mikielewicz, B. Sundén), 533–540, IFFM Publishers, Gdansk (Poland), 1999.

[17] A. Zukauskas, *High-performance single-phase heat exchangers*, Hemisphere Publ. Corp., Washington, USA, 1989.

[18] R.L. Webb, N.H. Kim, *Principles of enhanced heat transfer*, 2nd Ed., Taylor & Francis, New York, 2005.

[19] R.L. Webb, Enhancement of single-phase heat transfer, Ch. 17 in *Handbook of Single-Phase Convective Heat Transfer* (Eds. S. Kakac, R.K. Shah, W. Aung), J. Wiley & Sons, New York, 1987.

[20] S. Kakac, R.K. Shah, W. Aung, *Handbook of single-phase convective heat transfer*, Wiley-Interscience, New York, 1987.

[21] A. Zukauskas, Heat transfer from tubes in cross flow, *Advances in Heat Transfer*, Vol. 18, 87–159, 1987.

[22] A. Zukauskas, and R. Ulinskas, *Heat transfer in tube banks in cross flow*, Hemisphere Publ. Corp., New York, 1988.

[23] T. Tinker, Shell side characteristics of shell and tube heat exchangers, *Proc. Gen. Disc. on Heat Transfer*, Inst. Mech. Engn., London, 1951.

[24] L. Wang, R.M. Manglik, B. Sunden, *Plate heat exchangers: Design, Applications and performance*, WIT Press, Southampton, UK, 2007.

[25] W.M. Rohsenow, J.P. Hartnett, E.N. Ganic, *Handbook of heat transfer: Applications*, 2nd Ed., McGraw-Hill, New York, USA, 1985.

[26] B.R. Munson, D.F. Young, T.H. Okiishi, W.W. Huebsch, *Fundamentals of fluid mechanics*, 6th Ed., J. Wiley & Sons, New York, USA, 2009.

[27] F.M. White, *Fluid mechanics*, 7th Ed., McGraw-Hill, New York, USA, 2011.

[28] S. Mahjoob, K. Vafai, A synthesis of fluid and thermal transport models for metal foam heat exchangers, *Int. J. Heat Mass Transfer*, Vol. 51(15), 3701–3711, 2008.

[29] W.M. Lin, B. Sundén, J.L. Yuan, A performance analysis of porous graphite foam heat exchangers in vehicles, *Appl. Therm. Eng.*, 50(1), 1201–1210, 2013.

[30] A. Muley, C. Kiser, B. Sundén, R.K. Shah, Foam heat exchangers: a technology assessment, *Heat Transfer Eng.*, Vol 33(1), 42–51, 2012.

[31] I. Kaur, P. Singh, State-of-the-art in heat exchanger additive manufacturing, *Int. J. Heat Mass Transfer*, Vol. 178, 121600, 2021.

Chapter 2

Heat transfer augmentation

2.1 INTRODUCTION

Heat transfer augmentation (or enhanced heat transfer) is a topic that has developed to a mature stage being of great importance in many applications. For instance, the refrigeration, heating, and ventilation sector and automotive industries use enhanced surfaces in heat exchangers and related equipment. In the process industry, enhanced heat transfer surfaces are successfully employed in heat exchangers. In principle, all heat exchangers constitute potential candidates for heat transfer augmentation. Experimental and numerical investigations have been performed and significant achievements have been reached, see, for example, [1–7]. Criteria will be addressed later to evaluate the potential merits of any modified or enhanced heat transfer surface.

Initially, heat transfer equipment were used with smooth or plain surfaces. The aim of an augmented heat transfer surface with a specific surface structure is to provide a higher value of hA than for a corresponding plain surface (per unit base surface area). To quantify the advantage of using an augmented surface, a term enhancement or augmentation ratio (E) can be introduced as the ratio of hA of an augmented surface to hA of a plain surface. The ratio reads:

$$E = \frac{hA}{(hA)_0} \tag{2.1}$$

Consider a two-fluid counterflow heat exchanger as already described in Chapter 1. The heat transfer rate for a two-fluid heat exchanger is given as,

$$Q = UA\Delta T_m \tag{2.2}$$

The overall thermal resistance is the reciprocal of UA and is given below,

$$\frac{1}{UA} = \frac{1}{\varphi_1 h_1 A_1} + \frac{t_w}{k_w A_m} + \frac{1}{\varphi_2 h_2 A_2} \tag{2.3}$$

 DOI: 10.1201/9781003229865-2

where subscripts 1 and 2 mean fluids 1 and 2, respectively. The surface efficiency φ is introduced to reflect the employment of extended surfaces. For simplicity, Eqn. (2.3) does not include fouling resistances, which may be important (see Chapter 1, eqn. (1.14)). The heat exchanger performance will be improved if the value of UA is increased. An enhanced surface geometry can be employed to increase either of or both hA terms, compared to those of plain surfaces. Accordingly, the thermal resistance $1/UA$, will be reduced. This reduced thermal resistance can be used for any of the following objectives:

a. Size reduction: If the heat transfer rate is prescribed, the length of the heat exchanger can be reduced. A smaller heat exchanger is then achieved.
b. Increase of UA: This can be beneficial in either of two ways:
  1. Reduction of ΔT_m: If $\dot{Q}$ and the total tube length (L) are fixed, ΔT_m may be reduced. Then, an increase in the overall thermodynamic process efficiency can be reached, which yields savings of the operating costs.
  2. Increase of heat transfer rate. If L (the length of the equipment, e.g., tube length) is fixed, the increased value of UA/L will increase the heat transfer rate for the prescribed inlet temperatures of the fluids.
c. Reduction of pumping power to a fixed heat transfer duty.

The lesson to be learned is that the three different performance improvements can be achieved by an augmented heat transfer surface. Nevertheless, which one obtained depends on the objectives of the designers or end users.

The size reduction of objective (a) may be valued. Commonly, designers require that the size reduction is accompanied by cost reduction. Objective (b) in terms of refrigeration condensers and evaporators will enable reduction of the compressor power costs. Objective (c) is important for an existing heat exchanger if the desire is to increase its capacity.

Pressure drop (pumping power) is always a concern in heat transfer equipment. Accordingly, an augmented surface must deliver the desired heat transfer improvement at the required flow rate and satisfy the pressure drop constraints. Later in this chapter, a few evaluation criteria will be introduced to judge the performance and enable comparison of different augmentation techniques. However, first various techniques of augmentation will be presented.

2.2 TECHNIQUES FOR AUGMENTATION

The techniques to augment or enhance heat transfer are classified as passive and active ones. The passive techniques employ specific surface configurations (like ribs, dimples, inserts, etc.) or fluid additives for the augmentation.

For the active ones, external power, that is, electric, acoustic fields, or surface vibration, is required. In this section, brief descriptions of these techniques are provided.

2.2.1 Passive techniques

Extended surfaces like fins, ribs, protrusions, grooves, insertions, etc. are commonly employed in heat transfer equipment. From Eqn. (2.4), it is obvious that by increasing the heat transfer coefficient (h), or the surface area (A), or even both, the thermal resistance can be reduced. A plain fin provides mainly an increase of the area. Nevertheless, manufacturing of specifically shaped extended surfaces can in addition significantly enlarge the heat transfer coefficient. Extended surfaces for liquids typically use smaller fin heights than for gases. This is so because for liquids the heat transfer coefficients are higher than gases. Other chapters will give examples of applications of extended surfaces.

Rough surfaces appear either integral to the base surface or created by locating roughness elements adjacent to the heat transfer surface. The integral roughness can be achieved during the machine processing by various techniques to restructure the surface. Micro- and nanostructuring might be possible. In single-phase flow, the surface structure layout is often chosen to promote mixing in the boundary layers adjacent to the surface, but not to increase the heat transfer surface area.

Coated surfaces appear as a nonmetallic coating, for example, Teflon. Sintered porous metallic coatings are used as well. However, coating might be more suitable for condensation, evaporation, or boiling. However, in this book mainly single-phase convective heat transfer is considered.

Devices can be inserted into the flow channels to improve the energy exchange at the heat transfer surface. Such devices provide mixing in the bulk of the fluid in addition to the near-wall effects. Wire coils are examples of such devices.

Devices to create swirl flow appear as twisted tape inserts, as vortex generators, or as axial core inserts. These create a rotating or secondary flow pattern with high vorticity promoting the mixing and heat exchange process.

In heat exchangers, another passive technique is to apply coiled tubes. The secondary flow created in coiled tubes increases the convective heat transfer coefficient.

Additives are sometimes used particularly for liquids. In Chapter 8, the adding of nanoparticles to a base liquid, creating so-called nanofluids, will be discussed in detail.

2.2.1.1 Combined or compound techniques

Efforts have been made and are still carried out to combine various techniques with a hope to achieve better enhancement than the individual techniques. Then the phrasing fourth-generation heat transfer technology has been introduced as a complementary term to use compound or combined techniques, see, for example, Bergles [8].

Combined or compound techniques are exemplified by dimpled tubes with twisted tape inserts, helical ribbed tubes with double twisted tape inserts, wall roughness integrated with wavy strip inserts, and others. See, for example, Saha et al. [7].

2.2.2 Active techniques

Active techniques can be established by using mechanical means or rotating the surface to stir the fluid motion. In the chemical process industry, mechanical surface scrapers are extensively used for viscous liquids but are also applied for gas flows in ducts.

Vibration of the surface or fluid may be used but due to the high mass of the equipment, fluid vibration might be more practical. Then pulsations at low frequency to ultrasound vibration might be used.

Electrostatic fields have been applied in various ways to dielectric fluids. The idea is that the electrostatic field will create greater mixing of the fluid flow adjacent to the surface transferring the heat.

Another active technique is created by impinging jets which force a single-phase fluid to flow perpendicular toward a surface. Single jets as well as multiple jets appear in a variety of applications.

Injection and suction are utilized in some applications, but in general a porous heat transfer surface is needed.

The disadvantages of active techniques are costs, noise, safety, and reliability. Accordingly, passive techniques are dominating.

2.2.3 Judgment of the techniques

Enhancement or augmentation is applicable to heat and mass transfer processes. In this book, the modes of transfer considered are single-phase natural or forced convection, and convective heat transfer.

Most of the commercially applied augmentation techniques are of passive types. The limited usage of active techniques depends on the associated high cost, noise, safety issues, or reliability concerns of the specific technique. In this book, passive techniques are considered for single-phase applications.

The active techniques are popular in medical instrumentation, marine applications, and spacecraft engineering where the performance is the key feature and not the cost. However, low-cost solutions are highly desirable in heat exchangers to improve the thermal-hydraulic performance. Then the active techniques are not the first choice, but instead passive techniques are suitable. Thus, the flow field is altered favorably by the modified surfaces and/or insert devices.

2.2.4 Examples of augmented heat transfer surfaces

Here some augmented surfaces are visualized in Figures 2.1–2.5.

2.2.5 Opportunities by new technology

The design and possibilities of heat exchangers and other heat transfer equipment design are limited by current production processes. Additive manufacturing (AM) might provide freedom and create a potential for design and manufacturing of heat transfer equipment by advanced technology and then eliminate the current production methods and the associated constraints. It could then be possible to design, for example, heat exchanger flow passages with changing geometry or area aiming to accurately correspond to changing fluid flow conditions inside the passages. By traditional manufacturing methods, it would be hard to enable such designs. Nevertheless, AM may create problems like thermal stresses and deformations as well as uncontrolled surface roughness.

Currently, the powder bed laser fusion technology is regarded as suitable for AM of heat exchangers. The principle is that a laser melts and fuses a metal powder to solidify into the intended shapes, incrementally and vertically layer by layer. However, various cooling rates in different parts of the object may cause thermal stresses, which in turn can lead to deformed and warped structural parts.

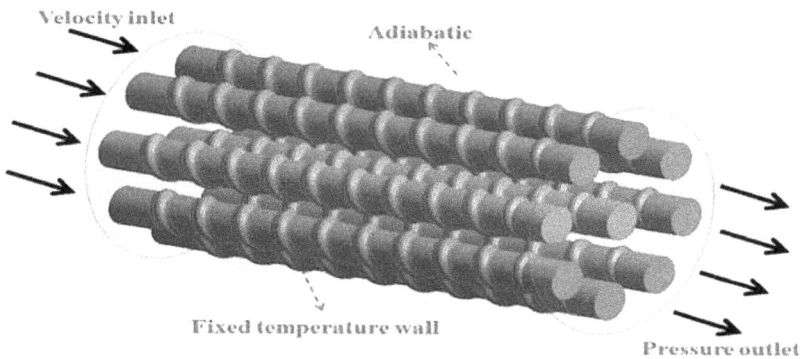

Figure 2.1 Illustration of a multi-tube heat exchanger with corrugated tubes [9].

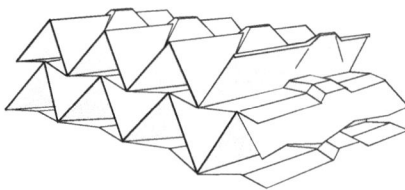

Figure 2.2 Partially corrugated triangular ducts using bumps [10].

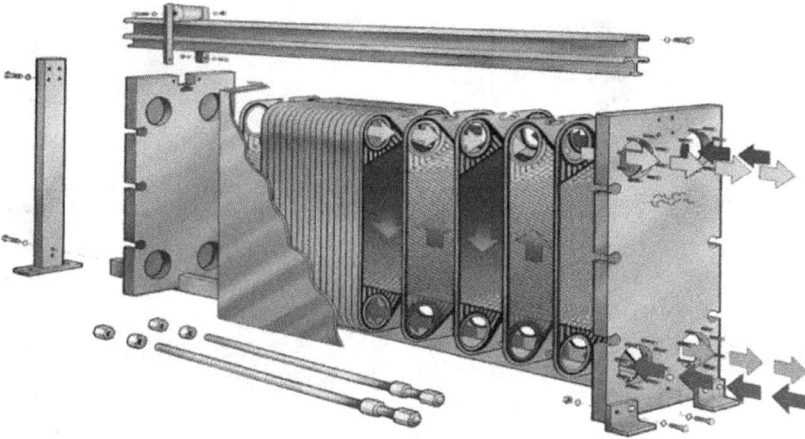

Figure 2.3 Plate-and-frame heat exchanger with corrugated plates. (Courtesy Alfa Laval AB.)

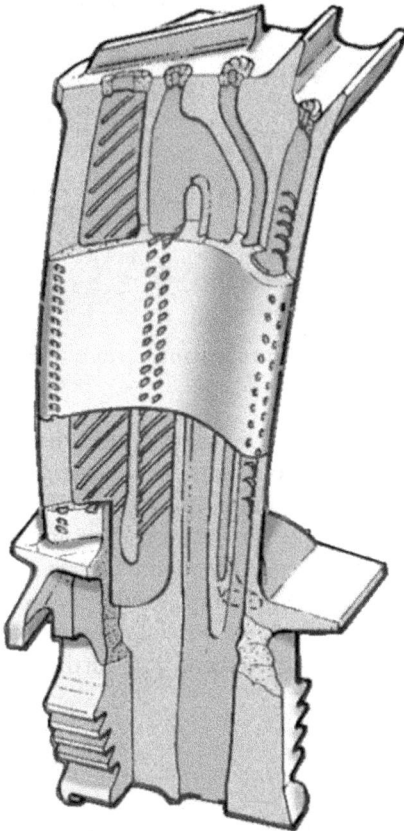

Figure 2.4 Gas turbine blade internal cooling ducts with periodic ribs [11].

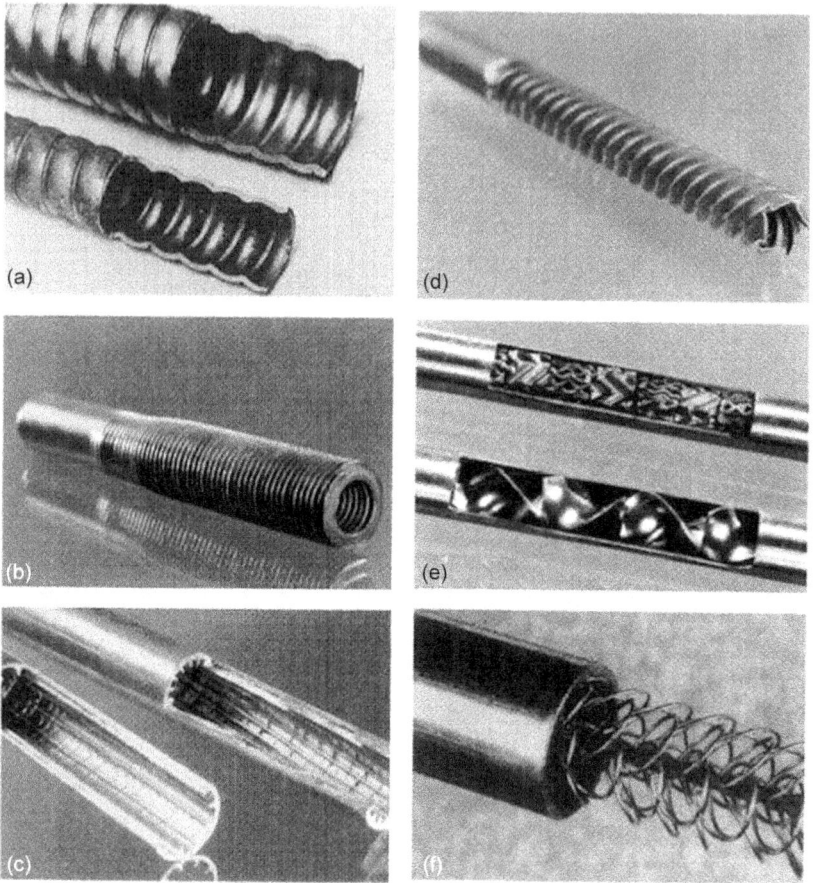

Figure 2.5 Examples of corrugated tubes and various inserts. From [12].

Surface roughness affects heat transfer efficiency and pressure drop. Roughness elements at proper locations are beneficial for heat transfer, while improper locations of the roughness may create obstruction to the flow, cause vibrations, or accumulate dirt or dust. The surface roughness depends on both surface orientation during the manufacturing and parameters in the process like layer heights, temperatures, and speed of the manufacturing process. To some extent, it would be possible to predict and control the end results toward a favorable achievement. For further details, see Kaur and Singh [13].

2.2.6 Nanofluids

Nanofluids are new types of heat transfer media prepared by dispersing nanoparticles (less than 100 nm) in a base heat transfer medium (e.g., water, ethylene glycol, heat transfer oil, and others) by ultrasonic and other technical means. Compared with traditional heat transfer media, nanofluids have excellent heat transfer characteristics and rheological properties, offering great potential in energy related applications, that is, heat exchangers, electronic cooling systems, and energy storage systems. Nanoparticles used to prepare nanofluids are classified as metals, metal oxides, carbon-based particles, as well as nitrides. Among these, graphene has become an ideal thermal conductivity enhancing filler in preparation of nanofluids because of its excellent thermal, mechanical, and electrical properties. In this book, Chapter 8 is devoted to nanofluids, and more details and reference to published literature will appear.

2.3 EVALUATION CRITERIA

Several measures of performance are possible, and such have been proposed in the literature [14–17]. It would be natural to compare the thermal-hydraulic performance of an augmented surface with a smooth surface. A straightforward way for evaluation the performance of various surfaces is to consider nondimensional parameters, such as Nusselt number, friction factor, and the j-factor (= $StPr^{2/3}$). Nevertheless, the final performance judgment of a certain heat transfer surface needs to be specified in at least four interrelated quantities, namely the heat transfer rate, pumping power of the fluid, size, and shape. Various performance evaluation criteria (PEC) exist to evaluate the performance of, for example, heat exchangers, and such have been classified into different forms. Many authors have presented different categories to classify the PECs. For instance, Shah [14] suggested (1) direct comparison of j and f factors (see definitions in Chapter 1), (2) comparison of convective heat transfer coefficients as the function of the fluid pumping power, (3) comparison of the performance with a known reference surface, and (4) miscellaneous direct methods of comparison. Yilmaz et al. [18] classified the PECs into two groups, one based on first law analysis and the second one on second law analysis. Webb [15] defined several criteria to define performance merits of tubular surfaces for augmentation of heat transfer primarily for single-phase flow and segregated the PECs by a few different geometrical constraints, namely FG = fixed geometry, FN = fixed flow area, and VG = variable geometry. For these geometry constraint groupings, three performance objectives were considered, namely reduction of surface area, increase of heat transfer rate or UA value, and reduction of pumping power. It is also known that minimization of entropy generation in any process implies energy conservation. In, for instance, heat exchangers, entropy is generated by heat transfer due to a temperature gradient and by irreversible dissipation of kinetic energy caused by the fluid friction.

The fundamental performance characteristics of augmented surfaces for a single-phase heat transfer process are the so-called j-factor (= $StPr^{2/3}$) and the friction factor f as functions of Reynolds number (Re). An opportunity to quantify the performance improvement is to calculate the ratios j/j_s and f/f_s, where the subscript s represents a smooth or plain surface at equal Re. Generally, for an augmented surface in single-phase flow, the friction factor is higher than for the smooth surface, when operation is at equal velocity or Reynolds number. Nevertheless, this method does not define the actual performance improvement, subject to specific operating constraints. The constraint on pressure drop is a very important consideration for determining the benefits of an augmented surface in single-phase flow.

PECs for revealing the benefits of enhancement or augmentation presented here are applicable to single-phase laminar or turbulent flows in various ducts and tubes as well as normal to tube banks. These are based on a first law analysis.

The performance ratios commonly being used by various researchers are (e.g., [14–16]) as follows:

$$PECS1 = \frac{Nu/Nu_s}{\left(f/f_s\right)^{1/3}} \tag{2.4}$$

$$PECS2 = \frac{St/St_s}{\left(f/f_s\right)^{1/3}} \tag{2.5}$$

$$PECS3 = \frac{Nu/Nu_s}{f/f_s} \tag{2.6}$$

$$PEC4 = \frac{q}{q_s} = \frac{h}{h_s} \tag{2.7}$$

PECS1 is sometimes referred to as the thermal performance factor (TPF) or thermal enhancement factor (TEF).

The pumping power P (W) can for a duct flow be expressed as (f is the Darcy friction factor) follows:

$$P = V \cdot \Delta p = \frac{\dot{m}}{\rho} \cdot f \frac{L}{D_h} \cdot \frac{\rho w^2}{2} \tag{2.8}$$

By introducing the duct cross-sectional area via the hydraulic diameter and using the definition of mass velocity $G = \rho w$, Eqn. (2.8) can be written as

$$P = \frac{\pi D_h L}{8\rho^2} f G^3 \sim f A G^3 \tag{2.9}$$

where $A = \pi D_h L$ $\pi D_h L$ is the heat transferring area.

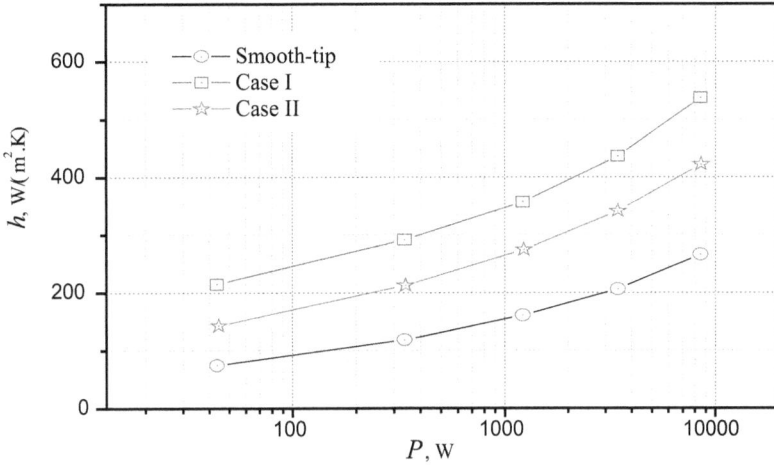

Figure 2.6 Example of heat transfer coefficient versus pumping power. Related to tip cooling with pin fins of a gas turbine blade. From Xie et al. [19].

As a comparison of various heat transfer surfaces is carried out, displaying the heat transfer coefficient versus the pumping power (from Eqn. (2.9)) is a common way to present the evaluation of the performance. Figure 2.6 shows an example.

Figure 2.6 presents an example of evaluation of the heat transfer coefficient versus pumping power. This example is related to the internal cooling of the tip of a gas turbine blade. Cases I and II represent various outline of pin fins placed on the internal tip surface of a U-bend duct. A high position in Figure 2.6 means efficient heat transfer and a smaller required volume. Figures 2.7 and 2.8 display the corresponding data for PEC3 and PECS1, respectively.

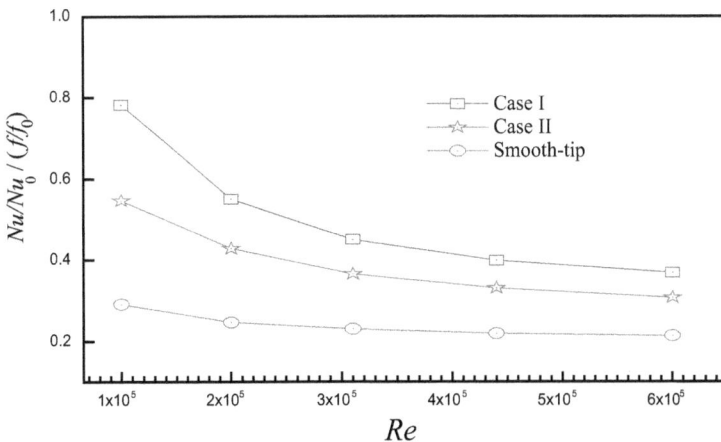

Figure 2.7 PECS3 for the same case as in Figure 2.6. From Xie et al. [19].

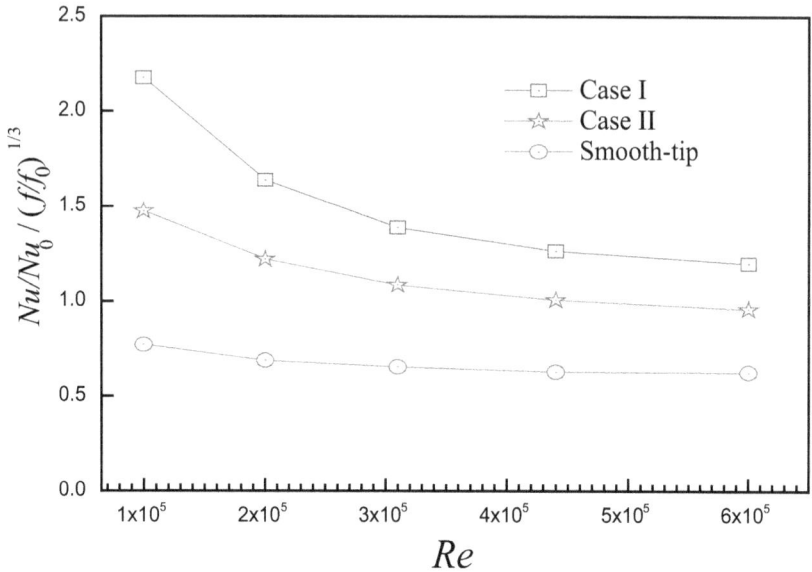

Figure 2.8 PECS I for the same case as in Figure 2.6. From Xie et al. [19].

From these three figures, it is obvious that augmented surfaces (in this case, using pin fins) provide favorable heat transfer performance. The advantage seems to be highest at low Reynolds numbers.

2.3.1 Entropy generation

The second law of thermodynamics can be used in evaluation of an augmented heat transfer surface, [20, 21]. The energy degradation during the process is then considered via entropy generation and work capability. This may be referred to as exergy analysis or minimization of entropy generation. Analysis of the entropy generation has been shown to be a suitable tool in the optimization and evaluation of complex thermodynamic systems [22–26]. The irreversible process of entropy generation in a heat exchanger is caused by the heat transfer and fluid friction, respectively. The total irreversibility is the sum of these two mechanisms. The entropy generation rate per unit length of a heat exchanger passage (s_{gen}) is given as [20]

$$S_{gen} = \frac{Q' \cdot \Delta T}{T^2} + \frac{\dot{m}}{\rho \Delta T}\left(-\frac{dp}{dx}\right) = S_{gen,\Delta T} + S_{gen,\Delta p} \tag{2.10}$$

where the heat amount per unit length is noted by Q', $\dot{m}$ is the mass flow rate, ΔT is the temperature difference between surface and fluid, $\dfrac{dp}{dx}$ is the pressure gradient, and ρ is the fluid density.

The irreversibility caused by the heat transfer is represented by the first term on the right-hand side of Eqn. (2.10), while the second one is caused by fluid friction. To discuss the advantage of an augmented surface, an entropy generation number is introduced as,

$$N_s = \frac{S_{gen}}{S_{gen,s}} \tag{2.11}$$

where $S_{gen,s}$ is the entropy generation rate of a smooth flow passage.

The quality level of the energy transfer is given by the amount of entropy generation. The higher the value of entropy generation, the poorer is the energy transfer quality. Minimization of entropy generation would therefore be suitable. A modified surface yielding a value of N less than unity reduces the irreversibility of the heat transfer equipment.

2.3.2 Synergy principle

To provide insight into the physical mechanisms of the enhancement or augmentation of convective heat transfer, the principle of field synergy has been introduced by Guo et al. [27]. It has then been used and further developed by others. The synergetic relation between velocity and temperature fields is the key and a higher synergy between the velocity and temperature fields implies that higher heat transfer intensity is achieved. The principle of field synergy has been proved for laminar and turbulent flows and its usefulness for the design of heat transfer equipment has been pointed out [28]. A gain in thermal performance is obtained if the intersection angle between the velocity and temperature gradient vectors is reduced. This implies that the higher the synergy of the velocity vector and temperature gradient is, the higher is the transferred heat amount. The field synergy equation for the transfer of heat and mass is derived from its energy equation. Results show that the transferred amount of heat is governed by the values of fluid velocity vector and enthalpy gradient as well as the synergy angle between the velocity vector and enthalpy gradient.

To qualitatively describe the essence of single-phase convective heat transfer augmentation, the field synergy principle has three criteria. Nevertheless, these three criteria are difficult to apply in practice for analysis of convective heat transfer. This is so as there are no available indicators to quantitatively describe them. A unified formula for the field synergy principle has recently been developed based on these three criteria using probabilistic techniques to overcome these defects, [29]. In both the laminar and turbulent flow regimes, the unified formula is applicable for incompressible flows with constant thermo-physical properties. It contains three categories of nondimensional indicators corresponding to the three criteria of the field synergy principle, respectively. These are the domain-averaged cosine of the synergy angle, the Pearson linear correlation coefficients between the scalar functions contained in the energy governing equation of convective heat transfer, and the variation coefficients of these functions. These indicators for the field synergy principle have physical meaning, and their connections with the known heat transfer augmenting mechanisms have been shown. Accordingly, an improved analytical system for the field synergy principle has been proposed. This allows for an efficient and quantitative analysis of all single-phase constant-property phenomena of convective heat transfer. The limitation of the conventional field synergy analytical system, which means that only analyzes the convective heat transfer mechanism based on the synergy angle, is eliminated.

In general, by considering the synergy principle of heat and mass transfer, methods can be established to improve the efficiency of heat and mass transport processes; see, for example, [30]. An illustration of the synergy principle is provided in Figure 2.9.

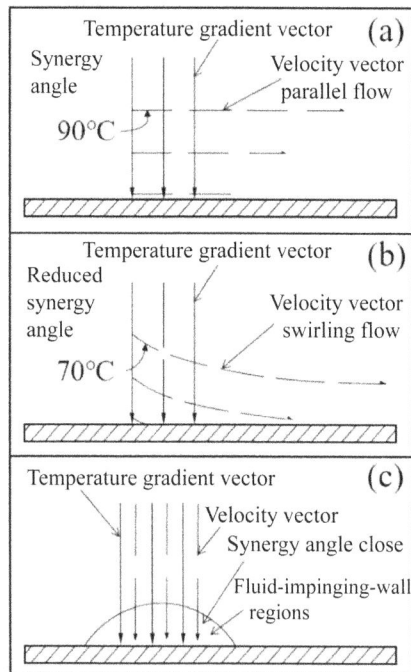

Figure 2.9 Illustration of the synergy principle: (a) weak synergy, (b) intermediate synergy, and (c) ideal synergy.

2.3.3 Comments on evaluation criteria

The application of various augmentation techniques in heat transfer equipment will result in some benefits. The equipment may enable a larger heat flux for fixed surface temperature or alternatively the equipment can operate at a lower temperature while keeping the same heat flux. The later fact is particularly relevant for electronic equipment as it enables improved service life for the components. The various evaluation criteria presented in the previous sections are useful as one likes to judge the benefits of augmented heat transfer surfaces.

2.4 PUBLISHED LITERATURE

The number of publications on augmented or enhanced heat transfer has grown considerably, and according to Manglik and Bergles [31], the rapid growth started around 1950. A figure was provided by Webb and Kim [6] covering the years 1850 to 2001. The total number of publications was 5,676. More recent reviews of the number of publications in this area were presented by Guo [32, 33]. In these reviews, the publications were structured in certain ways. In this section of the book, the status of the number of papers from 1989 to 2021 is presented in Figure 2.10. The information has been retrieved from Web of Science. However, only applications for single-phase convective heat transfer are included. The papers are separated into experimental ones and numerical ones. It is evident that in this narrowed topic area still an extensive number of publications are published every year. Thus, this means that augmented or enhanced heat transfer is important for research and development of heat transfer equipment. Also, the percentages

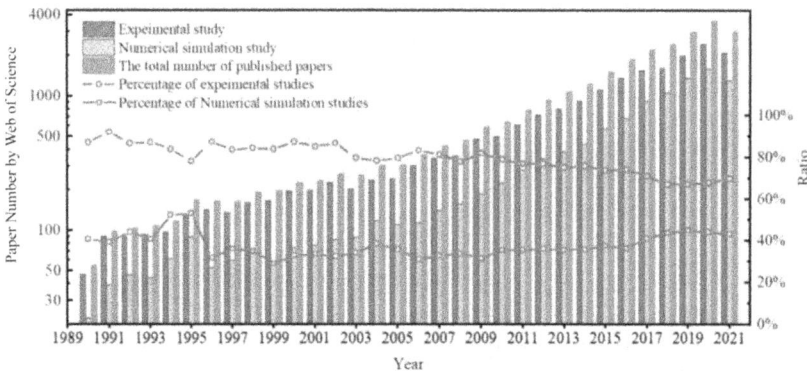

Figure 2.10 Published papers on augmentation of single-phase heat transfer 1989–2021.

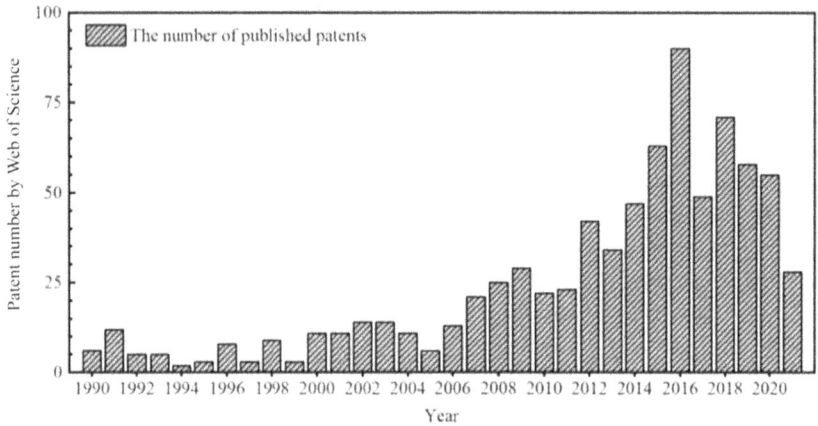

Figure 2.11 Patents on single-phase heat transfer augmentation.

of experimental and numerical papers are presented. It is obvious that experimental investigations still dominate.

In this literature survey, no deeper classification or analysis of the individual papers has been conducted. Basically, augmentation of heat transfer and single-phase convective heat transfer were used as keywords in the literature survey.

2.5 PATENTS

As augmentation or enhancement of heat transfer has strong engineering relevance, not the least in energy conservation, process industries, and heat recovery, many patents have been filed worldwide. The industries use patents to protect their proprietary interests. The classification of surfaces for providing enhanced heat transfer is different for patents and published journal papers. In this section, only the number of filed patents over the time1990 to 2021 is reported. Figure 2.11 presents a view of the progress in filed patents over the years. In Webb and Kim [6], US patents from 1989 were presented. From Figure 2.11, it is obvious that many patents are filed every year.

To create Figure 2.11, various data bases of patents were used.

2.6 CONCLUSIONS

The increase of the heat duty and demands on clean and sustainable technologies are forcing the developments of augmented or enhanced heat transfer techniques forward. This chapter introduced augmentation techniques and exemplified the topic for single-phase flows. The dominating passive techniques were mostly focus. In addition, evaluation criteria were discussed. Also, the importance of the subject was highlighted by considering the number of published articles in journals and the number of filed patents.

REFERENCES

[1] A.E. Bergles, Heat transfer enhancement-the maturing of the second-generation heat transfer technology, *Heat Transf. Eng.*, Vol. 18, 47–55, 1997.

[2] A.E. Bergles, Heat transfer enhancement-the encouragement and accommodation of high heat fluxes, *ASME J. Heat Transfer*, Vol. 119, 8–19, 1997.

[3] A.E. Bergles, Enhanced heat transfer: endless frontier, or nature or routine?, *J. Enhanc. Heat Transf.*, Vol. 6, 79–88, 1999.

[4] R.L. Webb, *Enhancement of single-phase heat transfer, Chapter 17 in Handbook of Single Phase Heat Transfer* (Eds. S. Kakac, R.K. Shah, W. Aung), J. Wiley & Sons, New York, USA, 1987.

[5] R.M. Manglik, *Heat Transfer enhancement, Chapter 14 in Heat Transfer Handbook* (Eds. A. Bejan, A.D. Kraus), Wiley, New York, USA, 2003.

[6] R.L. Webb, N.H. Kim, *Principles of enhanced heat transfer*, 2nd Ed., Taylor & Francis, New York, 2005.

[7] S.K. Saha, M. Tiwari, B. Sunden, Z. Wu, *Advances in heat transfer enhancement*, Springer Briefs in Applied Science and Technology, 2016.

[8] A.E. Bergles, ExHFT for fourth generation heat transfer technology, *Exp. Therm. Fluid Sci.*, Vol. 26, 335–344, 2002.

[9] W. Wang, Y. Shuai, L. Ding, B.X. Li, B. Sunden, Investigation of complex flow and heat transfer mechanism in multi-tube heat exchanger with different arrangement corrugated tube, *Int. J. Therm. Sci.*, 167, paper no. 107010, 2021.

[10] B. Sunden, Simulation of compact heat exchanger performance, *Int. J. Numer. Methods Heat Fluid Flow*, Vol. 20, 5, 551–569, 2010.

[11] S. Wang, S. Li, L. Luo, Z. Zhao, W. Du, B. Sundén, A high temperature turbine blade heat transfermultilevel design platform, *Numer. Heat Transf. A*, Vol. 79, 2, 122–145, 2021.

[12] A.E. Bergles, Some perspectives on enhanced heat transfer-second generation heat transfer technology, *ASME J. Heat Transfer*, Vol. 110 (4b), 1082–1096, 1988.

[13] I. Kaur, P. Singh, State-of-the-art in heat exchanger additive manufacturing, *Int. J. Heat Mass Transfer*, Vol. 178, paper no. 121600, 2021.

[14] R.K. Shah, Compact heat exchanger surface selection methods, *Proceedings of 5th Int. Heat Transfer Conference*, Vol. 4, 193–199, Hemisphere, New York, USA, 1974.

[15] R.L. Webb, Performance evaluation criteria for use of enhanced heat transfer surfaces in heat exchanger design, *Int. J. Heat Mass Transfer*, 24, 715–726, 1981.

[16] R.L. Webb, Performance evaluation criteria for air-cooled finned tube heat exchanger surface geometries, *Int. J. Heat Mass Transfer*, 25, 1770-, 1981.

[17] R. Karwa, C. Sharma, N. Karwa, Performance evaluation criterion at equal pumping power for enhanced performance heat transfer surfaces, *J. Sol. Energy*, Article ID 370823, 2013, Hindawi Publishing Corp.

[18] M. Yilmaz, D. Comakli, S. Vapici, O.N. Sara, Performance evaluation criteria for heat exchangers based on first law analysis, *J. Enhanc. Heat Transf.*, Vol. 12, 2, 121–157, 2005.

[19] G. Xie, B. Sunden, L. Wang, E. Utriainen, Augmented heat transfer of an internal blade tip by full or partial arrays of pin-fins, *Heat Transf. Res.*, Vol. 42, no. 1, 65–81, 2011.

[20] A. Bejan, *Entropy generation through heat and fluid flow*, Wiley, New York, USA 1982.

[21] A. Bejan, *Entropy generation minimization*, CRC Press, Boca Raton, Florida, USA, 1996.

[22] Y. Wang, X. Huai, Heat transfer and entropy generation analysis of an intermediate heat exchanger in ADS, *J. Therm. Sci.*, 27, 2, 175–183, 2018.

[23] W. Wang, Y. Zhang, J. Liu, Z. Wu, B. Li, B. Sunden, Entropy generation analysis of fully developed turbulent heat transfer flow in inward helically corrugated tubes, *Numer. Heat Transf. A*, Vol. 73, 11, 788–805, 2018.

[24] W. Wang, Y. Zhang, J. Liu, B. Li, B. Sunden, Numerical investigation of entropy generation of turbulent flow in a novel outward corrugated tube, *Int. J. Heat Mass Transfer*, Vol. 126, 836–847, 2018.

[25] W. Wang, K. Fu, Y. Zhang, Y. Yan, B. Li, B. Sunden, Entropy study on the enhanced heat transfer mechanism of the coupling of detached and spiral vortex fields in spirally corrugated tubes, *Heat Transf. Eng.*, 2020.

[26] V.D. Zimparov, N.L. Vulchanov, Performance evaluation criteria for enhanced heat transfer surfaces, *Int. J. Heat Mass Transfer*, Vol. 37, 12, 1807–1816, 1994.

[27] Z.Y. Guo, D.Y. Li, B.X. Wang, A novel concept for convective heat transfer enhancement, *Int. J. Heat Mass Transfer*, Vol. 41, 2221–2225, 1998.

[28] W.Q. Tao, Z.Y. Guo, B.X. Wang, Field synergy principle for enhancing convective heat transfer-its extension and numerical verifications, *Int. J. Heat Mass Transfer*, Vol. 45, 3849–3856, 2002.

[29] Y. Cui, Y. Zhang, W. Wang, B. Li, B. Sunden, Unified formula for the field synergy principle, *Numer. Heat Transf. B*, 77, 4, 287–298, 2020.

[30] W. Liu, P. Liu, Z.M. Dong, K. Yang, Z.C. Liu, A study on the multi-field synergy principle of convective heat and mass transfer enhancement, *Int. J. Heat Mass Transfer*, 134, 722–734, 2019.

[31] R. Manglik, A.E. Bergles, Enhanced heat and mass transfer in the new millennium: a review of the 2001 literature, *J. Enhanc. Heat Transf.*, Vol. 11, 87–118, 2004.

[32] Z. Guo, L. Cheng, H. Cao, H. Zhang, X. Huang, J. Min, Heat transfer enhancement-a brief review of literature in 2020 and prospects, *Heat Transf. Res.*, 52, 10, 65–92, 2021.

[33] Z. Guo, A review on heat transfer enhancement with nanofluids, *J. Enhanc. Heat Transf.*, 27, 1, 1–70, 2020.

Chapter 3

Using surface modification

3.1 INTRODUCTION

The surface modification method is the most widely used and maturely
enhanced heat transfer technology for single-phase convective heat transfer [1], which requires treatment of the heat transfer surface by welding
[2], extrusion [3], twisting [4], electrochemical machining, [5] micro milling [6], etc. Corresponding to the above manufacturing processes, one can
distinguish the enhanced heat transfer technology from four techniques:
finned, corrugated, coiled, and modified surfaces. These techniques mainly
involve breaking the flow and thermal boundary with a macro- or micro-
uneven surface to achieve the enhancement of heat transfer. In this way, an
enhanced heat transfer mechanism can be distinguished.

The finned surface technique involves welding various fin forms onto the
heat transfer surface. The types of fins can be divided into two-dimensional
(2D) flat and three-dimensional (3D) tilt fins, single-side and double-side
fins, continuous and discontinuous fins, and fins of various shapes and
arrangements [7]. The enhanced heat transfer mechanism of the finned surface technique mainly involves breaking the boundary layer development
with the production of secondary and inclined flows.

A corrugated surface is attained by extruding the heat transfer surface,
which produces an uneven surface. This technique is easily achieved and can
simultaneously improve the double-sided heat transfer performance. The
secondary flow can be produced by the sudden expansion and contraction
of the surface, breaking the boundary layer development and achieving heat
transfer augmentation [8]. Owing to the limitation of metal ductility, the
corrugation height cannot reach the height of the fins. Therefore, the heat
transfer augmentation of the corrugated surface technique is always smaller
than that of the finned surface technique. However, the corrugated surfaces
are always in a streamline and have a small flow resistance increment.

The coiled surface technique is mainly employed in tube (pipe) flow, and
straight tubes are twisted into coiled types. Owing to the large deformation

DOI: 10.1201/9781003229865-3

of the coiled tube, it is commonly used under low-pressure conditions with high compactness. The enhanced heat transfer mechanism is mainly due to the swirl flow caused by the centrifugal force and the uneven distribution of the flow and thermal fields [9].

The modified surface technique in the microscale has rapidly developed due to the high demand for electronic cooling. Coating, electrochemical machining, and micro milling techniques are employed to produce regular or irregular surface roughness in microchannels. The enhanced heat transfer mechanism is similar to the finned and corrugated surface techniques, but corresponds to microscale convection heat transfer [10].

The four types of surface modification techniques used for single-phase convective heat transfer introduced above have been prevalent in the literature from 1990 to 2020, with statistics from the Web of Science on publications shown in Figure 3.1. In the past 30 years, published works on modification techniques have continued to increase each year from 100 to nearly 5,000, nearly a 50-times increase, which indicates that the relative techniques have been widely applied and investigations are valuable. Among the four techniques, the finned surface technique is the most dominant, and the published number is almost one order of magnitude greater than the other three techniques, with a uniform growth rate. The coiled surface technique is also a mature technique, with levels of research similar to the finned surface technique in the early period. Corrugated and modified surfaces in microscale techniques were developed later, and few studies were published in 1990. The corrugated surface technique publication number rapidly reached a high level and then maintained a high growth rate. The modified surface in the microscale technique publication number increased slowly before 2003 and then quickly reached a level similar to that of the corrugated surface technique.

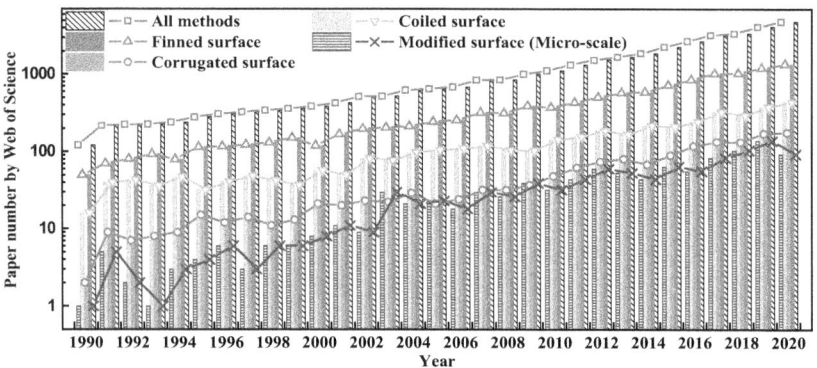

Figure 3.1 Statistics on published work for different surface modification techniques on single-phase convective heat transfer (1990–2020).

3.2 FINNED SURFACE

Ribs or fins were the earliest heat transfer enhancement techniques used in plate and tube heat exchangers. As relatively mature and simple methods, they have been widely used in refrigeration, petrochemicals, aerospace, vehicle manufacturing, power machinery, nuclear energy, and other industrial fields. In heat exchangers with additional fins, the fins can enhance the heat transfer performance by significantly disturbing the fluid movement and increasing the contact area between the fluid and the solid. To optimize the performance and satisfy different demands, various shapes, geometry parameters, and combinations and arrangements of ribs were developed as follows and summarized in Table 3.1.

The air flow in tubes with different arrangements of straight fins was numerically studied at a flow rate of 6–19 m/s [11]. The results showed that the heat transfer performance of a centrosymmetric eight-finned tube was 3.5–5.5 times that of a smooth tube, whereas the heat transfer coefficient of a straight rib tube arranged symmetrically was four to eight times that of a smooth tube.

The water flow in helical fin tubes was experimentally studied [12]. Measurements of friction factor and heat transfer performance implied that a tube with 45 fins, a 48° helix angle, and 0.38 mm height could be suggested for heat exchange applications because of its approximately 2.2 times higher j-factors and approximately 2.7 times higher f-factors compared with a smooth tube at Reynolds number (Re) in the range of 12,000–58,000.

The oil or water flow in helical micro-fin tubes was experimentally studied at Re in the range of 2,500–90,000 [13]. The results showed that once the Re reached a certain value, the heat transfer efficiency of the micro-fin tube was superior to that of a smooth tube. This critical Re was approximately 10,000 (for water) and 6,000 (for the higher Pr working medium, oil), and the heat transfer performance was twice higher than that of a smooth tube for Re > 30,000 (for water). This type of finned tube did not result in a large increase in pressure drop, which was approximately 40–50% over the smooth tube at Re > 30,000.

The air flow in 16 types of rib-roughened rectangular cooling channels was numerically studied at Re in the range of 5,000–50,000 [14]. The results showed that for different working conditions, the heat transfer and friction performance are strongly influenced by the cross-sectional rib shape. The slope of the front surface mainly affected the heat transfer performance, whereas the friction performance could not be generalized. Among the types of fins designed in this study, the new boot-shaped rib (Case 8) exhibited the most outstanding heat transfer performance, and the reverse-pentagonal rib (Case 11) exhibited the lowest fraction factor.

The water flow in seven types of rib-roughened channels was numerically studied [15]. The different dimensionless pitches (4–20), heights (0.02/ 0.05), and widths (0.5/2.0) of the ribs and the Re (20,000–60,000) were numerically analyzed. The results presented that the Nusselt number (Nu) of

Table 3.1 Summary of finned surface enhanced heat transfer studies [11–18]

Configurations	Method	Working conditions	Structure parameters	Performances
	Numerical [11]	Air flow; inlet velocity 6–19 m/s	Straight fins: centrosymmetric arrangement with 8 fins; left-right symmetric arrangement with 6–14 fins	$h/h_s \approx 4$–8 (left-right symmetric finned tubes); $h/h_s \approx 3.5$–5.5 (centrosymmetric finned tubes)
	Experimental [12]	Water flow; Re $\approx$ 12,000–58,000	Helical fins: helix angles (α): 25°–48°; number of fins (N_s): 10–45; height-to-diameter (e/D): 0.0199–0.0327	Best j-factor ≈ 0.005–0.0085 ($N_s = 45$, $\alpha = 48°$, $e/D = 0.0244$); lowest $f/f_s \approx 1.55$ ($N_s = 10$, $\alpha = 25°$, $e/D = 0.0243$)
	Experimental [13]	Oil flow (T_{in} = 40–90°C); Water flow (T_{in} = 15–45°C); Re = 2,500–90,000	Helical micro-fins in tubes	Nu $\approx$ Nu$_s$ (Re < 10,000 for water), Nu $\approx$ Nu$_s$ (Re < 6,000 for oil), Nu/Nu$_s$ $\approx$ 2 (Re > 30,000 for water); $f \approx f_s$ (Re < 10,000), $f/f_s \approx 1.4$–1.5 (Re > 30,000)
	Numerical [14]	Air flow; Re = 5,000–50,000	Sixteen rib shapes: pitch-to-hydraulic diameter ratio = 10; height-to-hydraulic diameter ratio = 0.047	Case 8 (boot-shaped) achieves the highest Nu; Case 11 (reverse-pentagonal shaped) achieves the lowest f

(Continued)

Numerical [15]	Water flow; Re = 20,000–60,000	Three rib shapes: square, triangular, and trapezoidal; pitch: 4–20, height: 0.02/0.05, width: 0.5/2.0	$Nu/Nu_s \approx 1.2–2.6$; $f/f_s \approx 1.5–13$; rectangular ribs achieve the lowest f
Experimental and Numerical [16]	Deionized water flow; Re = 100–400	Three fin shapes: circular, elliptical, and diamond	$h(W/m^2 \cdot K)$: circular (10,000–16,500) > diamond (8,000–11,500) > elliptical (7,000–9,500)
Numerical [17]	Air flow; Re = 20,000–160,000	Six rib arrangements: pitch of large ribs: 100 mm; height of large ribs: 5 mm; height of small ribs: 2.5 mm; Cases B/C/D: small ribs are 50/30/15 mm from the next large ribs, respectively	$Nu/Nu_s \approx 1.14–1.34$, Case B achieves the highest Nu; $f/f_s \approx 1.6–2.6$, Case D achieves the lowest f; $PEC \approx 0.92–1.02$
Numerical [18]	Laminar: helium flow; Turbulent: carbon dioxide flow; Re = 0–150,000	Fins in PCHE: zigzag shaped (15/40°), straight shaped, S shaped, and airfoil shaped	h (turbulent flow): airfoil shaped > zigzag shaped (40°) > straight shaped; h (laminar flow): airfoil shaped > zigzag shaped (15°) > straight shaped

Wall

Flow area

triangular rib-roughened channels reached up to approximately 2.6 times higher than that of channels without ribs (w/e = 2, Re = 60,000, p/e = 6), and the best performance evaluation criteria (PEC) was equal to 1.54 for triangular fins with p/e = 6, e/d = 0.02, and w/e = 2 at Re = 40,000.

The deionized water flow in the finned microchannel was studied using a combination of a microparticle image velocimetry system and numerical techniques, and three types of micro-pin fins (circular, elliptical, and diamond shaped) were compared at Re in the range of 100–400 [16]. The results showed that the microchannel with circular fins gave the best heat transfer coefficient, which was approximately 1.23–1.43 times that of diamond fins and 1.37–1.72 times that of elliptical fins.

The air flow in rectangular channels with different fin arrangements was numerically studied at Re in the range of 20,000–160,000 [17]. The fins consisted of large fins (h = 5 mm) and small fins (h = 2.5 mm). Small fins were added into the channel with large fins with a pitch-to-height ratio of 20 in different forms and compared the Nu and f in each case. The Nu of the channel in Case B (small ribs added in the middle of the large ribs) reached approximately 1.34 times the Nu of a smooth channel. The overall range of the Nu was about 1.14–1.34, and the range of the PEC was approximately 0.92–1.02.

The helium (laminar) or carbon dioxide (turbulent) flows in four types (zigzag shaped, straight shaped, S shaped, and airfoil shaped) of printed circuit heat exchangers (PCHEs) with different fins were compared with a numerical simulation for Re ranging from 0–150,000 [18], and a series of correlations were established to predict the Nu and heat transfer coefficient for each PCHE. The heat transfer coefficient of the airfoil PCHE achieved the best heat transfer performance with approximately 2.4–4.4 times that of the straight PCHE in laminar flow, and the PCHE with airfoil or zigzag fins had approximately 1.6–1.8 times better heat transfer coefficient than the straight PCHE in turbulent flow with a similar thermal performance.

It is evident that the addition of fins can enhance the heat transfer performance with an increase in the pressure drop. For the different combinations of fins and heat exchangers in the above-reviewed literature, the Nu of heat exchangers with fins are typically one to three times higher, and can be up to eight times higher than that of heat exchangers without fins. It is difficult to obtain the best shape and parameters of fins used in various application conditions. For example, in rib-roughened channels (250 × 10 × 1,000 mm) [15], the highest thermal performance was exhibited by triangular ribs with p/e = 6, e/d = 0.05, w/e = 2, and Re = 60,000, where the Nu was 2.6 times higher than that of channels without fins, and for the helically finned tubes ($\varnothing$ = 15.6 mm) [12], the highest thermal performance was exhibited by ribs with p/e = 2.577, e/d = 0.0244, and α = 45°, where the j-factor was approximately 2.2 times higher than that of smooth tubes.

Usually, several experimental or numerical studies are needed to determine the best parameters and fin shape of the heat exchangers design. Therefore, it is necessary to study the mechanism of heat transfer enhancement and provide a guideline for design work, which can significantly reduce the number of experiments required. The mechanism of heat transfer enhancement in some representative articles is summarized, and a brief explanation is provided below.

Most studies have contributed heat transfer enhancement to the broken boundary layers. In general, the turbulence generated by a vortex disturbs the development of laminar flow near the boundary layer, thus strengthening the heat conduction performance, which has a negative effect on the overall performance as the main heat transfer occurs at the boundary. The main purpose of vortex generators is to hinder the development of the laminar boundary layer, as shown in Figure 3.2.

It should be noted that the rib height should not exceed the boundary layer thickness because the distribution of turbulence is nearly useless for heat transfer improvement but increases the flow resistance significantly. From the expressions of the boundary layer thickness equations for a flat plate ($\delta/x = 5.0/Re^{0.5}$ for laminar, and $\delta/x = 0.385/Re^{0.2}$ for turbulent) [19], it can be found that the boundary thickness decreases with the increase in the Re. Therefore, one obtains guidance for the vortex generator parameter design, where a higher Re corresponds to a smaller rib height, and a smaller Re corresponds to a larger rib height.

Supported by these theories, many studies have further analyzed the effects and phenomena of the disturbed boundary layer caused by different fins on the model, in addition to studying the overall heat transfer and flow performance, in order to explore the mechanism of heat transfer enhancement by fins. Some important conclusions are as follows.

Some simple fins develop only in one direction (a 2D section can be approximately used to describe the overall flow situation). A simplified study can reveal some basic theories of the flow and heat transfer under the action of fins. A certain explanation of the relationship between vortex generation (destruction of boundary layer) caused by fins of different pitches and the change in heat transfer performance in a 2D mathematical model was provided in a previous study [15]. Figure 3.3 shows the changes in the shape and size of the vortices under different geometric conditions. When the fin pitch

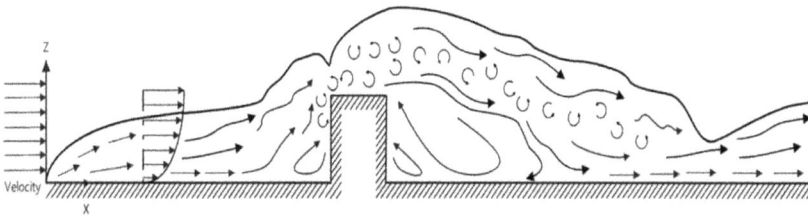

Figure 3.2 Enhanced heat transfer mechanism of ribs based on boundary layer theory.

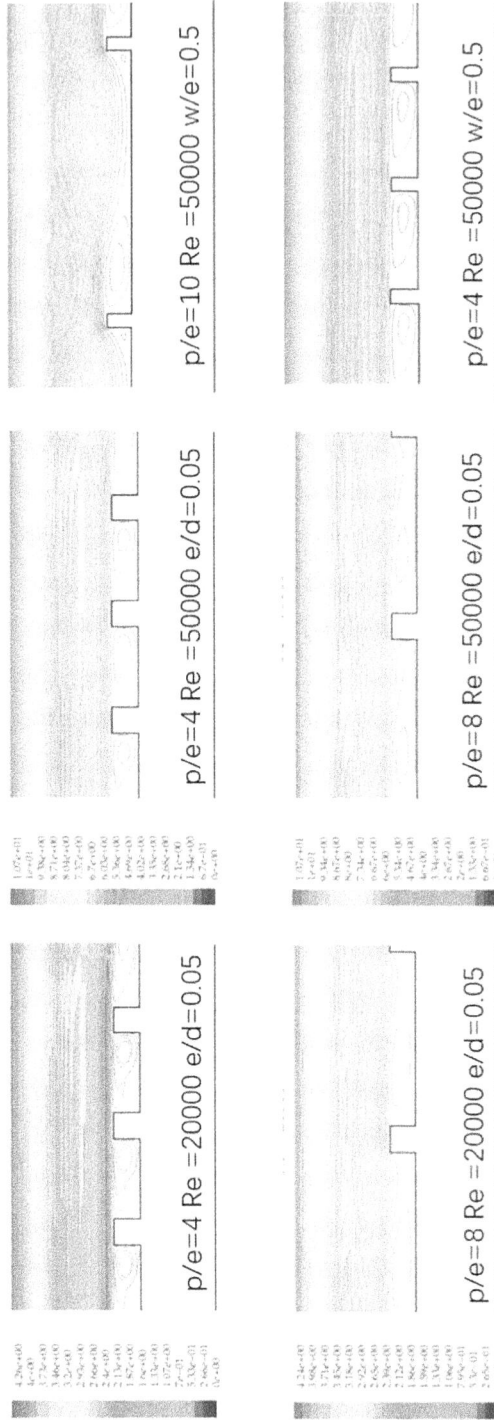

Figure 3.3 Part of stream function contours from Manca's study [15].

is too low, a d-type vortex is generated between the fins and is trapped by the next fin, resulting in a large increase in pressure drop and a small enhancement in heat transfer. An ideal flow state, or k-type behavior, at a proper pitch is achieved when a reattachment occurs as the flow distance increases (at about several times the height behind the rib) after the vortex generation, as shown in the top right in Figure 3.3. In this case, the Nu and friction factor are expected to be at their maximum values. Under stable flow conditions, the phenomena above are similarly repeated around each fin.

It is easy to imagine that the flow characteristics (mainly the separation vortex and reattachment phenomenon) will also change correspondingly with the fin shape, due to the change on the windward and leeward sides. Moon et al. explored the performance of 16 types of rib-roughened rectangular cooling channels with ribs at the bottom [14]. Parts of the streamlines on the streamwise-normal planes are shown below in Figure 3.4. It can be observed that in the cases having a 45° inclined front surface in common, an obviously large separation vortex produced downstream of the rib and caused relatively low heat transfer performance. It was pointed out that the slope of the front surface mainly affects the heat transfer performance, whereas the friction performance cannot be generalized.

The fin developed in the 3D direction can have a bigger influence on the overall flow in the heat exchanger, thereby playing a different role in enhancing the heat transfer. For example, helical fins cause a certain spiral flow, so the vortex occurrence and reattachment of the fluid do not necessarily occur in the plane of the main flow direction after the fins.

Research on helically micro-finned tubes provides a conclusion similar to the above results [20]. The contour of turbulent kinetic energy (TKE) of the tubes was explored, as shown in Figure 3.5. The authors pointed out that the augment of the heat transfer performance was due to the combination of TKE and separated/attached flow phenomena. In Figure 3.5, one can

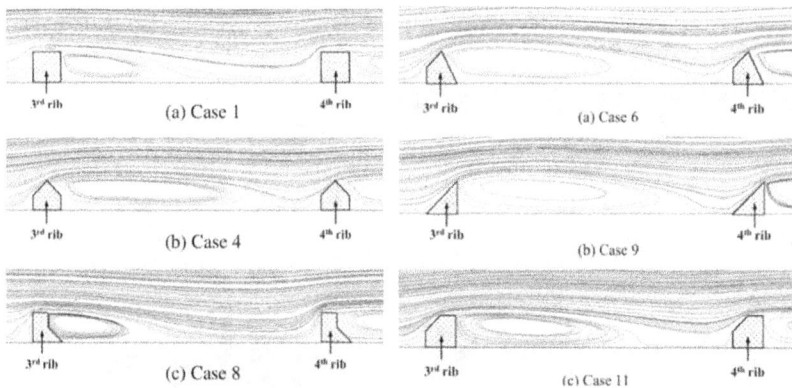

Figure 3.4 Part of stream function contours from Moon's study [14].

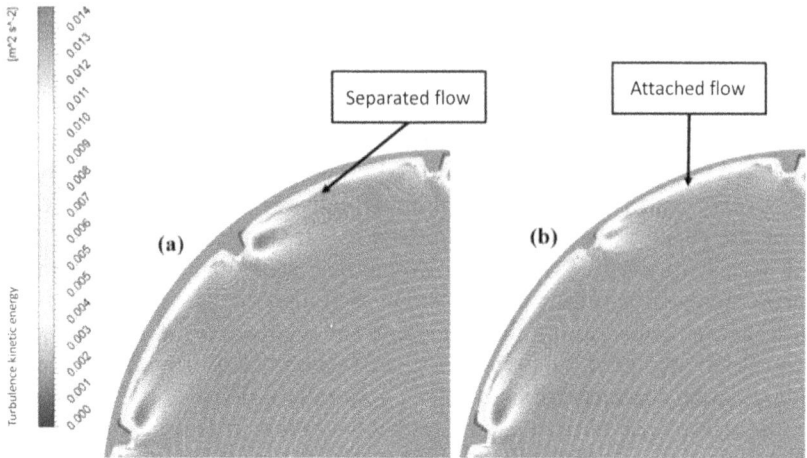

Figure 3.5 TKE contours for different fin height arrangements [20].

observe the distribution of TKE and the separated flow brought up by the 3D fins on the cross-section perpendicular to the main flow direction.

In addition, for fins developed in the 3D direction, a separate vortex (similar to the flow around the cylinder) can be generated on the plane parallel to the wall under certain conditions [16]. Qin studied the relationship between the heat transfer characteristics and vorticity distribution in a 0.5 mm high channel with micro-fin arrays of three different fins. When the Re was larger than a critical value, a vortex was generated by the fins (Figure 3.6). Circular fins had the lowest critical Re, followed by diamond and elliptical fins. Correspondingly, owing to the enhanced mixing of fluids, the heat transfer coefficient of the channel with circular fins was the highest, whereas the smaller vortex and stagnant zone at the end of the elliptical fins led to a decrease in the heat transfer performance (compared with the former).

Similarly, combinations of simple fins can result in complex effects. Zheng explored the influence of V-shape and P-shape arrangements of fins in heat

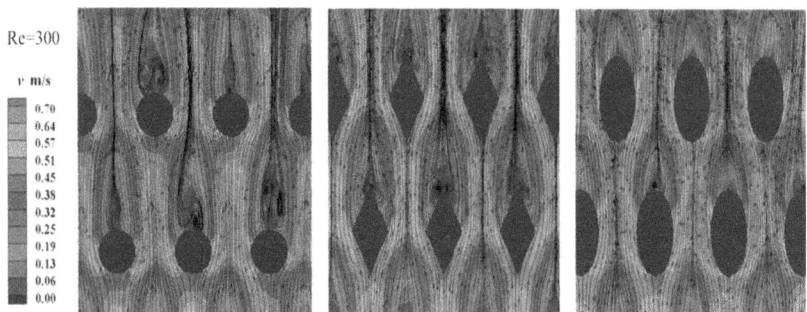

Figure 3.6 Part of streamline diagram for different shaped micro-pin fin heat sinks [16].

exchanger tubes [21]. Two vortices were induced as the fluid flowed over a rib, and the difference in the location where the two vortices interacted with each other and merged into a new one caused a macroscopic flow difference in the tubes. It was showed that a strong and single longitudinal swirl vortex was produced in the P-type ribbed tube, while three pairs of counter-rotating vortices and longitudinal swirl vortex were produced and coupled inside the V-type tube. It was pointed out that the longitudinal swirl vortex with multiple vortices induced in the V-type ribbed tube resulted in a longer flow path and obviously more severely turbulent mixing between the boundary and the core flow regions compared with the longitudinal vortex, thereby enhancing the heat transfer performance with a higher friction factor.

As the earliest method of surface modification, fins have a variety of types, processing methods, and applications, and new types are still emerging. As a method with high reliability and more research data, the addition of fins is suitable for a variety of structures and engineering applications. With the development of cutting-edge technologies in aerospace, shipbuilding, nuclear energy, and other fields, it will obtain a wider usage and have a more targeted application design.

3.3 CORRUGATED SURFACE

A corrugated surface can be easily added to a plate or tube heat exchanger as a type of vortex generator and is widely used in many industrial applications, such as the food industry, refrigeration engineering, and petrochemical industry. Compared with finned surfaces, some of the unique benefits of corrugated surfaces have led to increased research activities in recent decades. First, corrugated walls are easier to produce via extrusion, twisting, module bulging, etc. Second, corrugated walls create less resistance to the flow of fluid through them. In addition, corrugated walls can disturb the fluid on both sides and enhance the heat transfer. Many studies have been conducted on the influence of the arrangement, shape, and size of the corrugations on the enhanced heat transfer effect, and some of the studies on corrugated surfaces are briefly described below and summarized in Table 3.2.

The nitrogen (shell side) and helium (tube side) flows in outward-convex corrugated tubes were numerically studied using a 2D axisymmetric model [22]. Numerical simulation results for heat transfer and flow performance in two different tubes with symmetrical and asymmetrical corrugations were analyzed. The results showed the superiority of asymmetrical corrugations, which exhibited 3–4% less ΔP than symmetrical corrugations in the shell and tube sides with a similar (sometimes larger) Nu.

The oil or deionized water flow in alternating elliptical axis tubes was experimentally studied at Re in the range of 500–50,000 [23]. The surface variation of these tubes caused a multi-longitudinal vortex flow and significantly enhanced the heat transfer performance with a slight increase in flow

Table 3.2 Summary of corrugated surface enhanced heat transfer studies [22–29]

Configurations	Method	Working conditions	Structure parameters	Performances
	Numerical [22]	Shell sides: nitrogen flow; Re = 70,733, Tube sides: helium flow; Re = 16,9201	Symmetrical and asymmetrical corrugations: pitch = 20–60 mm; rs = 2–12 mm; rl = 5–30 mm; H = 1–3 mm	$Nu/Nu_s \approx 1.10$–1.33 (tube side); $\Delta P/\Delta P_s \approx 1.2$–$1.8$ (tube side); $Nu/Nu_s \approx 1.10$–1.35 (shell side); $\Delta P/\Delta P_s \approx 1.0$–$3.3$ (shell side)
	Experimental [23]	Oil and deionized water flow; Re = 500–50,000	Alternating elliptical axis tubes: ratio of the length of each alternate segment to the outer tube diameter: 2–3	$Nu/Nu_s = 2.0$–6.0 (500 < Re < 2,300); $f/f_s = 2.0$–4.5 (500 < Re < 2,300); $Nu/Nu_s = 1.35$–1.6 (10^4 < Re < 5×10^4); $f/f_s = 2.5$–3.0 (10^4 < Re < 5×10^4)
	Numerical [24]	Helium flow; Re = 3,800–43,800	Transverse and helical corrugations: pitch: 10–30 mm; height: 1–3 mm	$Nu/Nu_s \approx 1.05$–1.77; $f/f_s \approx 1.1$–4.0; PEC ≈ 0.9–1.4

Transverse corrugated tube

Helical corrugated tube

(Continued)

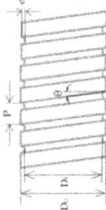

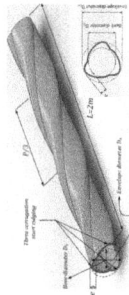

Experimental [25]	Ethylene glycol flow; Re = 90–800	Helical corrugations: pitch: 16–64 mm; depth: 1.5 mm	Nu: transverse corrugations ≈helical corrugations (in turbulent flow)
Experimental [26]	Water flow; Re = 5,500–60,000	Corrugations: pitch: 4.5–5.5 mm; height: 0.5–1.5 mm; helix angle: 15°	$Nu/Nu_s \approx 1.2$–3.05; $f/f_s \approx 1.46$–2.15; PEC ≈ 1.05–2.33
Numerical and Experimental [27]	Water flow; Re = 100–1,300	Corrugations: pitch: 10–50 mm; height: 2 mm	$Nu/Nu_s \approx 2.4$–3.7; $f/f_s \approx 1.7$–2.4; PEC: maximum 3.7

(Continued)

Inside-tube view

Outside-tube view

Table 3.2 (*Continued*) Summary of corrugated surface enhanced heat transfer studies [22–29]

Configurations	Method	Working conditions	Structure parameters	Performances
	Experimental [28]	Air flow; Re = 7,200	Four corrugation arrangements: conventional sinusoidal corrugation, primary corrugation with anti-phase/full-wave secondary corrugation	Heat transfer performance: almost no change; Flow performance: reduce pressure drop by 15% (anti-phase/full-wave secondary corrugation)
	Numerical [29]	Helium flow; Re = 7,200–36,000	Five corrugated tube arrangements: smooth tubes, transverse corrugated tubes (TCT), and helical corrugated tubes (HCT); arrangements: identical arrangement (ITA) and staggered arrangement (STA)	$Nu/Nu_s \approx 1.5$–2.5; STA-HCTs achieve the highest Nu; $f/f_s \approx 4.4$–6.2; ITA-HCTs achieve the lowest f; $PEC \approx 0.85$–1.4; STA-HCTs achieve the highest PEC

resistance. The enhancement of heat transfer was in the range of 35–60%, and the increase in flow resistance was in the range of 150–200% for Re ranging from 10,000 to 50,000, compared with a smooth tube.

The helium flow in outward corrugated tubes was numerically studied at Re values of 3,800–43,800 [24]. The results showed that helical corrugated tubes had similar heat transfer performance and lower resistance than transverse corrugated tubes of the same size ($pl/D = 1$, $hl/D = 0.1$). In addition, the maximum Nu of the helical corrugated tubes was 1.77 times that of the smooth tubes, with $hl/D = 0.1$, $pl/D = 0.5$, and Re = 6,260.

The ethylene glycol flow in a series of corrugated tubes with different parameters was experimentally studied at Reynolds numbers in the range of 90–800 [25]. The tubes with helical and transverse corrugations were compared, and the results showed that the tubes with transverse corrugation had a similar heat transfer performance to tubes with helical corrugation in turbulent flow.

The water flow in helical corrugated tubes was experimentally studied at Re in the range of 5,500–60,000 [26], and corrugations with different pitches and rib heights were compared. The results showed that the enhancement of the heat transfer was in the range of 1.2–3.05, and the increase in friction factor was in the range of 1.46–2.15, compared to smooth tubes. In addition, a maximum PEC of 2.33 was obtained for the enhanced tube with a pitch ratio of $P/D_h = 0.27$ and a rib height ratio of $e/D_h = 0.06$ at a low Re.

The water flow in a three-start spirally corrugated tube was studied using both experimental and numerical techniques at low Re in the range of 100–1,300 [27]. The enhancement of the heat transfer was in the range of 2.4–3.7, and the increase in the friction factor was in the range of 1.7–2.4. The maximum PEC reached 3.7.

The air flow in various cross-corrugated matrix unit cells was numerically studied [28], and different types of secondary corrugations added to primary corrugations were compared. The results showed that the anti-phase secondary corrugation and full-wave rectified trough corrugation addition produced a −15% pressure drop with unvaried heat transfer effect, compared to the cross-corrugated primary surface heat exchangers with only primary corrugations.

The helium flow in a multi-tube heat exchanger (MTHX) with transverse corrugated tubes and helical corrugated tubes was numerically studied at Re values in the range of 7,200–36,000 [29]. The results showed that among the four arrangements of tubes in the research, the MTHX with a staggered tube arrangement of helical corrugated tubes exhibited the best comprehensive performance compared to the other cases. The heat transfer index and friction factor of MTHX with CTs increase 1.61–2.46 and 5.86–5.25 times than those of MTHX with smooth tubes, respectively.

As mentioned above, the additional corrugated walls generally provide heat transfer enhancement with a Nu number that is one to two times higher than that of smooth walls; however, the impact of corrugated walls varies depending on where they are applied, such as the application of additional fins. Research on the mechanism of flow and heat transfer in various corrugations can help to design corrugations in heat exchangers, and some explanations for flow conditions and heat transfer enhancement of corrugated surfaces in some representative articles are summarized. A brief explanation is given below.

In Wang's research, the local heat transfer index was described by the ratio of the local Nu numbers, and the flow pattern was described by the velocity vectors of different tubes. Some of the results are shown below to describe the effect of corrugated walls on heat transfer enhancement [24]. Due to the sudden expansion and convergence of the corrugation walls, an adverse pressure gradient region was generated at each the front of corrugation. The beginning and ending points of the adverse pressure gradient region were labeled AP and AP', respectively. With the increase of the adverse pressure gradient, a secondary flow was produced. The separation point and reattachment point are labeled with SP and RP, respectively.

The corrugated walls did not provide heat transfer at every location of the tube. The adverse pressure gradient of the fluid entering the cross-sectional area resulted in a region of velocity decay (after AP in Figure 3.7), where the velocity was zero at the SP position, and this phenomenon led to a sudden

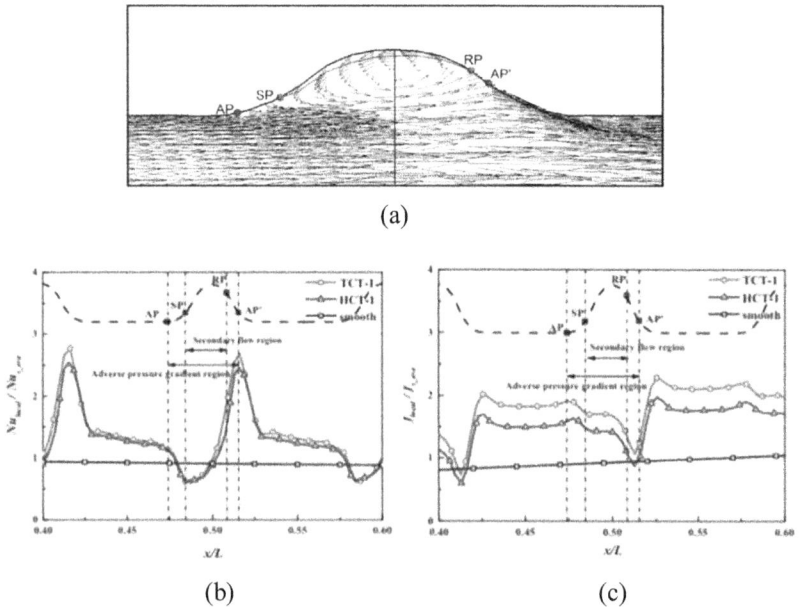

(a)

(b)

(c)

Figure 3.7 (a) Velocity vectors at longitudinal section; (b) local Nu for transverse and helical corrugated tubes; (c) local f for transverse and helical corrugated tubes [24].

decrease in heat transfer performance. At this point, the heat transfer performance was even lower than that of the smooth tube.

The area where the corrugated wall intensified the overall heat transfer was located beyond the middle of the corrugation. Because of the impact of the fluid on the windward side of the corrugations and the augmentation of the TKE, the heat transfer performance increased sharply. In addition, the authors pointed out that the rotational flow had little effect on the thermal performance. However, it inhibited the secondary flow and turbulent pulsation, thereby reducing the flow resistance.

The symmetry of corrugations also affected the flow field and heat transfer. The asymmetrical corrugations with different rl values (trough radius of leeward side) in Figure 3.8 (a) were compared by Han [22]. The results showed that the separation point was gradually ahead, and the vortex shedding was obviously suppressed with the increase in rl because the fluid flow was smoother along with the upstream of the corrugations, leading to a longer reversed boundary layer of the fluid in the tube. The TKE of the internal fluids had a distribution change (the TKE concentrated at the separation point decreased and increased at the attachment point), while the TKE of fluids outside the tube increased first and then decreased slightly. Overall, the increase in pressure and heat transfer slightly decreased with an increase in rl, and there was a significant peak value instead of a monotonic decrease in some cases with a proper pitch. The overall heat transfer performance of the corrugated tubes (including the shell and tube sides) was better than that of the smooth tubes.

As mentioned above, on the other side of the corrugated walls (convex side), the corrugations also produced a certain disturbance in the fluid. The fluid first reached the windward side of the corrugations and produced a strong impact on the wall, and then the accelerated fluid entered the leeward side to form a separation flow, producing an effect like that of additional fins to strengthen the heat transfer.

More compact corrugated walls of the heat exchangers also affect each other and lead to a complex effect on the performance (e.g., the interaction of the corrugated tubes in an MTHX, as shown in Figure 3.9 [29]). The results showed that the spiral flows of the ITA-HCTs clearly strengthened each other, whereas the spiral flows of the STA-HCTs weakened each other. The increased spiral flow inhibited the increase in heat transfer and friction factor and led to a better flow performance in ITA-HCTs compared to STA-HCT but a better heat transfer performance in STA-HCTs compared to ITA-HCTs. Meanwhile, the intensity of the backflow in the streamwise-normal planes for the HCTs was lower than that for the TCTs. The heat transfer performance of the staggered arrangement of the HCT was stronger than that of the two different arrangements of the TCT in that research. In general, as an important factor for heat transfer, the different tube arrangements can result in the spiral vortices with different directions, which affects the heat transfer and flow resistance, fundamentally.

(a)

(b)   (c)

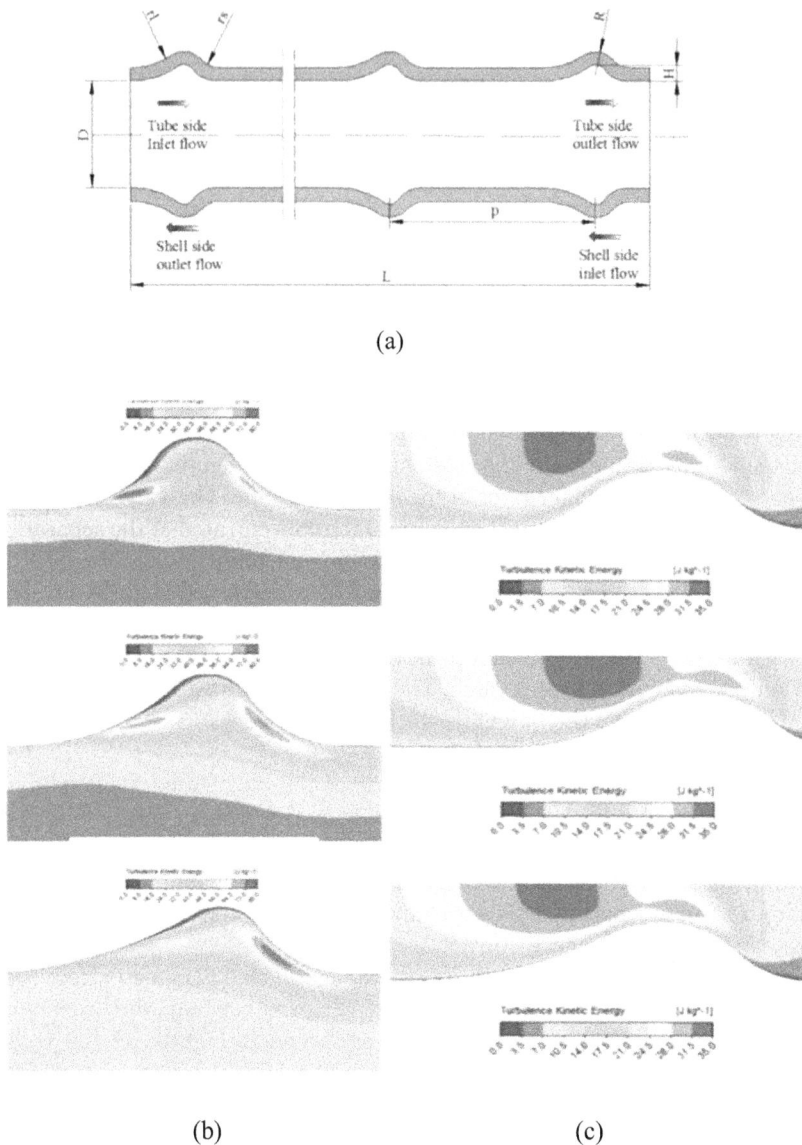

Figure 3.8 (a) Schematic view of asymmetrical corrugated tube; (b) TKE distribution for various *rl* at tube side; (c) TKE distribution for various *rl*
values at shell side [22].

Corrugated walls can be particularly useful in many situations because of
the complex effects of such walls on both sides of the fluid compared with
fins. As a surface modification method, corrugated walls are similar to finned
walls, but they can produce more varied effects in a simple form. The study

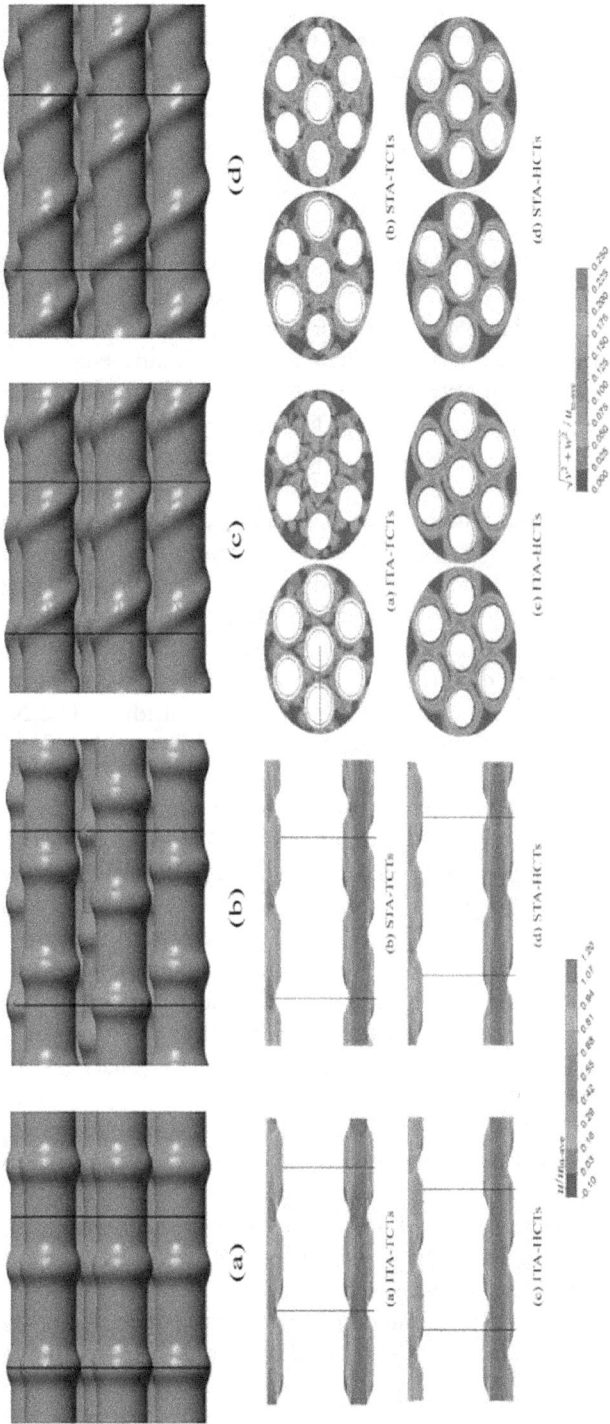

Figure 3.9 Flow features for the four types of helically corrugated tube arrangements [29]. (a) ITA-TCTs; (b) STA-TCTs; (c) ITA-HCTs; and (d) STA-HCTs.

of Doo showed that by adding some small corrugations on a wall, can further increase the heat transfer performance or decrease the flow resistance [28]. With further research on corrugated walls, this type of surface modification method can be more widely used.

3.4 COILED SURFACE

The coiled surface heat transfer technique is mainly directed toward helically coiled tube heat elements and heat exchangers. Helically coiled heat exchangers possess large heat transfer areas in a fixed volume with high heat transfer coefficients and are widely used in many engineering applications, such as food processing, nuclear reactors, heat recovery, chemical processing, refrigeration and air conditioning systems, etc. The fluid produces a complex flow effect due to the buoyancy and centrifugal forces in the tubes, which enhances the heat transfer efficiency. Like other enhanced heat transfer techniques, many types of helically coiled tube heat exchangers have been developed, and their characteristics and mechanisms have been investigated by both experimental and numerical methods. Representative studies are summarized as follows and in Table 3.3.

Engine oil was cooled in coiled tube heat exchangers by water [30], and the performance of three types of coiled tubes in different working conditions were experimentally and numerically studied. The Nu ranged from 16 to 50 with De = 35–200. A correlation was obtained, with an error of 0.79%.

A numerical study was performed on supercritical CO_2 in a helically coiled tube with various inclination angles (−90° to +90°) [31]. The effects of buoyancy and centrifugal forces on the enhanced heat transfer mechanism were discussed in detail. With a decrease in the inclination angles, the nonuniformity of the heat transfer index was improved, and the two centrifugal forces enhanced each other in the bottom half of the helically coiled tube but were inhibited in the top half. In the inclined helically coiled tube, the heat transfer index oscillated along the fluid flow owing to the buoyancy effect. However, no oscillation in the heat transfer coefficient was observed in the vertical helically coiled tube.

A coiled tube heat exchanger with a helical wire was experimentally studied with both water and air as internal fluids [32]. The results showed that the overall heat transfer coefficients for water flow, with and without a helical wire, were in the range of 470–1,660, and 430–1,260, respectively. The pressure drop for the coiled tube with a helical wire was nearly five times higher than that without a wire.

Helically coiled tubes interacting with spiral corrugations have been numerically studied [33]. The results showed that the Nu for a helically

Table 3.3 Summary of coiled tube enhanced heat transfer studies [30-35]

Configurations	Method	Working conditions	Structure parameters	Performances
	Experimental and numerical [30]	Engine oil inside ($T_{in} = 45$–$70°C$, $m = 0.013$–0.122 kg/s); Coolant water outside ($T_{in} = 11.9$–$16.2°C$, m $= 0.058$–0.062 kg/s)	Coiled tube inside diameter = 9 and 12 mm; coil pitch = 17.0, 21.4, and 26.7 mm; curvature ratio = 0.113 and 0.157	Nu = 16–50 with De = 35–200; A correlation was developed: $Nu_t = 0.554De^{0.496}\gamma^{-0.388}$ $Pr^{0.151}\Phi^{0.153}$
	Numerical [31]	Supercritical CO_2 ($T_{in} = 287.86$ K, $P_{in} = 8$ MPa, $m = 262$ kg/m²/s)	Inclination angle range of −90° to 90°; effect of buoyancy force and centrifugal force	The nonuniformity of h is enhanced with the decrease of the inclination angle; The inclined angle causes the h to oscillate along the flow direction.
	Experimental [32]	Both water (16°C, 1–3 L/min) and air (18°C, 1–5 L/min) for tube side; hot water for shell side (40°C, 1–5 L/min)	With and without helical wire	$U_w/U_{nw} = 1.09$–1.32; $\Delta P_{nf}/\Delta P_{bf} \approx 5$

(Continued)

Table 3.3 (Continued) Summary of coiled tube enhanced heat transfer studies [30-35]

Configurations	Method	Working conditions	Structure parameters	Performances
	Numerical [33]	Water flow (T_{in} = 288°C, Re ≈ 9,800–26,300)	Helical corrugation pitch h = 18.95, 7.59, 5.41 mm	Nu/Nu_s = 50–80%; fRe/fRe_s = 50–300%
	Numerical [34]	Water flow (Re = 15,000–40,000, T_{in} = 293.15 K, T_{wall} = 353.15 K)	Screw pitch (S = 31–130 mm); corrugation height (H = 1–7 mm); corrugation pitch (P = 35–57.5 mm)	Nu/Nu_s = 1.05–1.7; f/f_s = 1.01–1.28; maximum PEC = 1.56

(Continued)

| Numerical [35] | Nanofluids (0.025%, 0.05%vol); Water | Without helical ribs, co-current helical ribs, counter-current helical ribs | The Nu increases 12.45% with the helical ribs; The Nu increases 8.21% with the 0.05 vol% nanoparticles |

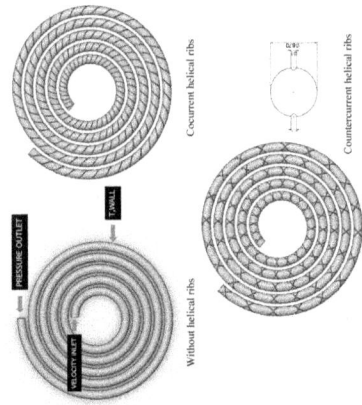

Without helical ribs

Cocurrent helical ribs

Countercurrent helical ribs

corrugated coiled tube was 50–80% higher than that of a smooth helically coiled tube, and the flow resistance was 50–300% higher.

A helically coiled tube with spherical corrugation was numerically studied [34], and configurations with different screw pitches, corrugation heights, and corrugation pitches were compared with those of a smooth helically coiled tube. The enhancement of heat transfer was in the range of 1.05–1.7, and the increase of friction factor was in the range of 1.01–1.28, with the maximum PEC reaching 1.56.

Nanofluids in spirally coiled tubes with and without helical ribs have been numerically studied [35]. The Nu for spirally coiled tubes with helical ribs was 12.45% higher than that without helical ribs. Adding 0.05 vol% nanoparticles enhanced the Nu by 8.21%.

The literature reviewed above has shown that employing different velocity inlets, curvature ratios, inclination angles, pitch coil diameters, corrugation surfaces, multi-tubes, insert wires, and nanofluids can result in different degrees of enhanced heat transfer performance. However, different evaluation criteria have been employed under different working conditions, and these are difficult to compare simultaneously. A comprehensive comparison of three types of spiral coiled tubes with straight tubes has also been performed [36] with different Re values and boundary conditions, and the numerical results are shown in Tables 3.4 and 3.5. The results indicated that the heat transfer efficiency and pressure drop for the three types of spiral coiled tubes were obviously higher than those of the straight tube, with a maximum ratio of 3.94 and 2.86, respectively. While the average Nu values for the three types of spiral coiled tubes were very close to each other, the

Table 3.4 Average Nu for various configurations

		Average Nusselt number (Nu_{avg})			
Re	Geometry	$T_{wall} =$ 323.15 K	$T_{wall} =$ 373.15 K	$T_{wall} =$ 473.15 K	$T_{wall} = 573.15$ K
100	Straight	3.32	3.76	4.50	5.26
	Conical spiral	4.91	4.90	5.54	6.11
	Helical spiral	4.54	4.82	5.44	5.99
	In-plane spiral	4.58	4.86	5.49	6.11
500	Straight	3.43	3.99	4.76	5.38
	Conical spiral	9.37	9.94	10.71	1.1.72
	Helical spiral	9.33	9.68	10.67	1.1.70
	In-plane spiral	9.66	9.79	10.80	1.1.71
1,000	Straight	3.64	4.26	5.03	5.61
	Conical spiral	14.32	14.96	16.32	17.46
	Helical spiral	14.26	14.86	16.27	17.45
	In-plane spiral	14.35	15.02	16.40	17.55

Table 3.5 Pressure drop for various configurations

		Pressure drop (Pa)			
Re	Geometry	$T_{wall} =$ 323.15 K	$T_{wall} =$ 373.15 K	$T_{wall} =$ 473.15 K	$T_{wall} = 573.15$ K
100	Straight	1.12	1.38	1.94	2.40
	Conical spiral	1.49	1.81	2.56	3.22
	Helical spiral	1.63	1.96	2.69	3.29
	In-plane spiral	1.47	1.79	2.54	3.18
500	Straight	6.56	7.68	11.18	14.32
	Conical spiral	12.16	14.73	20.87	26.44
	Helical spiral	13.92	16.49	23.02	28.66
	In-plane spiral	12.30	14.52	20.60	26.10
1,000	Straight	13.35	16.90	24.67	31.99
	Conical spiral	33.42	40.46	56.90	71.85
	Helical spiral	38.24	46.26	64.57	80.79
	In-plane spiral	33.71	40.38	56.40	71.25

pressure drop for the helical spiral coiled tube was obviously higher than that of the other two types.

Helically coiled corrugated tubes with different corrugation heights and pitches were compared with helically coiled smooth tube [34], and the results are shown in Figures 3.10 and 3.11. The augmentation factor of the Nu for different corrugation heights was approximately 1.05–1.7 as compared to the helically coiled smooth tube, corresponding to a friction factor of 1.01–1.24, and the PEC index was in the range of 1.07–1.54. For different corrugation pitches, the Nu and f increased by a factor in the range of 1.37–1.66 and 1.18–1.28, respectively, and the PEC range was 1.28–1.56, respectively. The PEC for a helical corrugated coiled tube with a corrugation height of 5 mm and pitch of 35 mm was higher than that of the other cases.

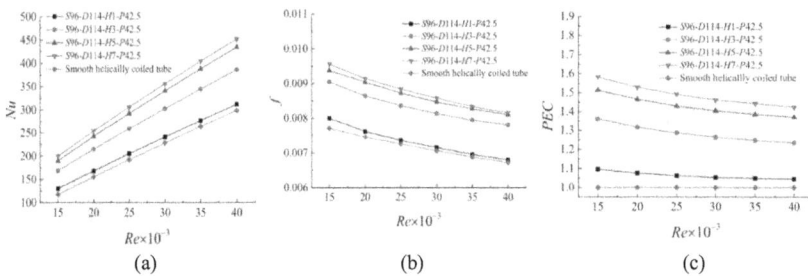

Figure 3.10 Various values of Nu, f, and PEC with Re for different corrugation heights [34]. (a) The variation of Nu; (b) The variation of f; and (c) The variation of PEC.

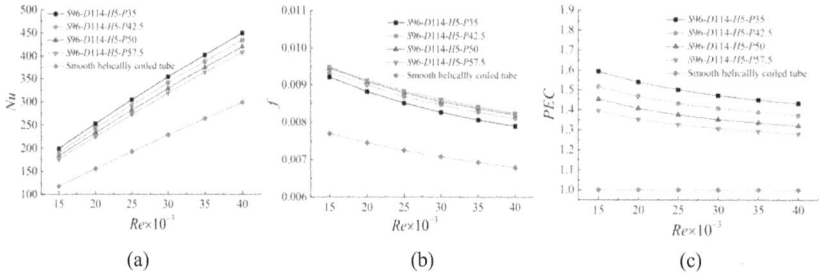

(a)   (b)   (c)

Figure 3.11 Various values of Nu, *f*, and PEC with Re for different corrugation pitches [34]. (a) The variation of Nu; (b) The variation of *f*; and (c) The variation of PEC.

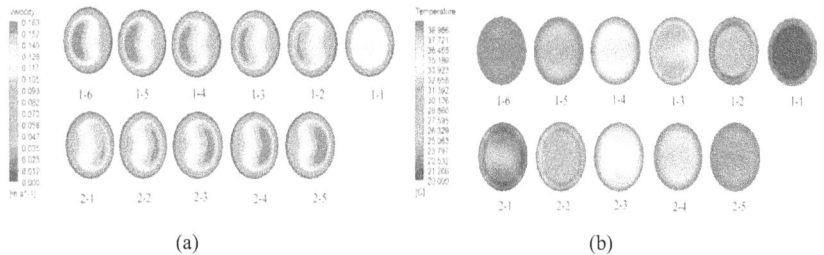

(a)   (b)

Figure 3.12 Velocity and temperature contours of nanofluids in helical coiled tube [35]. (a) Velocity distribution; and (b) Temperature distribution.

The enhanced heat transfer mechanism of the helically coiled tubes was mainly contributed by the centrifugal force, which significantly affected the velocity and temperature contours, as shown in Figure 3.12 [35]. The maximum flow velocity at different sections was more adjacent to the outward zone, the temperature contour was more adjacent to the outside of the curved tube, and the position of the minimum temperature was forced outward owing to the centrifugal force. Compared with the straight tube, the torsion of the helically coiled tube produced a secondary flow motion, and a nonuniform flow led to higher heat transfer efficiency.

3.5 MODIFIED SURFACE

In traditional macroscale fluid dynamics, the surface roughness effect on forced convection heat transfer is usually ignored when the relative roughness is less than 5% [37]. Therefore, for a long time in the past, the modified surface technique was not concerned with single-phase enhanced heat transfer, but was mostly employed in investigations of evaporation and condensation heat transfer [38].

However, the size of the heat transfer scale is significantly reduced owing to the rapid increase in micro electromechanical systems (MEMS) technology [39].

The hydraulic diameter for a heat transfer channel was defined and widely accepted by Kandlikar and Grande [40] as

1. For D_h > 3,000 μm: conventional channel
2. For 3,000 $\geq D_h$> 200 μm: mini-channel
3. For 200 $\geq D_h$> 10 μm: microchannel

The small absolute surface roughness, which may be ignored in large-scaled devices, has a significant effect on the flow dynamics and heat transfer characteristics in microchannels [41]. In microchannels, an increase in the relative surface roughness resulted in larger pressure drop, earlier transition to turbulence [37], and enhanced heat transfer in laminar and turbulent flows [42].

Therefore, it is necessary to clear the effect of surface characteristics on convective heat transfer. There are many types of micromachining methods that can be used to modify the heat transfer surface, such as electrical discharge machining (EDM), electrochemical machining (ECM), etching, and micro milling, which vary from 10^{-2} μm to 5 μm [43].

The heat transfer characteristics of a microchannel heat sink with different surface roughness values was numerically investigated [44]. A random surface roughness was generated using a prespecified exponential autocorrelation function. Six types of surfaces with different relative roughness (ε) from 0.5% to 3.0% are shown in Figure 3.13. The results showed that the average Nu and pressure drop increased linearly with an increase in ε,

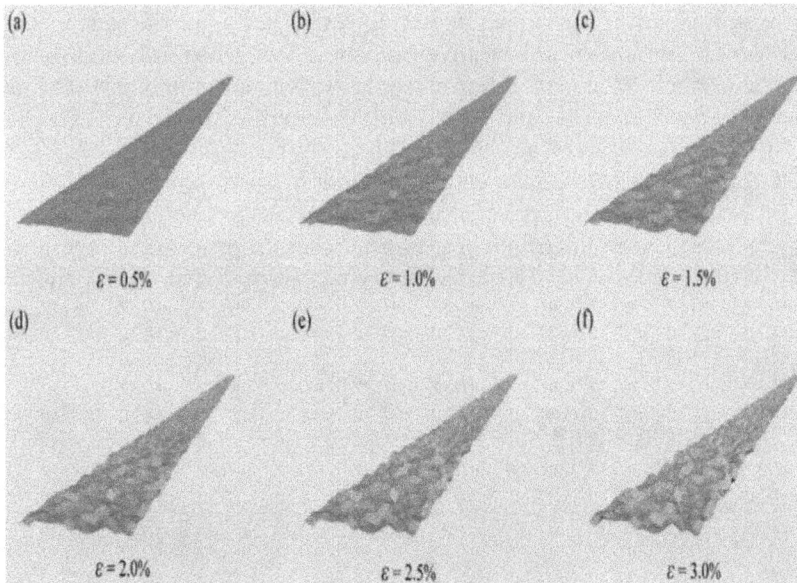

Figure 3.13 Isometric view of six types of surfaces with different relative roughness [44].

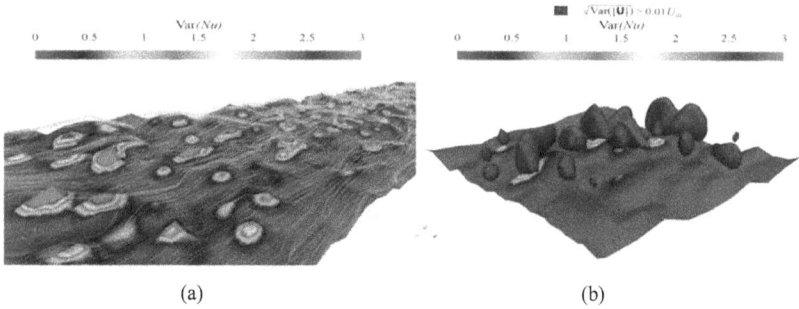

(a)   (b)

Figure 3.14 Local heat transfer and flow features on rough bottom surface [44]. (a) Contours for the variance of local Nu and streamlines; and (b) Contours for the variance of local Nu and standard deviation of velocity.

indicating that the relative roughness had a significant effect on the microchannel flow and heat transfer. The local heat transfer and flow features on the rough bottom surface are shown in Figure 3.14. The variance in the Nu on bottom surface reached high values at the peaks, whereas the variance in the valleys was relatively low. Consequently, the peaks of the rough surface can be considered as the key factor on convective heat transfer improvement. Therefore, the use of the surface modification technique to obtain peaks dominated by the surface wall was suggested.

The effect of surface roughness in three types of microchannels (square, wavy, and dimpled) was numerically studied [43], with the relative roughness ranging from 0 to 2%. From Figure 3.15, the ΔP, Nu, and PF are all increased with the relative roughness. However, the Nu for the square channel was constant when the relative roughness was below 1.2, leading to a decrease in PF. When the relative roughness was approximately 2%, ΔP increased by 3%, 5.3%, and 5.9%, and Nu increased by 10%, 6.5%, and 19.8% for square, wavy, and dimpled channels, respectively.

Rectangular microchannels with three kinds of micro-pins finned spacings (2, 4, and 8 µm) were fabricated using the MEMS technique, as shown in Figure 3.16. The laminar flow heat transfer characteristics were experimentally tested for Re = 70–250 [45]. The results showed that the Po and Nu

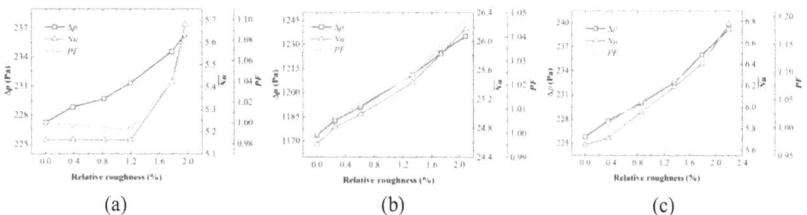

(a)   (b)   (c)

Figure 3.15 Pressure drop, average Nu, and PF versus relative roughness for three types of microchannels [43]. (a) Square channel; (b) Wavy channel; and (c) Dimpled channel.

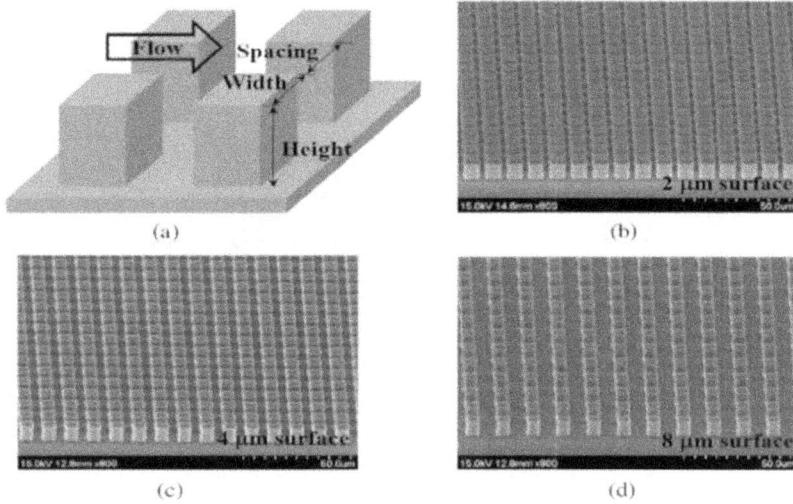

Figure 3.16 Photographs of micro-pin finned surfaces embedded in microchannels with three different spacings [45].

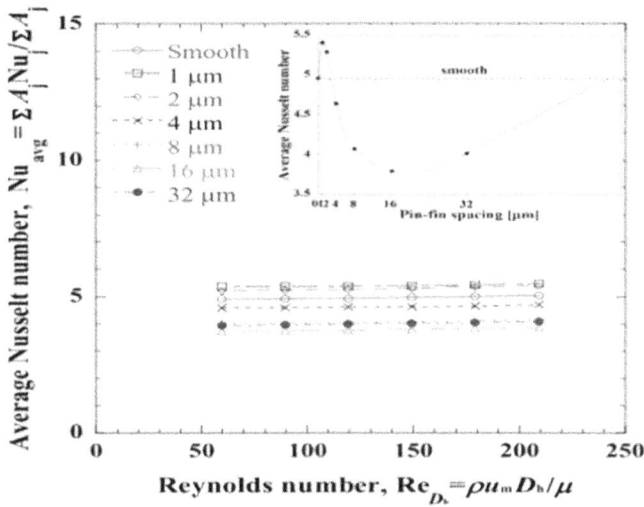

Figure 3.17 Average Nu for micro-pin finned surfaces with different spacings [45].

values decreased with increasing pin fin spacing. The local Nu was dramatically enhanced at the leading and trailing edges of the fins, but it decreased to the conduction level in the fin valleys owing to the weak recirculation flow. The average Nu values for different spacings are shown in Figure 3.17, where it is found that only Nu values for spacings below 4 μm are higher than those for the smooth surface.

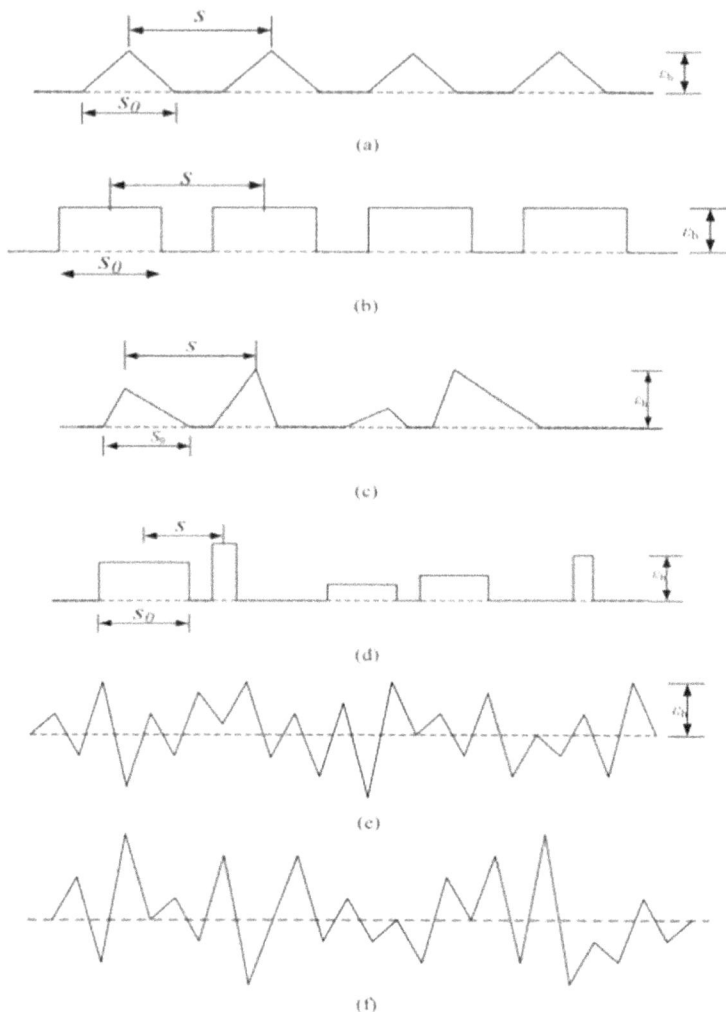

Figure 3.18 Six types of 2D rough surface models [37]. (a) Regular triangle model; (b) Regular rectangle model; (c) Random triangle model; (d) Random rectangle model; (e) 2D Gauss model; and (f) 2D Fractal model.

Six types of rough surface models (regular triangle, regular rectangle, random triangle, random rectangle, Gaussian, and fractal) were numerically compared, as shown in Figure 3.18 [37]. The average Po (Poiseuille number, $f \times$ Re) and Nu values for the six surface models under different relative roughness values are presented in Figure 3.19, and the results show that the heat transfer and friction factor for regular-roughness surface models were obviously higher than those of the other random-roughness surface models.

The surface topology and addition of a coating layer were both found to be effective in enhancing the heat dissipation in nanochannels [46].

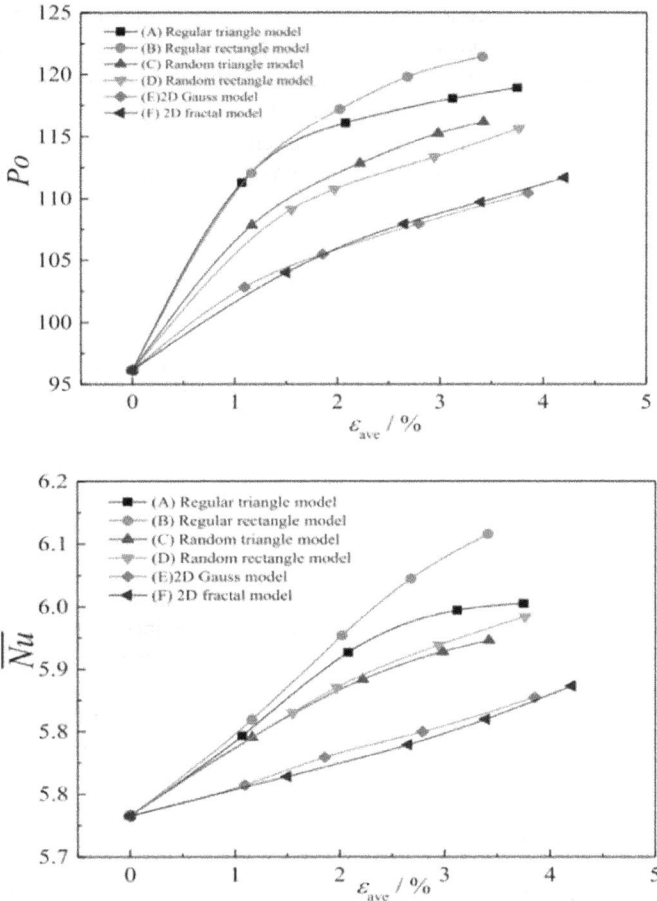

Figure 3.19 Flow and heat transfer performance for different 2D models [37].

3.6 SUMMARY AND OUTLOOK

The above four types of surface modification techniques are all maturely used and widely studied because of their unique manufacturing processes and enhanced heat transfer mechanisms. In this chapter, the various configurations of the four techniques with respect to the working conditions, structure parameters, and performance were reviewed. Enhanced heat transfer mechanisms have also been discussed in several studies. However, with plenty of geometric configurations, structural parameters, working fluids, and conditions, it is difficult to make a comprehensive comparison. The results obtained by different experiments and numerical methods exhibit certain differences. The mechanism analysis mostly focused on a single-heat transfer element and did not consider the interaction between the two strengthened structures.

For future studies, a benchmark work on performance tests for different techniques of surface modification using experimental and numerical methods is necessary. Then, one could obtain a more comprehensive and clear understanding of the different techniques. In addition, the mechanism analysis should be closer to the actual heat exchanger structure and provide a quantitative description of swirl flow, global heat transfer, and flow viscosity characteristics.

REFERENCES

[1] D.H. Nguyen, H.S. Ahn, A comprehensive review on micro/nanoscale surface modification techniques for heat transfer enhancement in heat exchanger, *Int. J. Heat Mass Transf.*, Vol. 178, 121601, 2021.

[2] R.L. Webb, Air-side heat transfer in finned tube heat exchangers, *Experi. Heat Transf.*, Vol. 1, 33–49, 1980.

[3] H. Zhang, X. Li, W. Tang, X. Deng, A.P. Reynolds, M.A. Sutton, Heat transfer modeling of the friction extrusion process, *J. Mater. Proces. Tech.*, Vol. 221, 21–30, 2015.

[4] S. Eiamsa-ard, P. Promthaisong, C. Thianpong, M. Pimsarn, V. Chuwattanakul, Influence of three-start spirally twisted tube combined with triple-channel twisted tape insert on heat transfer enhancement, *Chemi. Engineer. Proces: Proces. Internsifi.*, Vol. 102, 117–129, 2016.

[5] N.S. Qu, X.L. Fang, Y.D. Zhang, D. Zhu, Enhancement of surface roughness in electrochemical machining of Ti_6Al_4V by pulsating electrolyte, *Int. J. Advanc. Manufact. Techno.*, Vol. 69, 2703–2709, 2013.

[6] L. Chen, D. Deng, G. Pi, X. Huang, W. Zhou, Burr formation and surface roughness characteristics in micro-milling of microchannels, *Int. J. Advanc. Manufact. Techno.*, Vol. 111, 1277–1290, 2020.

[7] A. Sadeghianjahromi, C.-C. Wang, Heat transfer enhancement in fin-and-tube heat exchangers–A review on different mechanisms, *Renew. Sustain. Energ. Revie.*, Vol. 137, 110470, 2021.

[8] N. Kurtulmuş, B. Sahin, A review of hydrodynamics and heat transfer through corrugated channels, *Int. Commun. Heat Mass Transf.*, Vol. 108, 104307, 2019.

[9] S. Bhuvaneswari, G. Elatharasan, A study of the literature review on heat transfer in a helically coiled heat exchanger, in: *RTICCT-2019 Conference Proceedings*, Vol. 7, 1–3, 2019.

[10] C. Krishnamoorthy, R.P. Rao, A.J. Ghajar, Single-phase heat transfer in micro-tubes: A critical review, in: *Heat Transfer Summer Conference*, Vol. 42754, 979–989, 2007.

[11] Y. Wang, Q. Zhao, Q. Zhou, Z. Kang, W. Tao, Experimental and numerical studies on actual flue gas condensation heat transfer in a left–right symmetric internally finned tube, *Int. J. Heat Mass Transf.*, Vol. 64, 10–20, 2013.

[12] G.J. Zdaniuk, L.M. Chamra, P.J. Mago, Experimental determination of heat transfer and friction in helically-finned tubes, *Exp. Therm. Fluid Sci.*, Vol. 32, 761–775, 2008.

[13] X.-W. Li, J.-A. Meng, Z.-X. Li, Experimental study of single-phase pressure drop and heat transfer in a micro-fin tube, *Exp. Therm. Fluid Sci.*, Vol. 32, 641–648, 2007.

[14] M.-A. Moon, M.-J. Park, K.-Y. Kim, Evaluation of heat transfer performances of various rib shapes, *Int. J. Heat Mass Transf.*, Vol. 71, 275–284, 2014.

[15] O. Manca, S. Nardini, D. Ricci, Numerical analysis of water forced convection in channels with differently shaped transverse ribs, *J. Appl. Math.*, Vol. 2011, 1–25, 2011.

[16] L. Qin, J. Hua, X. Zhao, Y. Zhu, D. Li, Z. Liu, Micro-PIV and numerical study on influence of vortex on flow and heat transfer performance in micro arrays, *Appl. Therm. Eng.*, Vol. 161, 114186, 2019.

[17] G. Xie, S. Zheng, W. Zhang, B. Sundén, A numerical study of flow structure and heat transfer in a square channel with ribs combined downstream half-size or same-size ribs, *Appl. Therm. Eng.*, Vol. 61, 289–300, 2013.

[18] S.H. Yoon, H.C. No, G.B. Kang, Assessment of straight, zigzag, S-shape, and airfoil PCHEs for intermediate heat exchangers of HTGRs and SFRs, *Nucl. Eng. Des.*, Vol. 270, 334–343, 2014.

[19] Bengt Sundén, *Introduction to heat transfer*, WIT Press, Southampton, Boston, 2012.

[20] M. Ammar Ali, M. Sajid, E. Uddin, N. Bahadur, Z. Ali, Numerical analysis of heat transfer and pressure drop in helically micro-finned tubes, Vol. *Processes*, 9, 754, 2021.

[21] N. Zheng, P. Liu, F. Shan, Z. Liu, W. Liu, Effects of rib arrangements on the flow pattern and heat transfer in an internally ribbed heat exchanger tube, *Int. J. Therm. Sci.*, Vol. 101, 93–105, 2016.

[22] H.Z. Han, B.X. Li, B.Y. Yu, Y.R. He, F.C. Li, Numerical study of flow and heat transfer characteristics in outward convex corrugated tubes, *Int. J. Heat Mass. Transf.*, Vol. 55, 7782–7802, 2012.

[23] J.A. Meng, X.G. Liang, Z.J. Chen, Z.X. Li, Experimental study on convective heat transfer in alternating elliptical axis tubes, *Exp. Therm. Fluid Sci.*, Vol. 29, 457–465, 2005.

[24] W. Wang, Y. Zhang, B. Li, Y. Li, Numerical investigation of tube-side fully developed turbulent flow and heat transfer in outward corrugated tubes, *Int. J. Heat Mass. Transf.*, Vol. 116, 115–126, 2018.

[25] S. Rainieri, G. Pagliarini, Convective heat transfer to temperature dependent property fluids in the entry region of corrugated tubes, *Int. J. Heat Mass. Transf.*, Vol. 45, 4525–4536, 2002.

[26] S. Pethkool, S. Eiamsa-ard, S. Kwankaomeng, P. Promvonge, Turbulent heat transfer enhancement in a heat exchanger using helically corrugated tube, *Int. Commun. Heat Mass Transf.*, Vol. 38, 340–347, 2011.

[27] Z.S. Kareem, S. Abdullah, T.M. Lazim, M.N. Mohd Jaafar, A.F. Abdul Wahid, Heat transfer enhancement in three-start spirally corrugated tube: Experimental and numerical study, *Chem. Eng. Sci.*, Vol. 134, 746–757, 2015.

[28] J.H. Doo, M.Y. Ha, J.K. Min, R. Stieger, A. Rolt, C. Son, An investigation of cross-corrugated heat exchanger primary surfaces for advanced intercooled-cycle aero engines (Part-I: Novel geometry of primary surface), *Int. J. Heat Mass. Transf.*, Vol. 55, 5256–5267, 2012.

[29] W. Wang, Y. Shuai, B. Li, B. Li, K.-S. Lee, Enhanced heat transfer performance for multi-tube heat exchangers with various tube arrangements, *Int. J. Heat Mass. Transf.*, Vol. 168, 120905, 2021.

[30] M.R. Salimpour, Heat transfer characteristics of a temperature-dependent-property fluid in shell and coiled tube heat exchangers, *Int. Commun. Heat Mass Transf.*, Vol. 35, 1190–1195, 2008.

[31] X. Liu, X. Xu, C. Liu, J. Ye, H. Li, W. Bai, C. Dang, Numerical study of the effect of buoyancy force and centrifugal force on heat transfer characteristics of supercritical CO_2 in helically coiled tube at various inclination angles, *Appl. Therm. Eng.*, Vol. 116, 500–515, 2017.

[32] D. Panahi, K. Zamzamian, Heat transfer enhancement of shell-and-coiled tube heat exchanger utilizing helical wire turbulator, *Appl. Therm. Eng.*, Vol. 115, 607–615, 2017.

[33] Y.X. Li, J.H. Wu, H. Wang, L.P. Kou, X.H. Tian, Fluid flow and heat transfer characteristics in helical tubes cooperating with spiral corrugation, *Energy Procedia*, Vol. 17, 791–800, 2012.

[34] C. Zhang, D. Wang, S. Xiang, Y. Han, X. Peng, Numerical investigation of heat transfer and pressure drop in helically coiled tube with spherical corrugation, *Int. J. Heat Mass. Transf.*, Vol. 113, 332–341, 2017.

[35] P. Naphon, S. Wiriyasart, R. Prurapark, A. Srichat, Numerical study on the nanofluid flows and temperature behaviors in the spirally coiled tubes with helical ribs, *Case Studies in Thermal Engineering*, Vol. 27, 101204, 2021.

[36] J.C. Kurnia, A.P. Sasmito, A.S. Mujumdar, Thermal performance of coiled square tubes at large temperature differences for heat exchanger application, Vol. *Heat Transf. Eng.*, 37, 1341–1356, 2016.

[37] L. Guo, H. Xu, L. Gong, Influence of wall roughness models on fluid flow and heat transfer in microchannels, *Appl. Therm. Eng.*, Vol. 84, 399–408, 2015.

[38] Z.F. Zhou, Y.K. Lin, H.L. Tang, Y. Fang, B. Chen, Y.C. Wang, Heat transfer enhancement due to surface modification in the close-loop R410A flash evaporation spray cooling, Vol. *Int. J. Heat Mass. Transf.*, 139, 1047–1055, 2019.

[39] J. Jia, Q. Song, Z. Liu, B. Wang, Effect of wall roughness on performance of microchannel applied in microfluidic device, Vol. *Microsyst. Technol.*, 25, 2385–2397, 2018.

[40] S.G. Kandlikar, W.J. Grande, Evolution of microchannel flow passages—Thermohydraulic performance and fabrication technology, Vol. *Heat Transf. Eng*, 24, 3–17, 2003.

[41] H. Lu, M. Xu, L. Gong, X. Duan, J.C. Chai, Effects of surface roughness in microchannel with passive heat transfer enhancement structures, *Int. J. Heat Mass. Transf.*, Vol. 148, 119070, 2020.

[42] G.L. Morini, Single-phase convective heat transfer in microchannels: A review of experimental results, *Int. J. Therm. Sci.*, Vol. 43, 631–651, 2004.

[43] H. Lu, M. Xu, L. Gong, X. Duan, J.C. Chai, Effects of surface roughness in microchannel with passive heat transfer enhancement structures, *Int. J. Heat Mass. Transf.*, Vol. 148, 124501, 2020.

[44] B. Sterr, E. Mahravan, D. Kim, Uncertainty quantification of heat transfer in a microchannel heat sink with random surface roughness, *Int. J. Heat Mass. Transf.*, Vol. 174, 121307, 2021.

[45] K.Y. Heo, K.D. Kihm, J.S. Lee, Fabrication and experiment of micro-pin-finned microchannels to study surface roughness effects on convective heat transfer, *J. Micromech. Microeng.*, Vol. 24, 702–708, 2014.

[46] P. Chakraborty, T. Ma, L. Cao, Y. Wang, Significantly enhanced convective heat transfer through surface modification in nanochannels, *Int. J. Heat Mass. Transf.*, Vol. 136, 702–708, 2019.

Chapter 4

Heat transfer augumentation using vortex flow

4.1 INTRODUCTION

For decades, a vortex (or swirl) generator (VG) has been used as a passive heat transfer method to produce a vortex flow in heat exchangers (HX) [1–3]. The published work statistics depicted in Figure 4.1 imply that this technique has developed rapidly since 2000 and has maintained a high research interest in the last five years.

Various types of VGs have been developed based on the vortex flow direction, which can be classified into lateral [4, 5] and longitudinal VGs [6, 7]. The flow direction is similar to secondary [8, 9] and rotational flows [10, 11]. Furthermore, the generators can be categorized into surface vortex generators (S-VGs) [12, 13] and insert vortex generators (I-VGs) [14, 15], according to the applied locations. A vortex is produced against the surface or at the core of the fluid.

In this chapter, various types of S-VGs and I-VGs are summarized, and their structures, characteristics, and enhanced heat transfer mechanism are discussed in detail. A future outlook for the development of VG techniques is also provided.

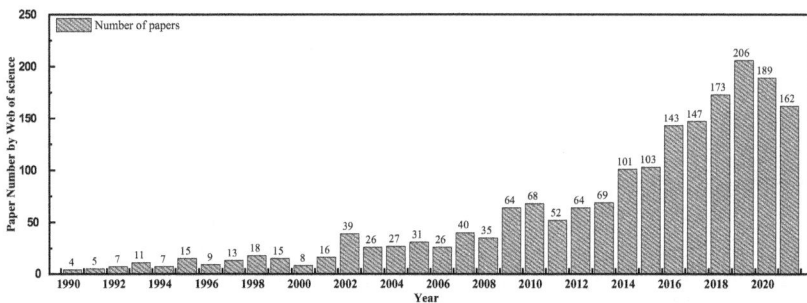

Figure 4.1 Published work statistics of using VG usage on enhanced heat transfer (1990–2020).

DOI: 10.1201/9781003229865-4

4.2 SURFACE VORTEX GENERATOR

Similar to the fin technique, the geometry structures, parameters, and arrangements of S-VGs have been extensively studied and widely used in plate HXs. Furthermore, S-VGs are generally combined with finned, corrugated, or dimpled walls. Studies on various S-VGs are summarized in Table 4.1, such as various-shape VGs [13], porous VGs [16], combination of pair VGs with flat-fins [17], variable-swept VGs [18], VGs with different slope angles [19], a combination of wavy fin and delta winglet VGs (WVGs) [20], curved triangular VGs [21], delta WVGs with various geometry parameters [22], pair of WVGs with different arrangements and attack angles [23], annular and inclined VGs embedded on tube bundle HXs [24], WVGs on flat plates [25], perforated WVGs [26], and punched curved WVGs [27].

Six different VG shapes (rectangular, rectangular trapezoidal, angular rectangular, wishbone winglets, intended, and waved VG) were numerically studied [13]. Compared to the other models, the heat transfer of the fin-plate HX was increased more (by 7%) by the simple rectangular VG. In addition, the increase in the height of the VGs increased the heat transfer index and the best slope angle for the VG arrangement was found to be 45°.

Porous VGs on laminar convective heat transfer have been studied [16] including porosity, Darcy number, VGs' width ratio, Re, and the heat transfer augment and vortex-induced vibration, which increase with increasing Re and width to height ratio. A porous VG with $B/h = 1.0$ when the Darcy number is 10^{-3} or 10^{-4} improves the overall heat transfer along the heated surfaces and weakens the vortex-induced vibration significantly.

A flat-fin heat sink with a pair of VGs was experimentally and numerically studied [17]. The effects of the horizontal and vertical distances between the VG and the heat sink, slope angle of the VG, height of the VG, configuration of the VG, and Re were considered. A slope angle of 30° was preferred to optimize the thermal resistance and pressure drop.

New triangular and rectangular VGs were punched from the longitudinal winglet with slope angles of 15°, 45°, and 75° [19]. The thermal dynamics in a channel with the new VGs was experimentally investigated. The result presented a 23–55% increase in heat transfer index owing to the new VGs.

A novel combination of a wavy fin and delta WVGs was developed and the effects of different corrugation angles of the wavy fin and different slope angles of the VGs were numerically analyzed to optimize the heat transfer and flow resistance performance of the combination [20]. When compared with the smooth wavy fin, the results for the wavy fin with $\theta = 20°$ and $\beta = 45°$ indicated that the span-averaged Nu and maximum thermal performance factor could increase up to 33% and 26.4%, respectively. Furthermore, a slope angle of 45° was found to be obviously more effective on the overall performance improvement.

Two types of delta winglets and rectangular-winglets on a flat plate were numerically compared [25], with different Re values (300, 600, and 900)

Table 4.1 Summary of surface vortex generator publications

Configuration	Method	Working conditions	Structure parameters	Performances
	Numerical [13]	Working fluid is air Re = 200 Inlet temperature = 300 K Wall temperature = 500 K	The space = 15 × 15 mm; Height of the VGs = 5 mm;	The simple rectangular VG increases the heat transfer significant with 7%; The best angle of VG is 45°.
	Numerical [16]	Unsteady, incompressible, laminar flow	The structure parameter is presented in the configuration.	A porous VG with B/h = 1.0 improves the overall heat transfer.

(Continued)

Table 4.1 (continued) Summary of surface vortex generator publications

Configuration	Method	Working conditions	Structure parameters	Performances
	Experimental and numerical [17]	Cooling airflow Maximum flow rate = 14 m³/min	Width W_{VG} = 5 mm; Heights and lengths, $H_{VG} = L_{VG}$ = 22.5, 45, and 67.5 mm; β =30°, 45°, 60°; B = 70 mm.	The performances are greater, and the increase in the pressure drop is lower at lower Reynolds numbers.
	Experimental [19]	Air flow fixed heat flux = 670 W/m² Re = 3,288–37,817	Attack angles = 15°, 45°, and 75°; Aspect ratio of rectangular channel = 2; Winglet pitch-to-height ratio = 0.59; Winglet height to channel height ratio = 0.6.	The enhancement factor of the punched triangular VG (~2.92) is observed to be the best.

Numerical [20]	Water flow $T_{in} = 300$ K $T_{wall} = 310$ K Re = 100–3,000	Corrugation angles are 10°, 20°, and 30°; Attack angles of VGs are 30°, 45°, 60°, and 75°.	The maximum heat transfer index increases by 26.4%; The optimum attack angle is 45°.
Numerical [25]	Water flow heated by solar energy Re = 300–900	Delta and rectangular WVGs Angle of attack = 15°, 30°, and 45°	The highest Nu is obtained for the rectangular WVG with attack angle = 45°, the best ratio of heat transfer and pressure drop is achieved for the delta-WVG with attack angle = 30°.
Experimental and numerical [26]	Air flow heated by solar energy Re = 4,100–25,500	Perforated rectangular and trapezoidal WVGs; Relative height = 0.2 and 0.48; Longitudinal pitch ratios = 1, 1.5, and 2; Punched hole diameters = 1, 3, 5, 7 mm.	The perforated rectangular WVG with hole diameter = 1 mm reach to the highest heat transfer and friction factor up to 6.78 and 84.32 times that of the smooth duct.

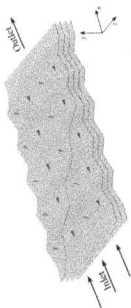

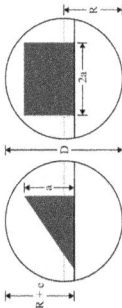

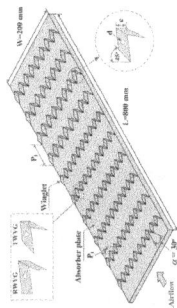

and angles of attack (15°, 30°, and 45°). The highest heat transfer index was obtained at a slope angle of 45° for both VGs, and was more significant for the rectangular WVG. However, the best equilibrium point between the heat transfer and pressure drop was obtained for the delta-WVG with a slope angle of 30°.

Two types of rectangular and trapezoidal WVGs with four punched hole diameters were experimentally and numerically studied [26], and the WVG height and pitch were analyzed. The results indicated that the perforated WVGs, and the perforated rectangular WVG (P-RWVG) at d = 1 mm yielded the highest heat transfer index and friction factors of up to 6.78 and 84.32 times higher than those of the smooth duct, respectively. However, the highest heat transfer performance of approximately 2.01 was observed for the (perforated trapezoidal WVG (P-TWVG) for d = 5 mm.

According to the literature review of SVGs mentioned in Table 4.1, the increase in the heat transfer performance ratio can reach 1.03–6.78 with various structures and parameters. The attack angle = 45° was found to be the optimum as mentioned in several works [13, 20, 23, 25]. A detailed parameter study was performed in Reference [26], which considered the relative height (B_R = 0.2 and 0.48), pitch ratios (PRs) (P_R = 1, 1.5, and 2), and punched hole diameters (d = 1, 3, 5, and 7 mm) for rectangular and trapezoidal WVGs at Re ranges from 4,100–25 and 500. As shown in Figures 4.2 and 4.3, a higher B_R leads to a higher Nu increase ratio (Nu_R), considerably higher friction increase ratio (f_R), and lower overall heat transfer performance (TEF). A lower P_R leads to a higher Nu_R and a considerably higher f_R, and P_R = 1.5 presents the optimum TEF. The trapezoidal WVG with B_R = 0.2 and P_R = 1.5 presents the efficient TEF with 1.84–1.44.

The punched WVGs with different hole diameters indicate that the punched hole can slightly decrease Nu_R and significantly decrease f_R. A larger d value indicates a more severe decline. The TEFs for all P-TWVGs were better than those of P-RWVGs and WVGs without holes. The best TEF could reach up to 2.0–1.55 for the P-TWVG with d = 5 mm.

The local Nu for the end wall of a protruding circular cylinder with a pair of WVGs was experimentally tested using steady-state liquid crystal thermography [23]. The local Nu values for pairs of WVGs with different orientations and streamwise and spanwise positions (S_x and S_z) are presented in Figures 4.4 and 4.5, respectively. The outward-orientation WVGs are harmful to heat transfer augment on both the leading and trailing edges of the circular cylinder, whereas inward-oriented WVGs can strengthen the overall local Nu. The smallest streamwise and spanwise positions (S_x = 0 and S_z = 0.1) showed a significant increase in the local Nu caused by the vortex between the interval of the pair of WCGs and the circular cylinder.

For a better relationship exploration between vortex flow and local heat transfer performance, a numerical method was employed for fin-tube HXs with different WVG arrangements [28] and attack angles [29]. Figure 4.6 depicts the local temperature and velocity contour distributions, which

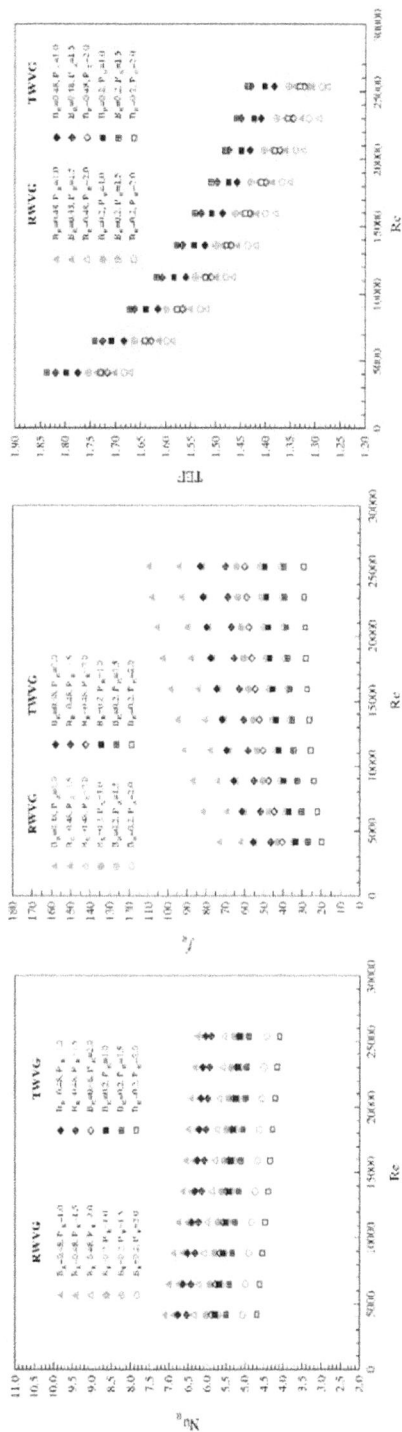

Figure 4.2 Thermo-hydraulic performance for rectangular and trapezoidal WVGs with different relative height and PRs [26].

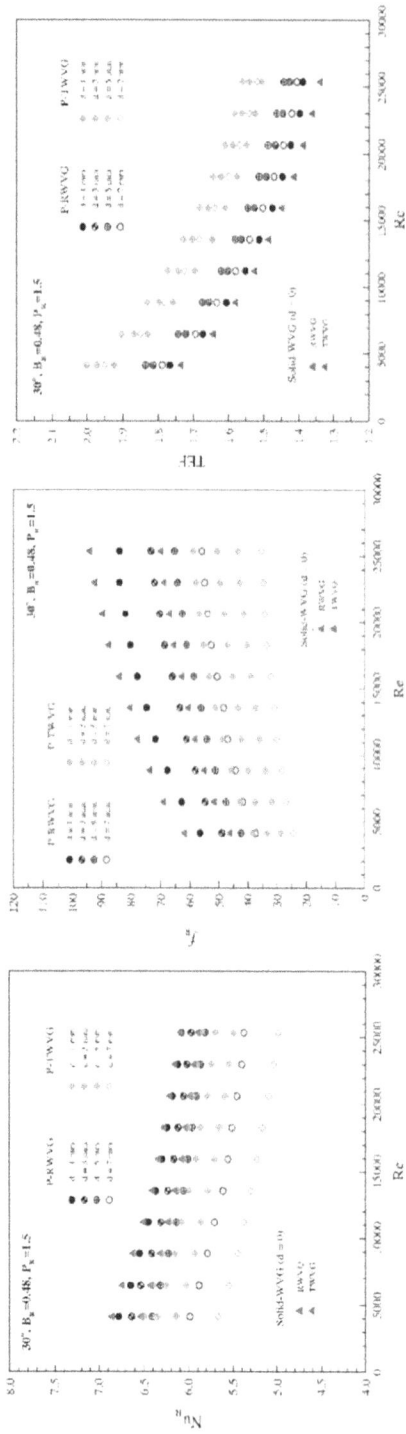

Figure 4.3 Thermo-hydraulic performance for punched rectangular and trapezoidal WVGs with different hole diameters [26].

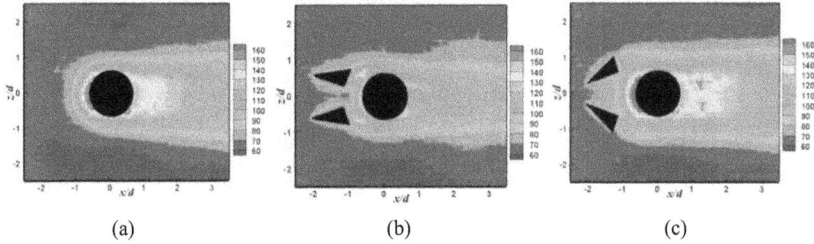

Figure 4.4 Local Nu for pair of WVGs with different orientations: (a) no WVG; (b) outward orientation; and (c) inward orientation [23].

demonstrates that the leeward side of the WVGs produces small vortices, which can effectively break the large secondary flow behind the circular tube. The small vortices increase the mixture of the core fluid elements, and the nozzle-like pair of WVGs accelerates the fluid flow. Under the combination of the abovementioned three effects, the temperature field of the staggered row tubes with WVGs is the most uniform, and the heat transfer performance is significantly improved. Based on the numerical method and evidential conclusions, one can conveniently and optimally design VGs on HXs.

4.3 INSERT VORTEX GENERATOR

I-VGs are commonly employed in tube or channel HXs. Various I-VGs have been developed, among which twisted tapes (TTs) [15, 30–35], coil inserts (CIs) [36–41], and swirl generators (SGs) [42–45] with different geometric shapes, parameters, and arrangements are most commonly used and studied. The representative I-VG values are summarized in Table 4.2.

Single TT, twin co-TT, and twin counter-TTs with different twist ratios were studied experimentally and numerically [15]. The results indicated that the heat transfer performance of the twin counter-TTs was approximately 12.5–44.5% and 17.8–50% higher than those of the twin co-TTs and single TT, respectively. The maximum thermal augment index (η) was 1.39.

TTs with wire nails and plain TTs were experimentally compared with Re ranging from 2,000 to 12,000 [30]. The Nu, f, and thermal augment factor of TTs with wire nails were 1.08–1.31, 1.1–1.75, and 1.05–1.13 times that of the plain TTs, respectively.

Serrated TTs with various serration widths and depths were experimentally studied and compared with typical TTs [31]. The heat transfer index of the serrated TTs was 4–27% higher than that of the TTs and the friction factor of the serrated TTs was 5–68% higher than that of the TTs.

TTs with triangular, rectangular, and trapezoidal wings in a round tube were experimentally studied with three types of wing-chord ratios [32]. The TTs with trapezoidal wings presented the highest Nu and f, followed by

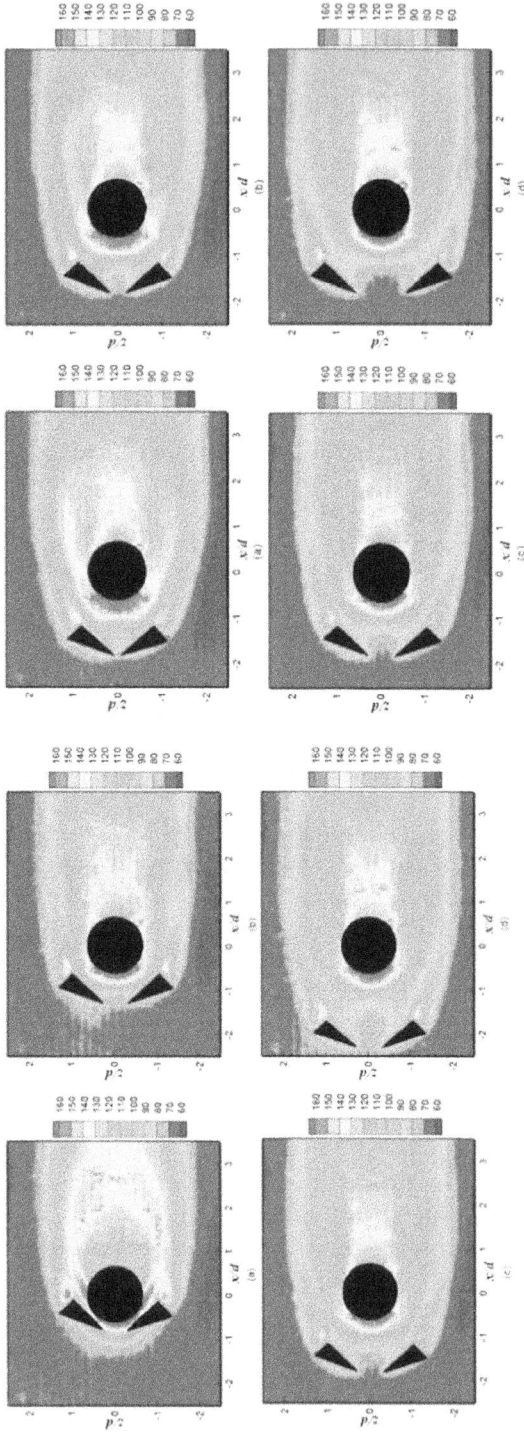

Figure 4.5 Local Nu for pair of WVGs with different positions: (a) S_x =0, 0.5, 1.0, and 1.5; (b) S_z =0.1, 0.25, 0.5, and 1.0 [23].

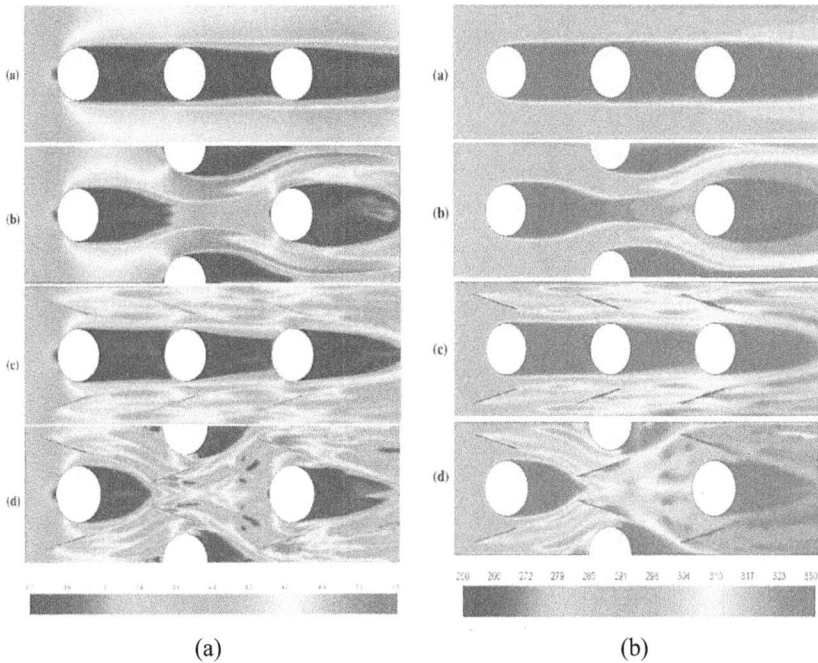

(a)   (b)

Figure 4.6 Temperature and velocity contours for fin-tube heat exchangers with different WVGs arrangements: (a) velocity contours and (b) temperature contours [28].

rectangular and triangular wings. The maximum heat transfer index was 1.42 for TTs with trapezoidal wings at a wing–chord ratio of 0.3. The corresponding heat transfer index and friction factor increase ratios were 2.84 and 8.02, respectively.

Perforated TTs with parallel wings were experimentally investigated, for Re between 5,500 and 20,500 [33]. Three types of hole diameter ratios ($d/W = 0.11, 0.33$, and 0.55) and wing depth ratios ($w/W = 0.11, 0.22$, and 0.33) were studied and compared with those of smooth tubes. The maximum heat transfer enhancements were 208% and 190% for perforated TTs with parallel wings and typical TTs, respectively. The maximum overall performance reached 1.32 for $d/W = 0.11$ and $w/W = 0.33$, at Re = 5,500.

TTs with uniform wire coils [46]and nonuniform wire CIs [34] were experimentally performed. The experimental data showed that the wire coils together with TTs led to a twice-increased ratio in heat transfer index compared to the wire coil or twisted tape alone. The TTs with uniform wire CIs could reach heat transfer augmentations of 3–6 and the best PEC = 1.55–1.05, corresponding to a spring PR of 4 and twist ratio of 4. The TTs with nonuniform wire CIs could reach heat transfer augmentations equal to 3–3.65 and the best PEC = 1.25–1.02, corresponding to the DI-coil and twist ratio = 3.

Table 4.2 Summary of insert vortex generator publications

Configuration	Method	Working conditions	Structure parameters	Performances
	Experimental and numerical [15]	Water flow $T_{in} = 27°C$ Re = 3,700–21,000	Single TT Co-TT Counter-TT Twist ratio = 2.5–4.0	The Nu of twin counter-TTs is approximately 12.5–44.5% and 17.8–50% higher than those of twin co-TTs and single TT, respectively. The maximum η index can reach to 1.39.
	Experimental [30]	Water flow Re = 2,000–12,000	TTs with wire nails Plain TTs Twist ratio = 2.0–6.0	The Nu, f, and thermal enhancement factor of TTs with wire nails are 1.08–1.31, 1.1–1.75, and 1.05–1.13 times than plain TTs, respectively.
	Experimental [31]	Air flow $T_{in} = 25°$ C Re = 4,000–20,000	TTs and TTs with serrated edge Serration depth ratio = 0.1–0.3 Serration width ratio = 0.1–0.3	The Nu of serrated TTs is 4–27% higher than that of TTs. The f of serrated TTs is 5–68% higher than that of TTs.

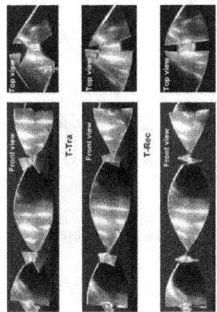

Experimental [32]	Water flow Re = 5,500–20,200	TTs with triangular, rectangular, and trapezoidal wings; wing-chord ratios = 0.1, 0.2, and 0.3	The maximum PEC = 1.42, corresponding Nu/Nu_0 = 2.84, and f/f_0 = 8.02 for TT with trapezoidal wings at wing-chord ratio = 0.3.
Experimental [33]	Water flow Re = 5,500–20,500	Perforated TTs with parallel wings Hole diameter ratio d/W = 0.11, 0.33, and 0.55 Wing depth ratio w/W = 0.11, 0.22, and 0.33	The maximum Nu/Nu_0 = 208% and 190% for perforated TTs with parallel wings and TTs, respectively. The maximum PEC = 1.32 for d/W = 0.11 and w/W = 0.33 at Re = 5,500.
Experimental [46]	Air flow T_{in} = 25° C Re = 3,000–18,000	Spring pitch ratio H/d = 4, 6, and 8; Twist ratios of TTs = 4 and 6	Nu/Nu_0 = 3–6. Best PEC = 1.55–1.05.

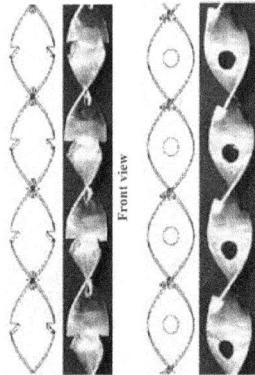

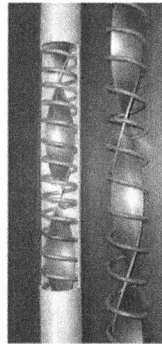

T-Tra

T-Rec

T-Tri

Front view

Top view

(Continued)

Table 4.2 (Continued)   Summary of insert vortex generator publications

Configuration	Method	Working conditions	Structure parameters	Performances
Decreasing/increasing coil pitch ratio arrangement	Experimental [34]	Air flow $T_{in} = 25°C$ Re = 4,600–20,000	Decreasing coil pitch ratio of spring (D-coil = 8:6:4:8:6:4) Decreasing/increasing coil pitch ratio of spring (DI-coil = 8:6:4:6:8…) Twist ratio of TTs = 3 and 4	Nu/Nu$_0$ = 3–3.65. Best PEC = 1.25–1.02.
	Experimental and numerical [35]	Air flow $T_{in} = 25°C$ Re = 4,000–20,000	Double-sided delta wing tape insert with alternate-axes, Pitch ratios = 0.75, 1.0, and 1.25, Wing width ratios = 0.5, 0.67, and 0.83.	The average Nu and f for T-W increase up to 165% and 14.8 times than the plain tube. The maximum PEC is 1.19.
	Experimental [36]	Air flow Re = 6,000–20,000	Twisted rings (TRs), and circular rings (CRs). Width ratios, W/D = 0.05, 0.1, and 0.15. Pitch ratios, p/D = 1, 1.5, and 2.	The Nu and f for most TRs is lower than those of CRs. The best PEC for TR = 1.24–1.02.

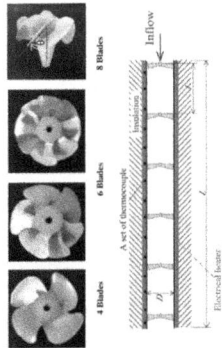

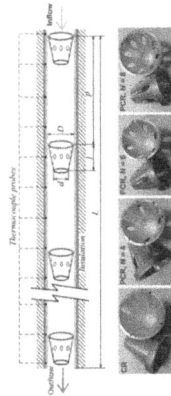

Experimental [42]	Air flow T_{in} = 25°C Re = 4,000–21,000	Blade angles, θ = 30°, 45°, and 60°. Pitch ratio = 5, 7, and 10. Number of blades = 4, 6, and 8.	Nu/Nu_0 = 2.07–2.18 times for blade angle = 60°, pitch ratio = 2, and blade number = 8. The maximum enhancement efficiency can reach up to 1.2.
Experimental [38]	Air flow Re = 4,000–20,000	Perforated conical ring. Pitch ratio, PR = 4, 6, and 12. Perforated hole number, N = 4, 6, and 8.	For N = 4, 6, and 8, Nu/Nu_0 = 1.9–3.39, 1.69–3.2, and 1.65–2.72, and f/f_0 = 1.57, 1.73, and 1.82, respectively. The maximum PEC is approximately 0.92 for PR = 4 and N = 8, with Re = 4,000.
Experimental[39]	Air flow T_{in} = 25°C Re = 6,000–26,000	Tube diameter ratios (d/D = 0.5, 0.6, and 0.7). Converging conical, diverging conical, and converging-diverging conical ring array (CR, DR, CDR).	The Nu for DR, CDR, and CR are 197–333%, 138–234%, and 91–175%, respectively, higher than that of smooth tube. The maximum PEC for DR and CR are 1.8 and 1.15 at d/D = 0.5.

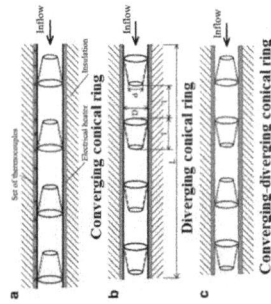

Converging conical ring

Diverging conical ring

Converging-diverging conical ring

(Continued)

Table 4.2 (Continued) Summary of insert vortex generator publications

Configuration	Method	Working conditions	Structure parameters	Performances
	Experimental [40]	Supercritical N_2 flow P_{in} = 35, and 40 bar m_{in} = 27.6, and 41.3g/min q_{wall} = 6.8, 8.0, and 9.3 kW/m²	hiTRAN™ wire matrix tube insert	The heat transfer index for matrix insert is enhanced by 1.42–2.35 times that of then on-insert tube, and the pressure drop increase is negligible.
	Experimental and numerical [41]	No. 45 transformer oil T_{in} = 313.15 K T_{wall} = 303.15 K Re = 400–1,800 u_{in} = 0.036–0.161 m/s	Twined coil insert	Nu/Nu_0 = 12.0–38.2 f/f_0 = 7.7–18.1 The maximum PEC can reach 2.16.

A double-sided delta wing TT (T-W) with alternate-axes was experimentally studied [35]. When using T-W, the increase in the average Nu and f reached up to 165% and 14.8 times that of the smooth tube, and the maximum PEC reach to 1.19. The T-W with the forward-wing presenting a higher heat transfer index than that with the back-wing at approximately 7%.

The twisted rings (TRs)and common circular rings (CRs) in the tube HX were experimentally compared [36]. The experimental results indicated that except for the biggest width ratio (W/D = 0.15) and smallest PR (p/D = 1.0), most TRs yielded lower Nu and f values than those of CRs. The best PEC was at 1.24–1.02 for TR with W/D = 0.05, and p/D = 1.0, at Re = 6,000–20,000.

Propeller type SGs in round tubes with various blade angles (θ = 30°, 45°, and 60°), PRs (5, 7, and 10), and numbers of blades (4, 6, and 8) were experimentally studied [42]. The heat transfer rate for propeller insert can exceed that of the plain tube around 2.07–2.18 times for blade angle = 60°, PR = 2, and blade number = 8, with a maximum enhancement efficiency of up to 1.2.

Perforated conical rings (PCRs) with three PRs (PR = 4, 6, and 12), three perforated hole numbers (N = 4, 6, and 8), and Re = 4,000–20,000 were experimentally studied [38]. The heat transfer index and friction factor of PCRs increased with both decreasing PR and N. For PR = 4, the average Nu/Nu_0 = 1.188, 1.539, and 2.37 times higher than those for PR = 6, and 12, and the plain tube, respectively. For N = 4, 6, and 8, Nu/Nu_0 = 1.9–3.39, 1.69–3.2, and 1.65–2.72, and f/f_0 = 1.57, 1.73, and 1.82, respectively. The maximum heat transfer index was approximately 0.92 for PR = 4 and N = 8, and Re = 4,000.

Conical ring inserts with three types of arrangements, namely converging conical ring array (CR), diverging conical ring array (DR), and converging-diverging conical ring array (CDR), were experimentally studied [39]. The mean heat transfer rates for DR, CDR, and CR with tube diameter ratios (d/D = 0.5, 0.6, and 0.7) were higher than those of the plain tube by 197–333%, 138–234%, and 91–175%, respectively. Furthermore, at d/D = 0.5, the maximum enhancement efficiencies for DR, CDR, and CR were 1.8, 1.4, and 1.15, respectively.

The hiTRAN™ wire matrix insert was experimentally studied with supercritical nitrogen (N_2) heated under different conditions [40]. The experimental results indicated that the supercritical N_2 presents similar heat transfer behavior when transiting across the pseudo-critical point as water flow; the heat transfer for the matrix insert was enhanced by 1.42–2.35 times that of the noninsert tube, while the increase of pressure drop was ignored.

The twined CIs were experimentally and numerically studied [41] with No. 45 transformer oil. The results showed that the twined CI could mix the flow complexity and significantly increase the intensity of the secondary flow. The ratio of Nu of the tubes with and without the insert was approximately

12.0–38.2, while the ratio of f with and without the insert was 7.7–18.1, with Re ranging from 400 to 1,800. The maximum heat transfer index could reach 2.16 at Re = 1,000.

Based on the literature review mentioned in Table 4.2, one may observe that the heat transfer improvement of various insert VGs can reach 1.1–4.6, and depending on the type and operation conditions, there is a corresponding increase in the friction factor that can reach 1.1–14.8. An abnormally excellent heat transfer increase ratio was obtained at 12.0–38.2 using a specially designed twined coil insert, see [41], and the friction factor increased by 7.7–18.1. Owing to the large variation range in thermal-hydraulic performance by various insert techniques, mechanism analysis has gained importance for the optimal design of insert structures.

The flow visualization study on helical TTs with and without rods [36] and TTs with alternate-axes and wings [32] clearly indicate that the TT insert can produce a violent longitudinal vortex flow and effectively disturb the core fluid flow, as depicted in Figure 4.7. For a tube heat transfer flow, the temperature contours were in a concentric circle arrangement; thus, it was found that only the rotational flow had little effect on the heat transfer improvement, and the horizontal vortex produced by the additional wings was more effective on the fluid mixture at different temperature contours.

According to the typical TTs, various modified TTs, such as winglet TTs, perforated TTs, TT arrays, wire coils, and conical ring inserts, have been developed, reviewed, and analyzed in Reference [47].

A delta winglet TT insert was experimentally studied [37], which could produce a main swirl flow and an additional fluid flow disturbance by delta winglets, as depicted in Figure 4.8(a). The results showed that the combined effects of the swirl flow and the additional vortex produced by winglets in the fluid near the wall assist in obtaining a higher heat transfer.

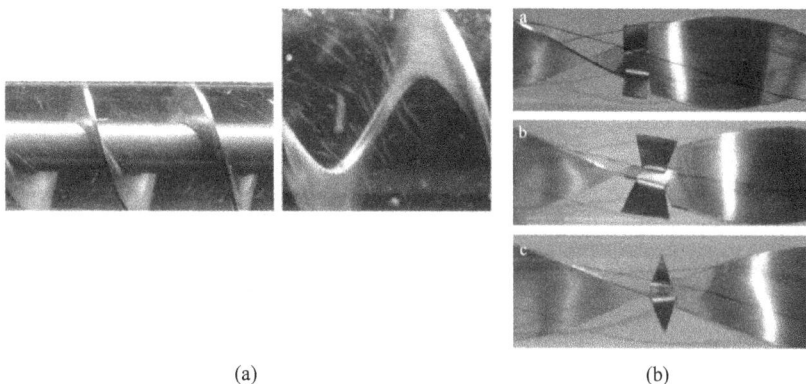

(a)   (b)

Figure 4.7 Flow visualization of twisted tape inserts: (a) helical tape with and without rods [36] and (b) twisted tapes with alternate-axes and wings [32].

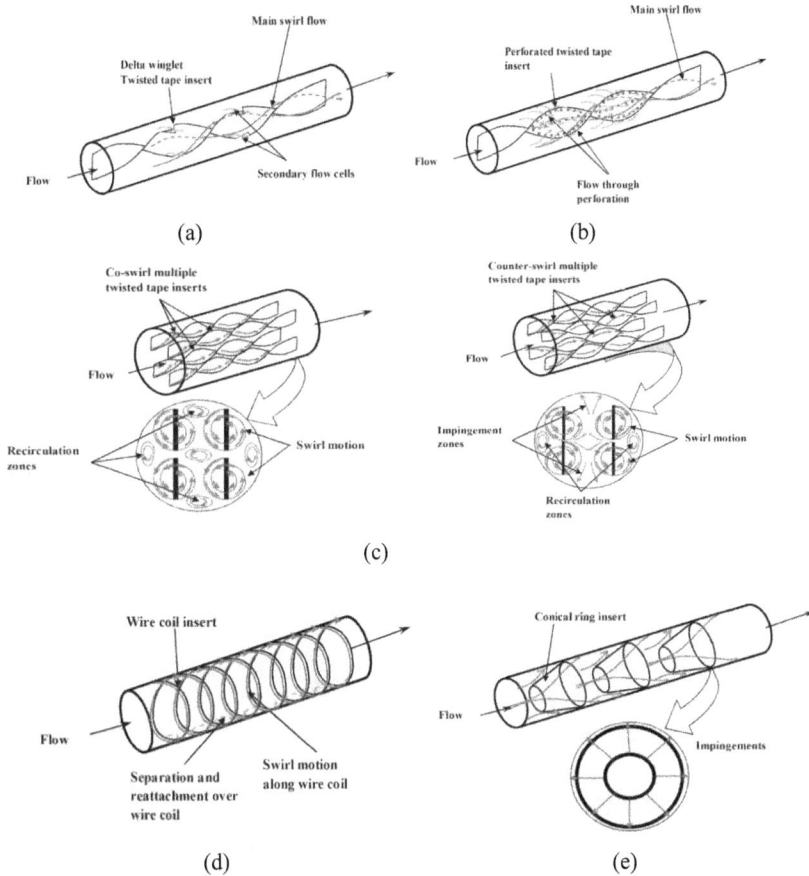

Figure 4.8 Fluid flow mechanism in various inserts: [47] (a) delta winglet twisted tape insert; (b) perforated twisted tape insert [20]; (c) four co-swirl and counter-swirl twisted tape inserts [38]; (d) wire coil insert [39]; and (e) conical ring insert [25].

A perforated TT insert was experimentally studied, and secondary flows were produced along with swirl flow, as shown in Figure 4.8(b), which elongated the path of fluid flow through the tube [20]. The friction factor was reduced owing to the increased open area, and the heat transfer rate increased because of the intense fluid mix by the radial and secondary swirls.

Co-swirl and counter-swirl TT inserts were experimentally compared [38], and multiple recirculation zones were produced in addition to swirl flow cells, as shown in Figure 4.8(c). The swirl motion revamped the boundary layer flow, whereas the impingement completely terminated the viscosity-dominated flow in the localized area, resulting in an increase in heat transfer and friction performance. In addition, the counter-swirl inserts caused higher enhancement ratios of the heat transfer and friction than those of the co-swirl inserts.

Wire CIs, which could induce separate flows along with the secondary flow, were experimentally studied [39], as shown in Figure 4.8(d). The synergy of the secondary flow vortex and the separated main flow resulted in significant heat transfer augment with a small increase in flow resistance.

Conical ring inserts were experimentally studied [25], which could cause better chaotic mixing between the core flow and the boundary flow and enhance the convective process, as shown in Figure 4.8(e). It was observed that the DR array provided better heat transfer performance than that of the converging and CDR arrays owing to the significant recirculation and larger contact surface area between the fluid and heating wall surface.

The abovementioned studies experimentally and theoretically analyzed the enhanced heat transfer mechanism of I-VGs. However, to better understand the mechanism and optimal structural parameters, numerical methods have been employed recently. These can be used to directly obtain the global flow and thermal characteristics. A unique twined coil insert was numerically studied, and the flow and thermal fields [41] are depicted in Figure 4.9.

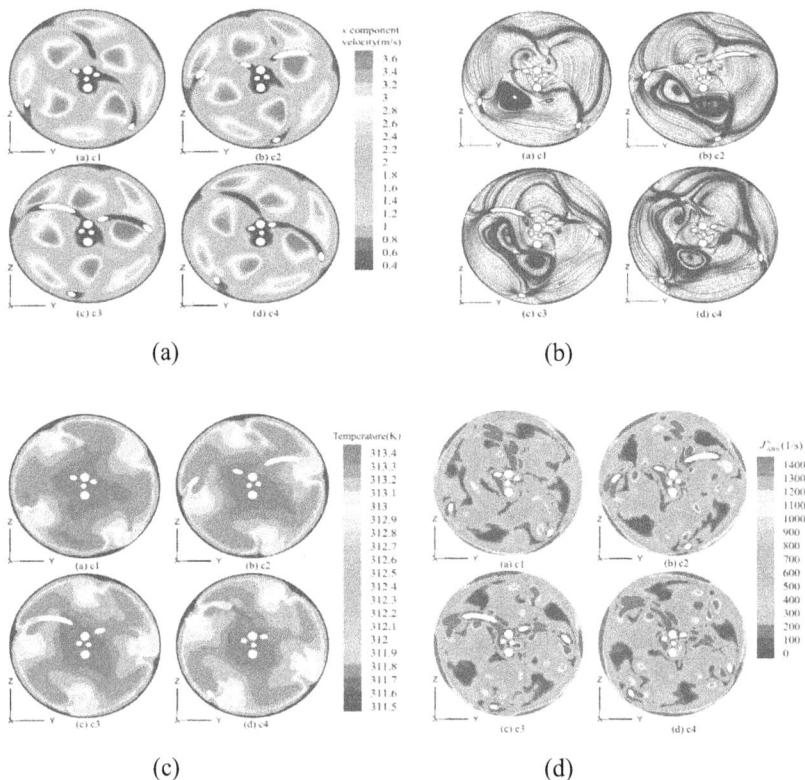

(a)   (b)

(c)   (d)

Figure 4.9 Mechanism analysis of twined coil insert [41]: (a) velocity component along the main flow direction; (b) streamlines; (c) temperature contours; (d) vorticity contours.

The channel flow was divided into several blocks by three-dimensional vortices, which accelerated the mixing of hot and cold fluids in the main flow region, increased the temperature gradient in the fluid at boundary region, and increased the heat transfer coefficient of the tube wall.

4.4 SUMMARY AND OUTLOOK

Based on the reviewed works, the use of a VG is effective in increasing the heat transfer performance. The surface and insert VGs can provide two strengthening mechanisms from the wall surface and core fluid, respectively. The structures, parameters, and arrangements have been investigated in detail using experimental, numerical, and theoretical methods, similar to surface modification techniques.

In the future, these two types of VGs can be used in conjunction with other surface modification techniques. In addition, air and water flows are mostly considered as working fluids in published VG studies. Employing his technique in various application fields must be considered. Because this technique was developed relatively late compared to surface modification techniques, industrial applications are expected to be significant in due course.

REFERENCES

[1] M. Veerabhadrappa Bidari, P.B. Nagaraj, G. Lalagi, Influence of different types of vortex generators (VGs) to enhance heat transfer performance in heat exchangers: A review, *Int. J. Ambient Energy*, Vol. 43 1–24, 2021.
[2] N. Zheng, F. Yan, K. Zhang, T. Zhou, Z. Sun, A review on single-phase convective heat transfer enhancement based on multi-longitudinal vortices in heat exchanger tubes, *Appl. Therm. Eng.*, Vol. 164, 114475, 2020.
[3] L. Chai, S.A. Tassou, A Review of Airside Heat Transfer Augmentation with Vortex Generators on Heat Transfer Surface, Vol. 11, 2737, 2018.
[4] G.P. Aravind, M. Deepu, Numerical studies on convective mass transfer augmentation in high-speed flows with lateral sweep vortex generator and dimple cavity, *Int. J. Therm. Sci.*, Vol. 153, 106379, 2020.
[5] J. Zhang, J. Kundu, R.M. Manglik, Effect of fin waviness and spacing on the lateral vortex structure and laminar heat transfer in wavy-plate-fin cores, *Int. J. Heat Mass Transf.*, Vol. 47, 1719–1730, 2004.
[6] A. Ebrahimi, B. Naranjani, S. Milani, F. Dadras Javan, Laminar convective heat transfer of shear-thinning liquids in rectangular channels with longitudinal vortex generators, *Chem. Eng. Sci.*, Vol. 173, 264–274, 2017.
[7] J.M. Wu, W.Q. Tao, Effect of longitudinal vortex generator on heat transfer in rectangular channels, *Appl. Therm. Eng.*, Vol. 37, 67–72, 2012.
[8] A. Stroh, K. Schäfer, P. Forooghi, B. Frohnapfel, Secondary flow and heat transfer in turbulent flow over streamwise ridges, *Int J Heat Fluid Flow*, Vol. 81, 108518, 2020.

[9] P. Forsyth, M. McGilvray, D.R.H. Gillespie, Secondary flow and heat transfer coefficient distributions in the developing flow region of ribbed turbine blade cooling passages, *Exp Fluids*, Vol. 58, 5, 2016.

[10] N.V.V. Krishna Chaitanya, D. Chatterjee, Influence of counter rotation on fluid flow and heat transfer around tandem circular cylinders at low Reynolds number, *J. Braz. Soc. Mech. Sci. Eng.*, Vol. 43, 357, 2021.

[11] S. Saleem, Heat and mass transfer of rotational flow of unsteady third-grade fluid over a rotating cone with buoyancy effects, *Math. Probl. Eng.*, Vol. 2021, 5544540, 2021.

[12] U. Kashyap, K. Das, B.K. Debnath, Effect of surface modification of a rectangular vortex generator on heat transfer rate from a surface to fluid: An extended study, *Int. J. Therm. Sci.*, Vol. 134, 269–281, 2018.

[13] M. Samadifar, D. Toghraie, Numerical simulation of heat transfer enhancement in a plate-fin heat exchanger using a new type of vortex generators, *Appl. Therm. Eng.*, Vol. 133, 671–681, 2018.

[14] M. Khoshvaght-Aliabadi, M.H. Akbari, F. Hormozi, An empirical study on vortex-generator insert fitted in tubular heat exchangers with dilute Cu–water nanofluid flow, *Chin. J. Chem. Eng.*, Vol. 24, 728–736, 2016.

[15] S. Eiamsa-ard, C. Thianpong, P. Eiamsa-ard, Turbulent heat transfer enhancement by counter/co-swirling flow in a tube fitted with twin twisted tapes, *Exp. Therm. Fluid Sci.*, Vol. 34, 53–62, 2010.

[16] S.W. Perng, H.W. Wu, T.C. Jue, Numerical investigation of heat transfer enhancement on a porous vortex-generator applied to a block-heated channel, *Int. J. Heat Mass Transf.*, Vol. 55, 3121–3137, 2012.

[17] H.Y. Li, C.L. Chen, S.M. Chao, G.F. Liang, Enhancing heat transfer in a plate-fin heat sink using delta winglet vortex generators, *Int. J. Heat Mass Transf.*, Vol. 67, 666–677, 2013.

[18] G.P. Aravind, M. Deepu, Effects of the Asymmetrical Vortex Interactions by a Variable Swept Vortex Generator (VSVG) on Heat Transfer Enhancement, in *7th Asian Joint Workshop on Thermophysics and Fluid Science AJWTF7 2018* (Eds. A. Suryan, D.H. Doh, M. Yaga, G. Zhang), 433–445, Springer, 2020.

[19] S. Caliskan, Experimental investigation of heat transfer in a channel with new winglet-type vortex generators, *Int. J. Heat Mass Transf.*, 78, 604–614, 2014.

[20] C. Luo, S. Wu, K. Song, L. Hua, L. Wang, Thermo-hydraulic performance optimization of wavy fin heat exchanger by combining delta winglet vortex generators, *Appl. Therm. Eng.*, Vol. 163, 114343, 2019.

[21] S. Liu, H. Jin, K. Song, L. Wang, X. Wu, L. Wang, Heat transfer and pressure drop characteristics of the tube bank fin heat exchanger with fin punched with flow redistributors and curved triangular vortex generators, *Heat Mass Transf.*, Vol. 53, 3013–3026, 2017.

[22] M.K. Nalawade, A. Bhati, R.P. Vedula, heat transfer and pressure drop characteristics for flow through square channel with delta wing vortex generator elements on two opposite walls, *J. Enhanc. Heat Transf.*, Vol. 26, 101–126, 2019.

[23] S. Hussain, J. Liu, L. Wang, B. Sunden, effects on endwall heat transfer by a winglet vortex generator pair mounted upstream of a cylinder, *J. Enhanc. Heat Transf.*, Vol. 23, 241–262, 2016.

[24] C.N. Lin, Y.W. Liu, J.S. Leu, Heat transfer and fluid flow analysis for plate-fin and oval tube heat exchangers with vortex generators, *Heat Transf. Eng.*, Vol. 29, 588–596, 2008.

[25] F.A.S. da Silva, D.J. Dezan, A.V. Pantaleão, L.O. Salviano, Longitudinal vortex generator applied to heat transfer enhancement of a flat plate solar water heater, *Appl. Therm. Eng.*, Vol. 158, 113790, 2019.
[26] S. Skullong, P. Promthaisong, P. Promvonge, C. Thianpong, M. Pimsarn, Thermal performance in solar air heater with perforated-winglet-type vortex generator, *Solar Energy*, Vol. 170, 1101–1117, 2018.
[27] G. Zhou, Z. Feng, Experimental investigations of heat transfer enhancement by plane and curved winglet type vortex generators with punched holes, *Int. J. Therm. Sci.*, Vol. 78, 26–35, 2014.
[28] A. Sinha, H. Chattopadhyay, A.K. Iyengar, G. Biswas, Enhancement of heat transfer in a fin-tube heat exchanger using rectangular winglet type vortex generators, *Int. J. Heat Mass Transf.*, Vol. 101, 667–681, 2016.
[29] Y.L. He, P. Chu, W.Q. Tao, Y.W. Zhang, T. Xie, Analysis of heat transfer and pressure drop for fin-and-tube heat exchangers with rectangular winglet-type vortex generators, *Appl. Therm. Eng.*, Vol. 61, 770–783, 2013.
[30] P. Murugesan, K. Mayilsamy, S. Suresh, Heat transfer and friction factor studies in a circular tube fitted with twisted tape consisting of wire-nails, *Chin. J. Chem. Eng.*, Vol. 18, 1038–1042, 2010.
[31] S. Eiamsa-ard, P. Promvonge, Thermal characteristics in round tube fitted with serrated twisted tape, *Appl. Therm. Eng.*, Vol. 30, 1673–1682, 2010.
[32] K. Wongcharee, S. Eiamsa-ard, Heat transfer enhancement by twisted tapes with alternate-axes and triangular, rectangular and trapezoidal wings, *Chem. Eng. Process. Process Intensif.*, Vol. 50, 211–219, 2011.
[33] C. Thianpong, P. Eiamsa-ard, P. Promvonge, S. Eiamsa-ard, Effect of perforated twisted-tapes with parallel wings on heat tansfer enhancement in a heat exchanger tube, *Energy Procedia*, Vol. 14, 1117–1123, 2012.
[34] S. Eiamsa-ard, P. Nivesrangsan, S. Chokphoemphun, P. Promvonge, Influence of combined non-uniform wire coil and twisted tape inserts on thermal performance characteristics, *Int. Commun. Heat Mass Transf.*, Vol. 37, 850–856, 2010.
[35] S. Eiamsa-ard, P. Promvonge, Influence of double-sided delta-wing tape insert with alternate-axes on flow and heat transfer characteristics in a heat exchanger tube, *Chin. J. Chem. Eng.*, Vol. 19, 410–423, 2011.
[36] C. Thianpong, K. Yongsiri, K. Nanan, S. Eiamsa-ard, Thermal performance evaluation of heat exchangers fitted with twisted-ring turbulators, *Int. Commun. Heat Mass Transf.*, Vol. 39, 861–868, 2012.
[37] E. Gholamalizadeh, E. Hosseini, M. Babaei Jamnani, A. Amiri, A. Dehghan Saee, A. Alimoradi, Study of intensification of the heat transfer in helically coiled tube heat exchangers via coiled wire inserts, *Int. J. Therm. Sci.*, Vol. 141, 72–83, 2019.
[38] V. Kongkaitpaiboon, K. Nanan, S. Eiamsa-ard, Experimental investigation of heat transfer and turbulent flow friction in a tube fitted with perforated conical-rings, *Int. Commun. Heat Mass Transf.*, Vol. 37, 560–567, 2010.
[39] P. Promvonge, Heat transfer behaviors in round tube with conical ring inserts, *Energ. Conver. Manage.*, Vol. 49, 8–15, 2008.
[40] Y. Wang, T. Lu, P. Drögemüller, Q. Yu, Y. Ding, Y. Li, Enhancing deteriorated heat transfer of supercritical nitrogen in a vertical tube with wire matrix insert, *Int. J. Heat Mass Transf.*, Vol. 162, 120358, 2020.
[41] W. Dang, L.-B. Wang, Convective heat transfer enhancement mechanisms in circular tube inserted with a type of twined coil, *Int. J. Heat Mass Transf.*, Vol. 169, 120960, 2021.

[42] S. Eiamsa-ard, S. Rattanawong, P. Promvonge, Turbulent convection in round tube equipped with propeller type swirl generators, *Int. Commun. Heat Mass Transf.*, Vol. 36, 357–364, 2009.

[43] L.H.K. Goh, Y.M. Hung, G.M. Chen, C.P. Tso, Entropy generation analysis of turbulent convection in a heat exchanger with self-rotating turbulator inserts, *Int. J. Therm. Sci.*, Vol. 160, 106652, 2021.

[44] İ. Kurtbaş, F. Gülçimen, A. Akbulut, D. Buran, Heat transfer augmentation by swirl generators inserted into a tube with constant heat flux, *Int. Commun. Heat Mass Transf.*, Vol. 36, 865–871, 2009.

[45] E. Taheran, K. Javaherdeh, Experimental investigation on the effect of inlet swirl generator on heat transfer and pressure drop of non-Newtonian nano-fluid, *Appl. Therm. Eng.*, Vol. 147, 551–561, 2019.

[46] P. Promvonge, Thermal augmentation in circular tube with twisted tape and wire coil turbulators, *Energ. Conver. Manage.*, Vol. 49, 2949–2955, 2008.

[47] B. Kumar, G.P. Srivastava, M. Kumar, A.K. Patil, A review of heat transfer and fluid flow mechanism in heat exchanger tube with inserts, *Chem. Eng. Process. Process Intensif.*, Vol. 123, 126–137, 2018.

Chapter 5

Heat transfer augmentation using pulsatile flows

5.1 INTRODUCTION

Many scientists and researchers are interested in studying the effects of pulsating flow on heat transfer. Various experimental and numerical investigations studied the heat transfer performance of pulsating flow across various conduits. Due to attributes of fluid pulsation like turbulence, dynamic change, and irregular mixing, etc., the thermal boundary layer is affected [1, 2]. The studies related to pulsating flow and heat transfer performance in channels, ducts, and tubes have gained prominence [1, 2]. To increase heat transmission, the flow of pulsating nature has been widely considered. With the advancement in electronic technology, components are continuously improved resulting in size compactness, low weight, and high thermal flux. As a result, finding effective and dependable cooling solutions to encounter the problem of high heat loads with constraints of volume and weight is quite critical. This is recognized and proven to promote heat transmission due to the advantage of fluid pulsatile behavior [1, 2]. As a result, pulsatile fluid flow research has improved heat transfer efficiency and received a lot of attention in recent years [3–7]. Pulsating flow possesses different forms that show various properties and methods affecting heat transfer phenomena. According to recent studies, a single-phase pulsating flow is helpful for industrial and medical applications [8–14]. Single-phase pulsating flow experiments that have already been done based on several limitations; hence, more experimental research should be done. Some researchers are interested in the study of prolonged pulsating flow on heat transfer due to boiling, while others are on long-period pulsating fields, however, short-period pulsating flows (also known as high-frequency pulsating flows) have recently received interest by certain researchers. When compared to a steady flow, most studies show that prolonged period pulsating flows reduce the critical heat flux and heat transfer coefficient of boiling [15].

In the case of intermittent impinging jets, previous research has shown that several characteristics, such as waveform, amplitude, and frequency of

DOI: 10.1201/9781003229865-5

intermittent impinging jets have an impact on heat transmission. In addition, the intermittent impinging jet heat transfer mechanism is examined. Air is the most studied working medium in this concern. The majority of studies show that an intermittent impinging jet improves heat transmission significantly.

Studying the mechanism and application of pulsating flow is both important and beneficial. At the moment, single-phase pulsating flow experimental work is limited, and high-frequency pulsating flow is receiving less attention. Exploring pulsating flow in other working fluids and under various experimental settings is especially important. Improvement in pulsating flow heat transfer is critical for its applications in various fields such as thermal, production and electrical engineering. As a result, improving heat transmission using pulsating flow is a topic worth studying and exploring.

5.2 IMPORTANT DIMENSIONLESS NUMBERS

To present the impact of pulsatile flow on heat transfer enhancement, some dimensionless numbers are introduced. A few of them are described in the following paragraphs:

5.2.1 Reynolds number

Reynolds number (Re) is a well-established parameter for predicting flow patterns. From airflow over an object to flow of liquid in channel/pipe, Reynolds number (Re) is used in a variety of contexts. Re predicts flow nature which can be laminar, transition, or turbulent flow and is also helpful in scaling of similar but different-sized flow conditions, for example a wind tunnel model and a full-scale aircraft. It is possible to foresee fluid behavior on a greater scale as well as associated meteorological and climatological consequences by being able to estimate scaling effects and anticipate the commencement of turbulence.

The interior bounding surface of pipe, for example, is a boundary layer where these forces change the flow behavior. The introduction of a stream of high-velocity fluid into a low-velocity fluid can have a similar result. This relative motion leads to the formation of fluid friction, which contributes to the development of turbulent flow. This is lessened by the viscosity of the fluid, which tends to suppress turbulence. The proportional importance of these two types of forces is quantified for certain flow circumstances by the Reynolds number, which also serves to foretell the occurrence of turbulent flow [16].

A high Reynolds number leads to turbulent flow, where the dominant nature of inertial forces is realized and is characterized by chaotic eddies and other flow instabilities. A low Re leads to laminar flow, where the effect of viscous forces is seen and is characterized by layered continuous fluid motion [17]. Re is mathematically defined as:

$$\text{Re} = \frac{\rho u L}{\mu} = \frac{u L}{v} \tag{5.1}$$

where ρ is fluid density, u is flow velocity, L a characteristic length, μ fluid dynamic viscosity, and v is kinematic viscosity.

5.2.2 Womersley number

In biofluid mechanics, Womersley number (α) is known as a dimensionless number. It is a one-dimensional representation of pulsatile flow frequency to viscous effects. It is named after John R. Womersley (1907–1958), who worked on arterial blood flow [18]. When scaling an experiment, the Womersley number is crucial for maintaining dynamic similarity. The Womersley number is useful for estimating the thickness of boundary layer and evaluating the entrance effects and whether it can be ignored or not. Womersley number is defined by the below relation and is generally given the notation α.

$$\alpha^2 = \frac{\text{transient inertial force}}{\text{viscous force}} = \frac{\rho \omega U}{\mu U L^{-2}} = \frac{\omega L^2}{v} \tag{5.2}$$

where ω is angular frequency, L a length scale, v (kinematic viscosity), μ (dynamic viscosity), and ρ (density) [19].

5.2.3 Strouhal number

Strouhal number (abbreviated as St or Sr to prevent confusion with Stanton number) is a dimensionless number which explains oscillating flow systems. V. Strouhal, a Czech physicist, experimented with wires experiencing vortex shedding in 1878 and called this parameter as named after him, [18, 19]. St number is expressed as,

$$\text{St} = \frac{f L}{U} \tag{5.3}$$

where f is frequency (vortex shedding), L is length (characteristic), and U is flow velocity. L is the amplitude of oscillation in some instances, such as heaving (plunging) flight. This length choice can be utilized to show the difference amid St and reduced frequency:

$$\text{St} = \frac{k A}{\pi c} \tag{5.4}$$

where k is reduced frequency, c is light velocity, and A is amplitude of heaving oscillation.

Viscosity dominates fluid flow for large Strouhal numbers (order of magnitude 1), causing a widespread oscillating motion similar to a fast-moving fluid plug flow. The quasi-steady-state component of motion dominates oscillation for low St (order of magnitude 10^{-4}) and lower. At intermediate Strouhal numbers, oscillation is characterized by fast accumulation and shedding of vortices [20].

There are two values of St in the Re range $8 \times 10^2 < \text{Re} < 2 \times 10^5$ for spheres in unvarying flow. Large-scale instability of wake flow is associated with lower frequency, which is not related to Re and is approximately 0.2. The St numbers of higher frequency are brought on by small-scale instabilities brought on by shear layer separation [21, 22].

5.2.4 Nusselt number

In a study of pulsating flow, Nusselt number (Nu) plays a vital role in quantifying heat transport. Normally the average Nu is utilized to evaluate heat transport characteristics. Nu over time is mathematically described as follows:

$$\text{Nu} = \int_0^L \int_0^T \text{Nu}_I(x,t)\,dt\,dx \qquad (5.5)$$

where Nu_I is local Nusselt number and T is time period of pulsating flow. Instantaneous local Nu varies depending on flow and operating conditions such as boundary conditions, geometry, etc.

5.3 PULSATING FLOW

Pulsating flows are categorized into a single-phase, two-phase, and intermittent jet flow as shown in Figure 5.1.

5.3.1 Single-phase pulsating flow

Improving the heat transfer rate of single-phase pulsating flow of laminar nature is one direction of the research of single-phase pulsating flow. Pulsating flow of laminar nature in a channel or tube is one of the most researched aspects of pulsating flow heat transfer [23–35]. There has also been research on pulsating flow over an object [62–73]. Additionally, there have been certain improvements made in the enhancement of heat transport [85–92].

There is numerous research on flow inside a channel [36–48] and external flow around objects [74–84]. It is worth debating whether the increase in

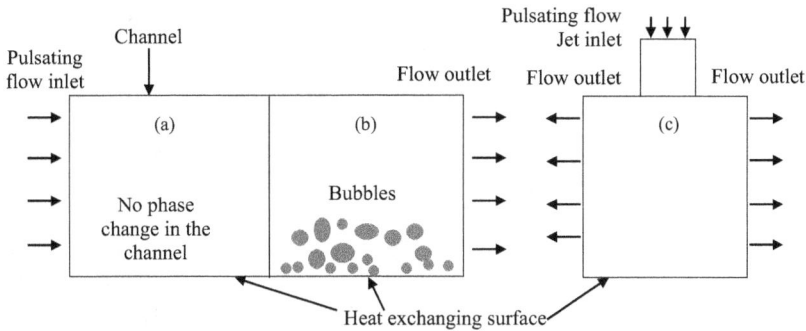

Figure 5.1 Pulsating flows types: (a) single-phase; (b) two-phase; and (c) intermittent jet flow [138].

the heat transfer rate of external flow or internal fluid flow is superior. Different heat transfer characteristics of exterior and interior flow can be compared. Pulsating flow heat transfer characteristics, such as heat transfer coefficient, Nu, and factors of heat transfer enhancement are the subject of investigations. In most circumstances, time-averaged Nusselt number for unidirectional laminar pulsating flows [49–61] doesn't change with respect to steady flow, and even thermal transmission impact might be weakened. Increases in Re, oscillation frequency, and amplitude promote heat transfer in reciprocating flow [86–93]. Pulsating flow is also frequently used with other heat transfer enhancement approaches, such as coupling porous medium with pulsating flow, with ribs, and with nanofluids, to boost heat transfer performance even further. The usage of porous material and pulsating flow within pipes was found to improve heat transmission [94–105]. A composite heat transfer enhancement method is single-phase nanofluid flow under pulsation. Many researchers have embraced the method of pulsating nanofluid flow to improve heat transmission. This chapter focuses on the influence of various-shaped particles and geometry on pulsating flow heat transfer [106–109]. The use of composite technology seems good for improving heat transfer. Furthermore, pulsating flow near fins has been shown to progress heat transfer [110–112].

5.3.2 Intermittent impinging jet

In reality, single-phase pulsation and two-phase pulsation are two types of intermittent impinging jet pulsation. On the other hand, intermittent jet research is included in a separate group because it has been described frequently as an effective heat transfer method. Previous research has examined the effect of intermittent impinging jet waveform, amplitude, and frequency [2–5]. Various research works have focused on various working media and most of the studies were conducted for air. The majority of studies show that an intermittent impinging jet improves heat transmission significantly.

The enhancement effect, according to the results, was significantly greater than a steady jet [113].

Effect of pulsating flow on heat transfer is unstable; it sometimes increases and sometimes decreases because of the complex nature of the flow and the variances in experimental geometry, operating circumstances, and waveforms. Studies on pulsing flow that are theoretical, experimental, and numerical have discrepancies. More studies have been done on applying pulsating flow to improve heat transfer, as well as its combinations with other heat transfer augmentation methods, for instance in examination of stable and pulsating flow in micro diffusers, as well as two types of flow in micro pumps [8]. In the field of medicine, a review of experimental and clinical research on intracranial pulsations has been completed, introducing the use of pulsing flow in the diagnosis of brain disorders [9]. Another study considered how periodic variation of nozzle flow can improve heat and mass transmission in impingement jets [10]. Utilizing a variety of experimental approaches, measurements of pulsing pressure and temperature have been explored [11].

5.4 SINGLE-PHASE PULSATING FLOW HEAT TRANSFER ENHANCEMENT

It can be categorized as unidirectional or reciprocating depending on flow direction. Figure 5.2 depicts many single-phase pulsating flows that have been investigated. Theoretical, numerical, and experimental methodologies are employed in these studies. Single-phase pulsating flow experiments that have already been done have several limitations; hence, more experimental research should be done.

5.4.1 Theoretical analysis

Analysis of the pulsing flow (laminar) forced convection heat transfer in a circular tube and a parallel-plate channel led to the identification of the link between velocity, frequency, and heat transfer [24]. The fluid velocity in the working environment is extremely low, resembling slug flow (not the same as pulsating flow). The pulsating flow can be explained as the inflow and outflow in a tube/channel are fluctuating. In a circular tube, a pulsing flow problem with constant heat flux boundary conditions was explored [25]. The Green's function approach is used to obtain the complex temperature fluctuation equation. The solution did not meet the necessary differential equation describing thermal energy conservation. To derive an analytical formula for temperature distribution, the Green's function method was applied [26]. Findings indicated that pulsation has a major impact on heat transmission, with heat transfer increasing as pulsating amplitude increases but decreasing as frequency and Prandtl number increase. Numerical simulation has been

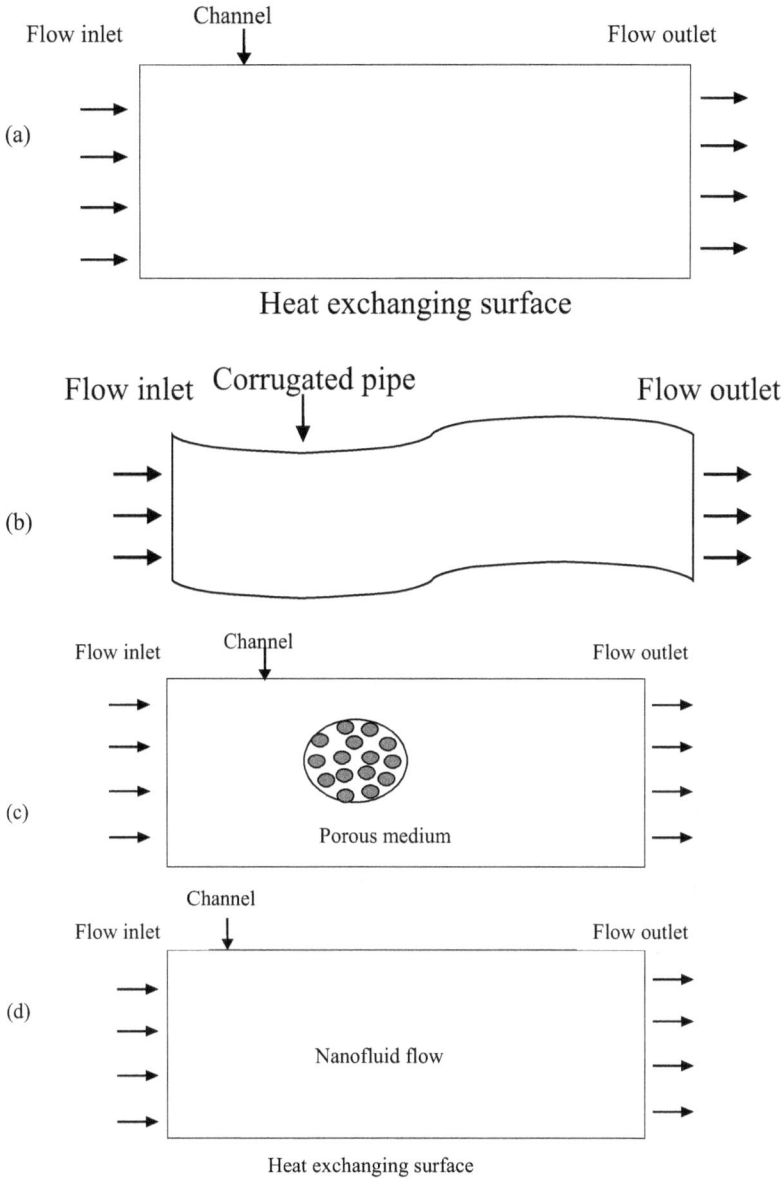

Figure 5.2 Single-phase pulsating flow types: (a) straight channels; (b) curved channels; (c) porous medium channels; (d) nanofluid based pulsing flow [138].

(*Continued*)

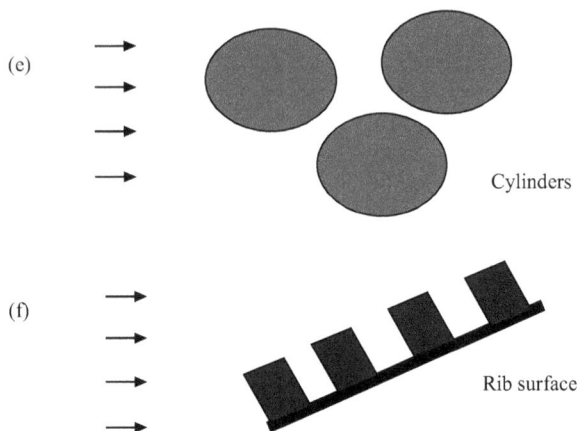

Figure 5.2 (Continued) Single-phase pulsating flow types: (e) flow over cylinder surface; and (f) flow around ribs.

used to calculate Nusselt number in a thermally developing region. Yuan et al. [30] also theoretically studied pulsating flow with nonvarying heat flux boundary state. Time-average Nu was found to be unaffected by pulsation and higher pulsation frequency. With increasing Prandtl number, the peak value drops. In a theoretical investigation of pulsing flow (laminar) between parallel plates, the impact of wall heat capacity under nonvarying heat flux boundary state was investigated [30]. In contrast to pulse heat transfer without paying attention to wall heat capacity, fluid temperature fluctuates less and mean temperature difference is larger. Whether wall thermal inertia is taken into account or not, pulsating flow is not conducive to heat transmission. The flow filled of the pipe component of a refrigeration system was successively approximated using an analytical method [31]. Both the temperature oscillation range and the rate of heat transmission diminish when the pulse in tube is shifted from cold to hot end. Pulsating flow in a pipe was studied theoretically for heat transmission properties. The difference in heat transfer between pulsating and constant flows decreases throughout the pipe, and this variation affects thermal developing region more [32].

5.4.2 Numerical research

A numerical and theoretical investigation has been performed to analyze heat transfer rate in pulsating flow [26]. Heat transfer pulsation in thermal developing zone is suppressed by wall thermal resistance, according to numerical simulation [26]. Turbulent pulsation has a stronger heat transfer enhancing effect than laminar pulsation. Nusselt number effects were theoretically evaluated in a study [33]. At varying pulsation frequencies and small amplitudes, heat transfer increases and decreases, but at large amplitudes it improves. The flow and heat transmission in a circular tube with laminar pulsing flow were numerically analyzed [34]. Findings show not only that pulsation has little impact on time-average heat transfer but also that pulsation causes the Nu number to fluctuate close to the pipeline intake. Craciunescu and Clegg [35] analyzed the effect of pulsed blood velocity on bioheat transmission. The results show that axial velocity causes temperature changes.

Energy equation and boundary layer equation of pulsating flow in a tube were numerically solved in a study [36]. Results reveal that, depending on pulsation frequency, Nusselt number might grow or decrease in comparison to steady flow in a fully developed zone. When the Prandtl number is low, below unity, the Nusselt number rises as the amplitude rises. The influence of pulsing flow in a thermal developing zone was studied statistically by Kim et al. [37]. At Re = 50 and Pr = 0.7, the findings were simulated in a relatively sluggish laminar flow. When the frequency is minimal, the impact of frequency on Nu is significant, according to the findings. The impact of pulsation on heat transport is notably reduced at higher pulsation frequencies. Wang et al. [39] developed a three-dimensional and transient model for fluid dynamics and heat transport in a channel using LVG (longitudinal vortex generator). Findings revealed that both amplitude and period are critical parameters that have a significant impact on flow and heat transfer efficiency. Heat transmission in wavy channels has been widely investigated due to the large number of wavy channels in heat exchangers. Heat transfer is aided by the synergetic impact of wavy channels and pulsating flow. Heat transfer of corrugated tubes, however, is restricted by the stable flow. Wavy channel research is primarily focused on numerical methods. Reference [40] looked into how fuel cells are affected by pressure changes in wavy cathode channels. The study considered the impact of pressure oscillation amplitude and frequency on the efficiency of PEM fuel cells. PEM fuel cell efficiency boosted by about 7% when amplitude is 0.7 times pressure loss in a wavy channel, but the influence of pulsating frequency on power is minimal. The heat transfer rate achieves its greatest value for a crucial frequency.

5.4.3 Experimental study

Researchers have also shown inconsistent results when it comes to experiments. Experimental results from laminar flow pulsation [47–51] showed that velocity pulsation can lower or increase Nu. The effect of pulsation with different amplitudes in a heat exchanger (with coiled wire around the inner tube) has been investigated in [53]. A reciprocating device produced pulse frequencies. Compared to a smooth annulus with no pulsation, the maximum Nusselt number enhancement ratio was 12.7, while the friction factor ratio was 8.7 for pitch ratio 10 and stroke ratio 4.2. In a study, constant heat flux boundary conditions were used to examine pulsating turbulent air in a tube [54]. Frequency-based investigations were done. The local

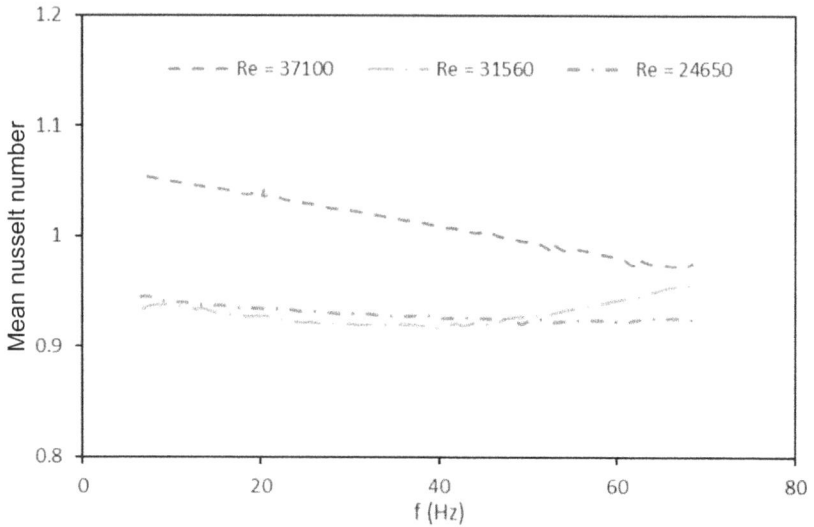

Figure 5.3 Mean Nusselt number variation trend with pulsating frequency at different Reynolds number [54].

value varies and averaged Nu appears to increase or decrease depending on the frequency range. Figure 5.3 shows the relation between the averaged Nusselt number and pulse frequency. The maximum measurable heat transfer augmentation is around 9% at Re = 37,100, f = 13.3 Hz, and x^* of roughly 0.5. This is attributed to the interaction of predicted burst with frequency of pulse.

Reactors, heat exchangers, and fuel cells contain microchannels. In a narrow rectangular pipe, friction characteristics in a low-frequency oscillating turbulent-laminar transition zone were studied [55]. Velocity oscillation affects friction mostly in transition region. Complex frequency and oscillating amplitude affect the time-average friction. Figure 5.4 shows how pulse frequency affects friction ratio.

5.5 PULSATING FLOW AROUND A CYLINDER

Several researchers have studied heat transfer and flow process surrounding a cylinder in pulsating cross-flow, and it has been suggested that unsteady flow behavior at the cylinder surface can improve heat transfer [63–68].

5.5.1 Experimental study

Due to the complexity of the phenomena, the majority of studies on this topic focuses on experimental studies [70]. A rectangular pipe with pulsating flow has been examined to evaluate heat transfer properties experimentally.

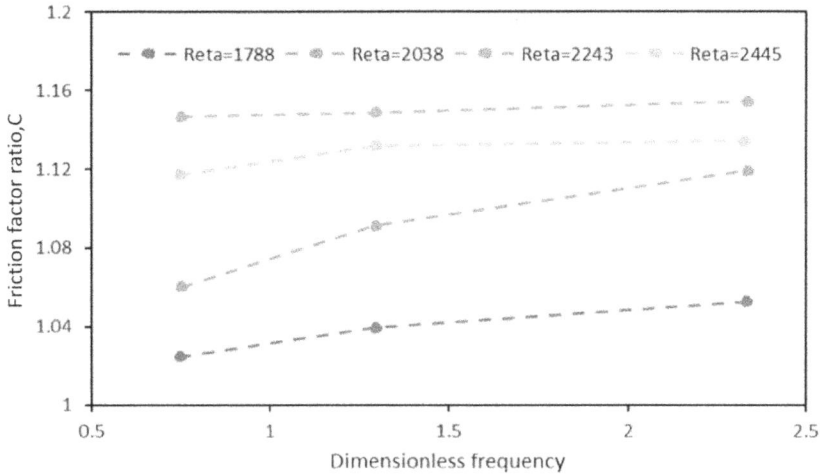

Figure 5.4 Friction coefficient ratio, amplitude, and dimensionless frequency relation image [55].

As pulsation amplitude and frequency rise, the rate of heat transmission increases. Experimental investigation of pulsing flow around a cylinder has been conducted to find laws of vortex shedding and heat transmission [77]. The results suggest that forcing external flow to pulse can improve average heat transfer on cylinder surface. The pulsating cross-flow of type IV exhibits the greatest improvement in average heat transfer on the cylinder surface. Figure 5.5 depicts vortex shedding pattern. The largest rise in average heat transfer coefficient within the range of analyzed unsteady state parameters is roughly 15%.

Finally, pulsating cross flow outcomes [63–72] present that forced pulsations can boost the rate of heat transmission around a cylinder. Flow pattern affects the cylinder heat transfer. Pulsation amplitude and frequency can cause multiple flow phenomena that affect heat transfer.

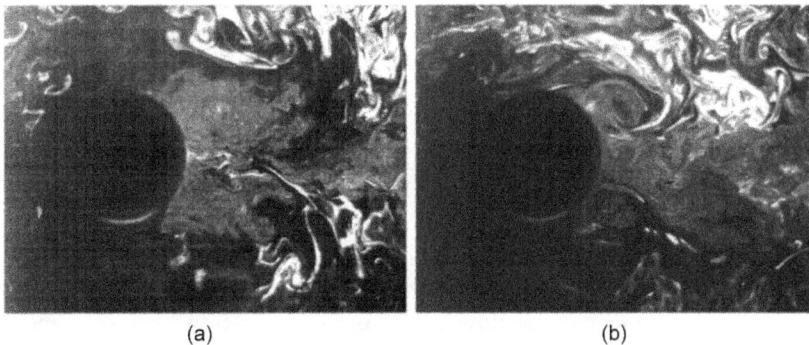

(a)   (b)

Figure 5.5 Pulsating flow picture around cylinder at (a) St = 0.29 and (b) St = 0.69 [78].

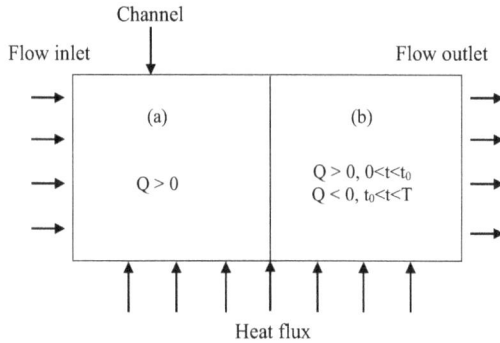

Figure 5.6 (a) Pulsating flow of unidirectional nature and (b) reciprocating fluid flow [138].

5.6 RECIPROCATING FLOW

Reciprocating flow, in light of unidirectional flow, is classified individually due to its superior heat transmission capabilities. As shown in Figure 5.6, reciprocating flow is a type of pulsating flow in which flow rate is intermittently positive and negative. In this field, numerous studies have been done. The discovered "annular effect" using experimental procedures and following studies [114–116] confirmed this numerically and experimentally. Single-phase heat transfer characteristics of reciprocating flow have been investigated in a piston–cylinder mechanism. It is noted that increasing pulsation amplitude and frequency improves heat transfer rate. Using air as the working medium was considered in a numerical simulation of pulsating flow problems [88]. It was pointed out that as flow amplitude and frequency increase, the effect of increasing Reynolds number on heat transfer enhancement is favorable. Zhao and Cheng [89] numerically simulated the incompressible periodic flow of laminar nature with constant wall temperature. The Reynolds number, pulsation amplitude, Prandtl number, and aspect ratio are found to be the parameters governing heat transfer performance in this situation similarly to what has been found by other researchers also [90, 91].

5.7 SINGLE-PHASE PULSATING FLOW AND POROUS MEDIA

To remove more heat from chips in high-power electronic devices, a porous medium has been investigated as a heat sink in a restricted channel with pulsating flow. For creeping flow condition, investigation of porous media pulsating flow has been analyzed analytically [94]. The geometric characteristics of the porous media and the Stokes number are shown to be related to the impact of flow on heat transmission. Pulsating air flow in aluminum foam was experimentally studied to analyze heat transfer characteristics [95–97]. The distribution of pressure drop, surface temperature through the aluminum foam, and fluid flow velocity within the medium were investigated. The mean Nusselt number of pulsating flows was found to be higher than that for constant flow velocity. They also discovered that when the dimensionless frequency and amplitude grew, the heat transfer enhancement by pulsating flow was increased. Bayomy and Saghir [93] studied heat transfer rate of aluminum foam radiators using experimental and numerical methods for pulsating water flow. The experimental data revealed that when the heat flux declines and the magnitude and frequency of flow increase, the cyclic mean local temperature decreases. In comparison to stationary flow, the average Nu of pulsing flow is 14% higher. They also discovered that pulsating water flow showed a 73% lower uniformity than steady-state water flow.

5.8 PULSATING NANOFLUID FLOW

Because of the superior performance of nanofluids, such as pairing with magnetic and electrical fields [119–120] and reducing the friction drag [122–123], studies on the pulsating flow of nanofluids have become increasingly popular. Many studies have considered nanofluid properties and found that they have better thermal properties than the base fluids [125–130]. Many researchers have embraced the method of pulsating nanofluid flow to improve heat transmission. The heat transmission and flow parameters of nanofluids under a magnetic field in a micro-finned tube were investigated [131]. Heat transfer increased as magnetic field intensity, particle concentration, and frequency increased.

The composite heat transfer improvement method has the potential to improve heat transfer in the energy sector and chemical equipment. Figure 5.7 shows the effect of frequency and Re on heat transfer. High-frequency pulsing flow has a larger Nusselt number in the absence of a magnetic field.

5.9 PULSATING FLOW AROUND RIBS

The heat transfer characteristics of oscillating airflow behind ribs in an experimental study were investigated in the studies [110, 111]. To further assess the rate of heat transfer, they utilized copper wires to detect wall temperatures. The experimental findings reveal that under constant wall temperature and varying temperature conditions, heat transfer augmentation of pulsing flow is 25% maximum as shown in Figure 5.8 [132]. Figure 5.9 displays the positions (a, b, and c) of the acceleration phase of the period.

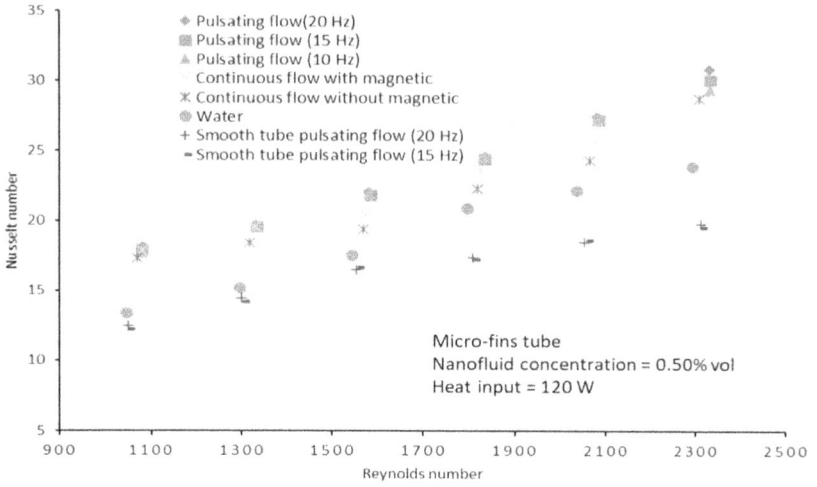

Figure 5.7 Nu versus Re number variation under magnetic field for micro-finned and smooth pipes [133].

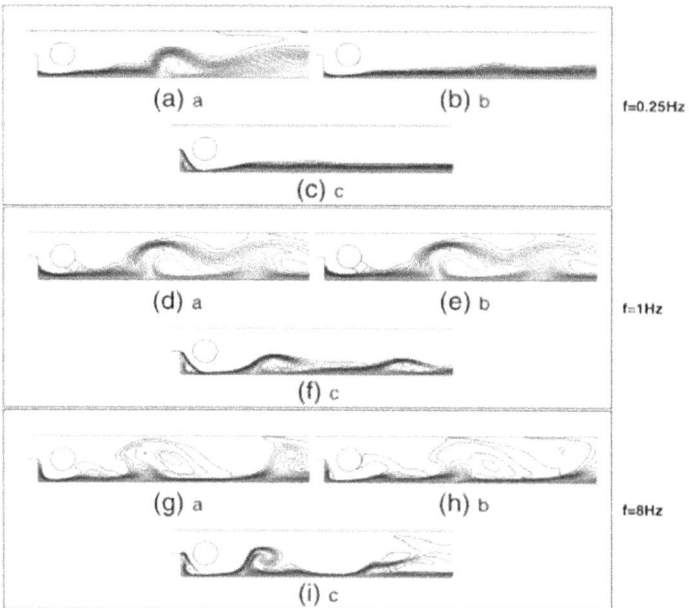

Figure 5.8 Pulsating flow temperature contours under different frequencies and under three conditions (Figure 14) at ϕ = 0.002 and Re = 100 [134].

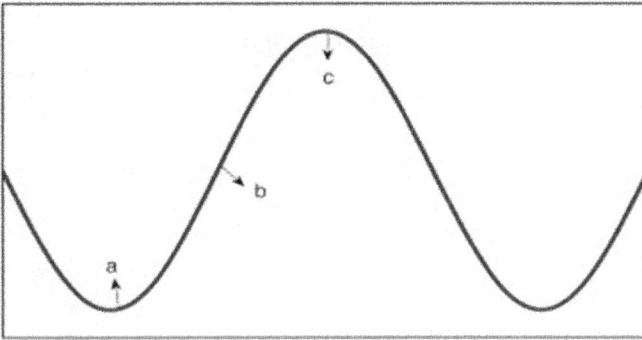

Figure 5.9 Points a, b, and c during the acceleration phase of a period [134].

Table 5.1 Effect of the pulsating period on HTC

Period	Average HTC	Reference
0.05 to 1 s	Improved by 26%	Davletshin et al. [113]
2 to 20 s	Improved by 27%	Eibeck et al. [114]
20 to 120 s	Insignificant	Richardson and Tyler [115]
10 to 20 s	Insignificant	Shokouhmand et al. [116]
1.04 to 10.6 s	Insignificant	Naphon and Wiriyasart [133]
20 to 120 s	Insignificant	Selimefendigil and Öztop [134]
3 to 20 s	Insignificant	Zhao et al. [135]
1 to 9 s	Improved initially then decreased	Wang et al. [136]

Heat transmission of constant and fluctuating air flows via a spanwise rib was examined [112]. The distributions of transverse velocity component and heat transfer coefficient are highly correlated.

Because of the contradicting results of different research studies, it is tough to find an appropriate judgment on the impact of pulsing flow duration on mean heat transfer coefficient as shown in Table 5.1. However, past research studies have revealed that the condition of pulsating flow can boost heat transmission when used for a short time period. It can also be deduced that high-frequency flow helps to improve heat transmission.

5.10 CONCLUSIONS

Because pulsating flow is employed in many industrial applications and in the medical field, researchers are studying the mechanism of pulsating flow and how it is used. Pulsating flows have been shown to improve heat transfer under certain conditions in both numerical and experimental investigations. Despite that numerous investigations have been conducted on pulsating flow, the mechanism of pulsating flow remains unknown. In this chapter, theory and research related to pulsating flow have been properly consolidated to guide future research. This chapter describes single-phase pulsatile flow in various systems. Below are some of the most important findings:

1. Theoretical study on pulsating flow is further hampered by the complexity of turbulence theory. Only single-phase laminar pulsating flow has been explored theoretically using Green's function approach so far. Furthermore, Reynolds number of laminar flow is found to be very small based on features of the pulsating flow. As a result, theoretical research is extremely limited.

2. Current single-phase pulsating flow in a channel produces paradoxical outcomes, and flow and heat transmission may be strengthened or diminished. According to earlier studies, single-phase pulsating flow has a marginal effect on heat transmission, but combined with micro-surface modification technologies, curved channels, and other techniques, a stronger heat transfer effect can be accomplished.

3. Pulsating cross-flow data revealed that forcing pulsations could improve average heat transfer from a cylinder. While there are still some inconsistencies in the results of inner flow in a channel, the majority of studies suggest that outer pulsating flow can improve heat transmission. According to available data, it is found that internal or external pulsating flow improves the heat transmission by less than 20%. External pulsating flows provide a more stable enhancement of heat transfer.

4. Pulsating flow research results for unidirectional laminar pulsating flows demonstrated that pulsating flow gives no more than 20% increase in heat transfer under specified experimental settings. When compared to a stable flow, the time-averaged Nusselt number rarely changes, and, in some situations, it even weakens the heat transfer. Increases in Reynolds number, oscillation frequency, and amplitude promote heat transfer in reciprocating flow.

5. When using nanofluids, heat transfer improvement of pulsating flow is more substantial. Pulsating flows have been found in previous studies to improve heat transmission in a majority of circumstances. Most researchers feel that combining nanofluids and pulsating flows can improve heat transfer. The heat transfer enhancement effect was dramatically boosted as nanoparticle concentration, pulse amplitude, and pulse frequency were increased.

6. Pulsating flow heat transfer in porous media channels is influenced by the material, shape, frequency, and amplitude of pulsating flow. According to the major studies mentioned above, there is a critical frequency for achieving maximal heat transfer enhancement. Heat transport in porous channels can be improved by using materials with

low permeability and high thermal conductivity. Heat transmission is improved when pulsating flow and porous media are combined.

7. According to some research, pulsating flow near the fins improves heat transmission. It is advantageous to combine pulsating flow and nonlinear boundary conditions to improve heat transfer.

8. When compared to traditional heat transfer methods, jet impingement is a better option when combined with pulsation, which has been shown to increase heat transmission by 2.5 times. Intermittent jet impingement boosts or reduces heat transmission depending on circumstances. Existing research has concluded that a square wave pulse impinging jet is superior to a sine wave for heat transmission. More theoretical research is needed in the future to reveal the mechanism. In addition, more experimental studies are needed to validate and complement numerical simulation results.

REFERENCES

[1] E.C. Mladin, D.A. Zumbrunnen, Alterations to coherent flow structures and heat transfer due to pulsations in an impinging air-jet, International *J. Therm. Sci.*, Vol. 39, 236–248, 2000.

[2] H.S. Sheriff, D.A. Zumbrunnen, Effect of flow pulsations on the cooling effectiveness of an impinging jet, *J. Heat Transfer*, Vol. 116, 886–895, 1994.

[3] E.C. Mladin, D.A. Zumbrunnen, Local convective heat transfer to submerged pulsating jets, *Int. J. Heat Mass Transf.*, Vol. 40, 3305–3321, 1997.

[4] D.J. Sailor, D.J. Rohli, Q.L. Fu, Effect of variable duty cycle flow pulsations on heat transfer enhancement for an impinging air jet, *Int. J. Heat Fluid Flow*, Vol. 20, 574–580, 1999.

[5] H.J. Poh, K. Kumar, A.S. Mujumdar, Heat transfer from a pulsed laminar impinging jet, *Int. Commun. Heat Mass Transf.*, Vol. 32, 1317–1324, 2005.

[6] C.A. Chen, T.F. Lin, W.M. Yan, Experimental study on time periodic evaporation heat transfer of R-134a in annular ducts due to wall heat flux oscillation, *Int. J. Heat Mass Transf.*, Vol. 106, 1232–1241, 2016.

[7] C.A. Chen, T.F. Lin, H.M. Ali, et al., Bubble dynamics in evaporation flow of R-134a in narrow annular ducts due to flow rate oscillation, *Int. Commun. Heat Mass Transf.*, Vol. 100, 27–34, 2019.

[8] M. Nabavi, Steady and unsteady flow analysis in microdiffusers and micropumps: A critical review, *Microfluid. Nanofluid.*, Vol. 7, 599–619, 2009.

[9] M.E. Wagshul, P.K. Eide, et al., The pulsating brain: A review of experimental and clinical studies of intracranial pulsatility, *Fluids Barriers CNS*, Vol. 8, 5, 2011.

[10] V. Tesa, Z. Tr´avníek, Pulsating and Synthetic Impinging Jets, *J. Vis.*, Vol. 8, 201–208, 2005.

[11] M. Nabavi, Invited review article: Unsteady and pulsating pressure and temperature: A review of experimental techniques, *Rev. Sci. Instrum.*, Vol. 81, 031101, 2010.

[12] M. Metwally, Review of compressible pulsating flow effects on system performance, *13th International Conference on Aerospace Sciences & Aviation Technology*, ASAT-13, 2009.

[13] M. Hemmat Esfe, et al., A critical review on pulsating flow in conventional fluids and nanofluids: Thermo-hydraulic characteristics, *Int. Commun. Heat Mass Transf.*, Vol. 120, 104859, 2021.

[14] N.A. Cumpsty, E.M. Greitzer, A simple model for compressor stall cell propagation, *Trans. ASME J. Eng. Gas Turb. Power*, Vol. 104, 170–176, 1982.

[15] B. Yuan, Y.H. Zhang, L. Liu, J.J. Wei, Y. Yang, Experimental research on heat transfer enhancement and associated bubble characteristics under high-frequency reciprocating flow, *Int. J. Heat Mass Transf.* Vol. 146 (2020), 118825.

[16] M.A. Moyers-Gonzalez, R.G. Owens, J. Fang, A nonhomogeneous constitutive model for human blood: Part 1. Model derivation and steady flow, *J. Fluid Mech.*, Vol. 617, 327–354, 2008.

[17] R.G. Owens, M. Moyers-Gonzalez, J. Fang, On the simulation of steady and oscillatory blood flow in a tube using a new nonhomogeneous constitutive model, *Biorheology*, Vol. 45, 83–84, 2008.

[18] G.T. Liu, X.J. Wang, B.Q. Ai, et al., Numerical study of pulsating flow through a tapered artery with stenosis, *Chin. J. Physiol.*, Vol. 42, 401–408, 2004.

[19] J.R. Buchanan, C. Kleinstreuer, S. Hyun, et al., Hemodynamics simulation and identification of susceptible sites of atherosclerotic lesion formation in a model abdominal aorta, *J. Biomech.*, Vol. 36, 1185, 2003.

[20] R. Botnar, G. Rappitsch, M. Scheidegger, et al., Hemodynamics in the carotid artery bifurcation, *J. Biomech.*, Vol. 33, 137, 2000.

[21] L. Moreno, F. Calderas, G. Sanchez-Olivares, et al., Effect of cholesterol and triglycerides levels on the rheological behavior of human blood, *Korea Aust. Rheol. J.*, Vol. 27, 1–10, 2015.

[22] M.A. Moyers-Gonzalez, R.G. Owens, Mathematical modelling of the cell-depleted peripheral layer in the steady flow of blood in a tube, *Biorheology*, Vol. 47, 39–71, 2010.

[23] R. Siegel, Transient heat transfer for laminar slug flow in ducts, *ASME J. Appl. Mech. Series E*, Vol. 26, 140–142, 1959.

[24] R. Siegel, M. Perlmutter, Heat transfer for pulsating laminar duct flow, *ASME J. Heat Transf.*, Vol. 84, 111–123, 1962.

[25] T. Moschandreou, M. Zamir, Heat transfer in a tube with pulsating flow and constant heat flux, *Int. J. Heat Mass Transf.*, Vol. 40, 2461–2466, 1997.

[26] H.N. Hemida, M.N. Sabry, H. Mansour, Theoretical analysis of heat transfer in laminar pulsating flow, *Int. J. Heat Mass Transf.*, Vol. 45, 1767–1780, 2002.

[27] J.C. Yu, Z.X. Li, T.S. Zhao, An analytical study of pulsating laminar heat convection in a circular tube with constant heat flux, *Int. J. Heat Mass Transf.*, Vol. 47, 5297–5301, 2004.

[28] D.A. Nield, A.V. Kuznetsov, Forced convection with laminar pulsating flow in a saturated porous channel or tube, *Int. J. Therm. Sci.*, Vol. 46, 551–560, 2007.

[29] B.H. Yan, L. Yu, et al., Heat transfer with laminar pulsating flow in a channel or tube in rolling motion, *Int. J. Therm. Sci.*, Vol. 49, 1003–1009, 2010.

[30] H. Yuan, et al., Theoretical analysis of wall thermal inertial effects on heat transfer of pulsating laminar flow in a channel, *Int. Commun. Heat Mass Transf.*, Vol. 53, 14–17, 2014.

[31] M. Azadi, A. Jafarian, M. Timaji, Analytical investigation of oscillating flow heat transfer in pulse tubes, *Scientia Iranica B*, Vol. 20(3), 483–491, 2013.

[32] G. Hafez, O. Montasser, A theoretical study on enhancing the heat transfer by pulsation, in *11th International Mechanical Power Engineering Conference*, H128–H137, Cairo, February 5–7, 2000.

[33] Z. Guo, H.J. Sung, Analysis of the Nusselt Number in Pulsating Pipe Flow, *Int. J. Heat Mass Transf.*, Vol. 40, 2486–2489, 1997.

[34] H. Chattopadhyay, F. Durst, S. Ray, Analysis of heat transfer in simultaneously developing pulsating laminar flow in a pipe with constant wall temperature, *Int. Commun. Heat Mass Transf.*, Vol. 33, 475–481, 2006.

[35] O.I. Craciunescu, S.T. Clegg, Pulsatile blood flow effects on temperature distribution and heat transfer in rigid vessels, *J. Biomech. Eng.*, Vol. 123, 500–505, 2001.

[36] H.W. Cho, J.M. Hyun, Numerical solution of pulsating flow and heat transfer characteristics in a pipe, *Int. J. Heat Fluid Flow*, Vol. 11, 321–330, 1990.

[37] S.Y. Kim, B.H. Kang, J.M. Hyun, Heat transfer in the thermally developing region of a pulsating channel flow, *Int. J. Heat Mass Transf.*, Vol. 36, 4257–4266, 1993.

[38] P.K. Papadopoulos, A.P. Vouros, Pulsating turbulent pipe flow in the current dominated regime at high and very-high frequencies, *Int. J. Heat Fluid Flow*, Vol. 58, 54–67, 2016.

[39] Y. Wang, Y.L. He, W.W. Yang, et al., Numerical analysis of flow resistance and heat transfer in a channel with delta winglets under laminar pulsating flow, *Int. J. Heat Mass Transf.*, Vol. 82, 51–65, 2015.

[40] H.R. Ashorynejad, K. Javaherdeh, H.E.A. Van den Akker, The effect of pulsating pressure on the performance of a PEM fuel cell with a wavy cathode surface, *Int. J. Hydrogen Energy*, Vol. 41, 14239–14251, 2016.

[41] M. Jafari, M. Farhadi, K. Sedighi, Pulsating flow effects on convection heat transfer in a corrugated channel: A LBM approach, *Int. Commun. Heat Mass Transf.*, Vol. 45, 146–154, 2013.

[42] T.K. Nandi, H. Chattopadhyay, Numerical investigations of simultaneously developing flow in wavy microchannels under pulsating inlet flow condition, *Int. Commun. Heat Mass Transf.*, Vol. 47, 27–31, 2013.

[43] M. Shu, D. Qing, H. Wang, Study on heat transfer performance of pulsating flow in convergent-divergent tube, *IOP Conf. Ser.: Earth Environ. Sci.*, Vol. 153, 032023, 2018.

[44] V.Q. Hoang, T.T. Hoang, C.T. Dinh, et al., Large eddy simulation of the turbulence heat and mass transfer of pulsating flow in a V-sharp corrugated channel, *Int. J. Heat Mass Transf.*, Vol. 166, 120720, 2021.

[45] D. Zheng, X. Wang, Q. Yuan, The effect of pulsating parameters on the spatiotemporal variation of flow and heat transfer characteristics in a ribbed channel of a gas turbine blade with the pulsating inlet flow, *Int. J. Heat Mass Transf.*, Vol. 153, 119609, 2020.

[46] S.K. Gupta, T.R.D. Patel, R.C. Ackerberg, Wall heat/mass transfer in pulsatile flow, *Chem. Eng. Sci.*, Vol. 37(12), 1727–1739, 1982.

[47] R. Lemlich, Vibration and pulsation boost heat transfer, *Chem. Eng.*, Vol. 68, 171–176, 1961.

[48] T. Niida, T. Yoshida, R. Yamashita, S. Nakayama, The influence of pulsation on laminar heat transfer in pipes, *Chem. Eng.*, Vol. 38, 47–53, 1974.

[49] M.A. Habib, A.M. Attya, A.I. Eid, A.Z. Aly, Convective heat transfer characteristics of laminar pulsating pipe air flow, *Heat Mass Transf.*, Vol. 38(3), 221–232, 2002.

[50] M. Mackley, G. Tweddle, I. Wyatt, Experimental heat transfer measurements for pulsatile flow in baffled tubes, *Chem. Eng. Sci.*, Vol. 45(5), 1237–1242, 1990.

[51] V. Mamayev, V. Nosov, N. Syromyatnikov, Investigation of heat transfer in pulsed flow of air in pipes, *Heat Transf. Sov. Res.*, Vol. 8(3), 111–116, 1976.

[52] G.B. Darling, Heat transfer to liquids in intermittent flow, *Petroleum*, Vol. 180, 177–180, 1959.

[53] A.E. Zohir, A.A. Abdel Aziz, et al., Heat transfer characteristics and pressure drop of the concentric tube equipped with coiled wires for pulsating turbulent flow, *Exp. Therm Fluid Sci.*, Vol. 65, 41–51, 2015.

[54] E.A.M. Elshafei, M.S. Mohamed, et al., Experimental study of heat transfer in pulsating turbulent flow in a pipe, *Int. J. Heat Fluid Flow*, Vol. 29, 1029–1038, 2008.

[55] N. Zhuang, S. Tan, et al., The friction characteristics of low-frequency transitional pulsatile flows in narrow channel, *Exp. Therm Fluid Sci.*, Vol. 76, 352–364, 2016.

[56] R. Blythman, T. Persoons, N. Jeffers, et al., Localised dynamics of laminar pulsatile flow in a rectangular channel, *Int. J. Heat Fluid Flow*, Vol. 66, 8–17, 2017.

[57] M. Simonetti, C. Caillol, P. Higelin, et al., Experimental investigation and 1D analytical approach on convective heat transfers in engine exhaust-type turbulent pulsating flows, *Appl. Therm. Eng.*, Vol. 165, 114548, 2020.

[58] C.S. Wang, T.C. Wei, P.Y. Shen, et al., Lattice Boltzmann study of flow pulsation on heat transfer augmentation in a louvered microchannel heat sink, *Int. J. Heat Mass Transf.*, Vol. 148, 119139, 2020.

[59] S. Kato, K. Okuyama, T. Ichikawa, S. Mori, A single, straight-tube pulsating heat pipe (examination of a mechanism for the enhancement of heat transport), *Int. J. Heat Mass Transf.*, Vol. 64, 254–262, 2013.

[60] L.J. Guo, X.J. Chen, Z.P. Feng, et al., Transient convective heat transfer in a helical coiled tube with pulsatile fully developed turbulent flow, *Int. J. Heat Mass Transf.*, Vol. 41(19), 2867–2875, 1998.

[61] N. Kurtulmus, B. Sahi, Experimental investigation of pulsating flow structures and heat transfer characteristics in sinusoidal channels, *Int. J. Mech. Sci.*, Vol. 167, 105268, 2020.

[62] H. Khosravi-Bizhaem, A. Abbassi, A.Z. Ravan, Heat transfer enhancement and pressure drop by pulsating flow through helically coiled tube: An experimental study, *Appl. Therm. Eng.*, vol. 160, 114012, 2019.

[63] M.R.H. Nobari, J. Ghazanfarian, Convective heat transfer from a rotating cylinder with inline oscillation, *Int. J. Therm. Sci.*, Vol. 49, 2026–2036, 2010.

[64] M.R.H. Nobari, J. Ghazanfarian, A numerical investigation of fluid flow over a rotating cylinder with cross flow oscillation, *Comput. Fluids*, Vol. 38, 2026–2036, 2009.

[65] M.R.H. Nobari, H. Naderan, A numerical study of flow past a cylinder with cross flow and inline oscillation, *Comput. Fluids*, Vol. 35, 393–415, 2006.

[66] F.M. Mahfouz, H.M. Badr, Forced convection from a rotationally oscillating cylinder placed in a uniform stream, *Int. J. Heat Mass Transf.*, Vol. 43, 3093–3104, 2000.

[67] S.-J. Lee, J.-Y. Lee, Temporal evolution of wake behind a rotationally oscillating circular cylinder, *Phys. Fluids*, Vol. 19(10), 105104, 2007.

[68] S. Choi, H. Choi, S. Kang, Characteristics of flow over a rotationally oscillating cylinder at low Reynolds number, *Phys. Fluids*, Vol. 14(8), 2767, 2002.

[69] D. Jian, S. Xue-ming, R. An-lu, Vanishing of three-dimensionality in the wake behind a rotationally oscillating circular cylinder, *J. Hydrodyn. Ser. B*, Vol. 19(6), 751–755, 2007.

[70] S.J. Lee, J.Y. Lee, PIV measurements of the wake behind a rotationally oscillating circular cylinder, *J. Fluids Struct.*, Vol. 24, 2–17, 2008.

[71] W. Cooper, V. Nee, K. Yang, An experimental investigation of convective heat transfer from the heated floor of a rectangular duct to a low frequency, large tidal displacement oscillating flow, *Int. J. Heat Mass Transf.*, Vol. 37, 581–592, 1994.

[72] G.S. Jones, C. Barbi, D.P. Telionis, Natural and forced vortex shedding, *IUTAM Unsteady Turbulent Shear Flows Symposium*, 228–247, Toulouse, France, 1981.

[73] C. Barbi, D. Favier, C. Maresca, Vortex shedding from a circular cylinder in oscillatory flow, *IUTAM Unsteady Turbulent Shear Flows Symposium*, 248–261, Toulouse, France, 1981.

[74] C. Barbi, D.P. Favier, C.A. Maresca, D.P. Telionis, Vortex shedding and lock-on of a cylinder in oscillatory flow, *J. Fluid Mech.*, Vol. 170, 527–544, 1986.

[75] E. Konstantinidis, S. Balabani, Symmetric vortex shedding in the near wake of a circular cylinder due to streamwise perturbations, *J. Fluids Struct.*, Vol. 23(7), 1047–1063, 2007.

[76] Y. Kikuchi, H. Suzuki, M. Kitagawa, K. Ikeya, Effect of pulsating Strouhal number on heat transfer around a heated cylinder in pulsating cross-flow, *JSME Int.l J. Series B*, Vol. 43(2), 250–257, 2000.

[77] E. Konstantinidis, D. Bouris, Vortex synchronization in the cylinder wake due to harmonic and non-harmonic perturbations, *J. Fluid Mech.*, Vol. 804, 248–277, 2016.

[78] N.I. Mikheev, V.M. Molochnikov, A.N. Mikheev, Hydrodynamics and heat transfer of pulsating flow around a cylinder, *Int. J. Heat Mass Transf.*, vol. 109, 254–265, 2017.

[79] G. Li, Y. Zheng, G. Hu, et al., Experimental investigation on heat transfer enhancement from an inclined heated cylinder with constant heat input power in infrasonic pulsating flows, *Exp. Therm Fluid Sci.*, Vol. 49, 75–85, 2013.

[80] I.A. Davletshin, N.I. Mikheev, et al., Separation of a pulsating flow, *Dokl. Phys.*, Vol. 52, 695–698, 2007.

[81] G. Narayana, S. Selvaraj, Transient heat transfer measurements on pulsating and oscillating flows in a shock tunnel, *Aerosp. Sci. Technol.*, Vol. 104, 105879, 2020.

[82] D. Palfreyman, R.F. Martinez-Botas, The pulsating flow field in a mixed flow turbocharger turbine: An experimental and computational study, *J. Turbomach.*, Vol. 127, 144–155, 2005.

[83] P. Sellappan, T. Pottebaum, Vortex shedding and heat transfer in rotationally oscillating cylinders, *J. Fluid Mech.*, Vol. 748, 549–579, 2014.

[84] T.I. Mulcahey, M.G. Pathak, S.M. Ghiaasiaan, The effect of flow pulsation on drags and heat transfer in an array of heated square cylinders, *Int. J. Therm. Sci.*, Vol. 64, 105–120, 2013.

[85] G. Mishra, R.P. Chhabra, Influence of flow pulsations and yield stress on heat transfer from a sphere, *App. Math. Model.*, Vol. 90, 1069–1098, 2021.

[86] U. Akdag, M. Ozdemir, Heat removal from oscillating flow in a vertical annular channel, *Heat Mass Transf.*, Vol. 44, 393–400, 2008.

[87] U. Akdag, A.F. Ozguc, Experimental investigation of heat transfer in oscillating annular flow, *Int. J. Heat Mass Transf.*, Vol. 52, 2667–2672, 2009.

[88] K.C. Leong, L.W. Jin, Characteristics of oscillating flow through a channel filled with open-cell metal foam, *Int. J. Heat Fluid Flow*, Vol. 27, 144–153, 2006.

[89] M. Ghafarian, D. Mohebbi-Kalhori, J. Sadegi, Analysis of heat transfer in oscillating flow through a channel filled with metal foam using computational fluid dynamics, *Int. J. Therm. Sci.*, Vol. 66, 42–50, 2013.

[90] T. Zhao, P. Cheng, A numerical solution of laminar forced convection in a heated pipe subjected to a periodically reversing flow, *Int. J. Heat Mass Transf.*, Vol. 38, 3011–3022, 1995.

[91] Z. Chen, Y. Utaka, Y. Tasaki, Measurement and numerical simulation on the heat transfer characteristics of reciprocating flow in microchannels for the application in magnetic refrigeration, *Appl. Therm. Eng.*, Vol. 65, 150–157, 2014.

[92] H.-W. Wu, R.-F. Lay, C.-T. Lau, W.-J. Wu, Turbulent flow field and heat transfer in a heated circular channel under a reciprocating motion, *Heat Mass Transf.*, Vol. 40, 769–778, 2003.

[93] G. Xiao, C. Chen, B. Shi, K. Cen, M. Ni, Experimental study on heat transfer of oscillating flow of a tubular Stirling engine heater, *Int. J. Heat Mass Transf.*, Vol. 71, 1–7, 2014.

[94] A.M. Bayomy, M.Z. Saghir, Heat transfer characteristics of aluminum metal foam subjected to a pulsating/steady water flow: Experimental and numerical approach, *Int. J. Heat Mass Transf.*, Vol. 97, 318–336, 2016.

[95] J. Khodadadi, Oscillatory fluid flow through a porous medium channel bounded by two impermeable parallel plates, *J. Fluids Eng.*, Vol. 113, 509–511, 1991.

[96] K. Leong, L. Jin, An experimental study of heat transfer in oscillating flow through a channel filled with aluminum foam, *Int. J. Heat Mass Transf.*, Vol. 48, 243–253, 2005.

[97] K. Leong, L. Jin, Heat transfer of oscillating and steady flows in a channel filled with porous media, *Int. Commun. Heat Mass Transf.*, Vol. 31, 63–72, 2004.

[98] H. Fu, K. Leong, X. Huang, C. Liu, An experimental study of heat transfer of a porous channel subjected to oscillating flow, *ASME J. Heat Transf.*, Vol. 123, 162–170, 2001.

[99] J. Paek, B. Kang, J. Hyun, Transient cool-down of a porous medium in pulsating flow, *Int. J. Heat Mass Transf.*, Vol. 42, 3523–3527, 1999.

[100] G. Al-Sumaily, M. Thompson, Forced convection from a circular cylinder in pulsating flow with and without the presence of porous media, *Int. J. Heat Mass Transf.*, Vol. 61, 226–244, 2013.

[101] M. Sozen, K. Vafai, Analysis of oscillating compressible flow through a packed bed, *Int. J. Heat Fluid Flow*, Vol. 130–136(2), 12, 1991.

[102] S.Y. Kim, B.H. Kang, J.M. Hyun, Heat transfer from pulsating flow in a channel filled with porous media, *Int. J. Heat Mass Transf.*, Vol. 14(37), 2025–2033, 1994.

[103] T.C. Jue, Analysis of flows driven by a torsionally-oscillatory lid in a fluid saturated porous enclosure with thermal stable stratification, *Int. J. Therm. Sci.*, Vol. 41(8), 795–804, 2002.

[104] P.C. Huang, C.F. Yang, Analysis of pulsating convection from two heat sources mounted with porous blocks, *Int. J. Heat Mass Transf.*, Vol. 51(25–26), 6294–6311, 2008.

[105] P. Forooghi, M. Abkar, M. Saffar-Avval, Steady and unsteady heat transfer in a channel partially filled with porous media under thermal non-equilibrium condition, *Transp. Porous Media*, Vol. 86(1), 177–198, 2011.

[106] Z.X. Guo, S.Y. Kim, H.J. Sung, Pulsating flow and heat transfer in a pipe partially filled with a porous medium, *Int. J. Heat Mars Transf.*, Vol. 40, 4209–4218, 1997.

[107] A.A. Al-Haddad, N. Al-Binally, Prediction of heat transfer coefficient in pulsting flow, *Int. J. Heat Fluid Flow*, Vol. 10, 131–133, 1989.

[108] M.A.A. Hamad, I. Pop, Unsteady MHD free convection flow past a vertical permeable flat plate in a rotating frame of reference with constant heat source in a nanofluid, *Heat Mass Transf.*, Vol. 47, 1517–1524, 2011.

[109] M. Rahgoshay, A.A. Ranjbar, et al., Laminar pulsating flow of nanofluids in a circular tube with isothermal wall, *Int. Commun. Heat Mass Transf.*, Vol. 39, 463–469, 2012.

[110] T.G. Elizarova, et al., Simulation of separating flows over a backward-facing step, *Comput. Math. Model.*, Vol. 15(2), 167–193, 2004.

[111] N. Mikheev, I. Gazizov, and I. Davletshin, Heat transfer in pulsating flow behind a rib in the channel inlet region. *J. Phys. Conf. Ser.*, Vol. 1105, 012018, 2018.

[112] I.A. Davletshin, A.N. Mikheev, N.I. Mikheev, et al., Data on distribution of heat transfer coefficient and profiles of velocity and turbulent characteristics behind a rib in pulsating flows, *Data Brief* 33 (2020), 106485.

[113] I.A. Davletshin, A.N. Mikheev, N.I. Mikheev, et al., Heat transfer and structure of pulsating flow behind a rib, *Int. J. Heat Mass Transf.*, Vol. 160, 120173, 2020.

[114] P.A. Eibeck, J.O. Keller, T.T. Bramlette, D.J. Sailor, Pulse combustion: Impinging jet heat transfer enhancement, *Combust. Sci. Technol.*, Vol. 94, 147–165, 1993.

[115] E.G. Richardson, E. Tyler, The transverse velocity gradient near the mouths of pipes in which an alternating or continuous flow of air is established, *Proc. Phys. Soc.*, Vol. 42, 1–15, 1929.

[116] H. Shokouhmand, A. Mosahebi, B. Karami, Numerical simulation and optimization of heat transfer in reciprocating flows in two dimensional channels, in *Proceedings of the World Congress on Engineering*, vol. II, London, UK, 2008.

[117] R. Blythman, T. Persoons, N. Jeffers, K.P. Nolan, D.B. Murray, Localised dynamics of laminar pulsatile flow in a rectangular channel, *Int. J. Heat Fluid Flow*, Vol. 66, 8–17, 2017.

[118] O.J. Lobo, D. Chatterjee, Development of flow in a square mini-channel: Effect of flow oscillation, *Phys. Fluids*, Vol. 30(42001–42003), 42003–42013, 2018.

[119] U. Akdag, M. Ozdemir, Heat transfer in an oscillating vertical annular liquid column open to atmosphere, *Heat Mass Transf.*, Vol. 42, 617–624, 2006.

[120] M. Sheikholeslami, S. Soleimani, D.D. Ganji, Effect of electric field on hydrothermal behavior of nanofluid in a complex geometry, *J. Mol. Liq.*, Vol. 213, 153–161, 2016.

[121] M. Sheikhbahai, M.N. Esfahany, N. Etesami, Experimental investigation of pool boiling of Fe3O4/ethylene glycol-water nanofluid in electric field, *Int. J. Therm. Sci.*, Vol. 62, 149–153, 2012.

[122] D. Liu, Q. Wang, J.J. Wei, Experimental study on drag reduction performance of mixed polymer and surfactant solutions, *Chem. Eng. Res. Des.*, Vol. 132, 460–469, 2018.

[123] A.R. Pouranfard, D. Mowla, F. Esmaeilzadeh, An experimental study of drag reduction by nanofluids through horizontal pipe turbulent flow of a Newtonian liquid, *J. Ind. Eng. Chem.*, Vol. 20, 633–637, 2014.

[124] A. Steele, I.S. Bayer, E. Loth, Pipe flow drag reduction effects from carbon nanotube additives, *Carbon*, Vol. 77, 1183–1186, 2014.

[125] M. HemmatEsfe, et al., Estimation of thermal conductivity of CNTs-water in low temperature by artificial neural network and correlation, *Int. Commun. Heat Mass Transf.*, Vol. 76, 376–381, 2016.

[126] L. Yang, J. Xu, K. Du, X. Zhang, Recent developments on viscosity and thermal conductivity of nanofluids, *Powder Technol.*, Vol. 317, 348–369, 2017.

[127] T. Yiamsawasd, A.S. Dalkilic, S. Wongwises, Measurement of the thermal conductivity of titania and alumina nanofluids, *Thermochim Acta*, Vol. 545, 48–56, 2012.

[128] M. Bahiraei, S. Heshmatian, Electronics cooling with nanofluids: A critical review, *Energ. Conver. Manage.*, Vol. 172, 438–456, 2018.

[129] M. Bahiraei, H. Kiani Salmi, M.R. Safaei, Effect of employing a new biological nanofluid containing functionalized graphene nanoplatelets on thermal and hydraulic characteristics of a spiral heat exchanger, *Energ. Conver. Manage.*, Vol. 180, 72–82, 2019.

[130] Z. Hajjar, A. Rashidi, A. Ghozatloo, Enhanced thermal conductivities of graphene oxide nanofluids, *Int. Commun. Heat Mass Transf.*, Vol. 57, 128–131, 2014.

[131] M. Amani, P. Amani, O. Mahian, P. Estelle, Multi-objective optimization of thermophysical properties of eco-friendly organic nanofluids, *J. Clean. Prod.*, Vol. 166, 350–359, 2017.

[132] M.H. Esfe, M.K. Amiri, M. Bahiraei, Optimizing thermophysical properties of nanofluids using response surface methodology and particle swarm optimization in a non-dominated sorting genetic algorithm, *J. Taiwan Inst. Chem. Eng.*, Vol. 103, 7–19, 2019.

[133] P. Naphon, S. Wiriyasart, Experimental study on laminar pulsating flow and heat transfer of nanofluids in micro-fins tube with magnetic fields, *Int. J. Heat Mass Transf.*, Vol. 118, 297–303, 2018.

[134] F. Selimefendigil, H.F. Öztop, Identification of forced convection in pulsating flow at a backward facing step with a stationary cylinder subjected to nanofluid, *Int. Commun. Heat Mass Transf.*, Vol. 45, 111–121, 2013.

[135] D.W. Zhao, G.H. Su, Z.H. Liang, Y.J. Zhang, W.X. Tian, S.Z. Qiu, Experimental research on transient critical heat flux in vertical tube under oscillatory flow condition, *Int. J. Multiphase Flow*, Vol. 37, 1235–1244, 2011.

[136] S.L. Wang, C.A. Chen, T.F. Lin, Oscillatory subcooled flow boiling heat transfer of R-134a and associated bubble characteristics in a narrow annular duct due to flow rate oscillation, *Int. J. Heat Mass Transf.*, Vol. 63, 255–267, 2013.

[137] S.L. Wang, C.A. Chen, Y.L. Lin, T.F. Lin, Transient oscillatory saturated flow boiling heat transfer and associated bubble characteristics of FC-72 over a small heated plate due to heat flux oscillation, *Int. J. Heat Mass Transf.*, Vol. 55, 864–873, 2012.

[138] Qianhao Ye, Yonghai Zhang, Jinjia Wei, A comprehensive review of pulsating flow on heat transfer enhancement, *Appl. Therm. Eng.*, Vol. 196, 117275, 2021.

Chapter 6

Heat transfer augmentation using ultrasound and magnetic forces

6.1 INTRODUCTION

Higher frequency sound waves than those humans can hear are referred to as ultrasound (i.e., above 16 kHz). They are categorized as high-power ultrasound or low-frequency ultrasound (20 kHz < frequency < 100 kHz) and low-power ultrasound or high-frequency ultrasound (frequency 1 MHz). High-powered ultrasound can change the medium in which it is transmitted. At liquid–gas, liquid–liquid, and liquid–solid interfaces, high-power ultrasound causes cavitation, heating, micro-streaming, and surface instability. These characteristics are appealing to heat exchanger performance because high-power ultrasound waves can provide two benefits, namely reduced heat exchanger fouling and increased heat transfer. As a result, high-powered ultrasound is used in a variety of procedures, including cleaning, sonochemistry, etc. On the other hand, low-power ultrasound has no impact on the propagation medium. As a result, it is frequently employed in medical diagnostics and nondestructive material testing. Various utilizations of ultrasound are depicted in Figure 6.1 based on their frequency or power.

Using ultrasonic waves to improve heat and mass transmission is a fascinating technology that has been rising in popularity more recently [1]. Ultrasonic waves can also be utilized to make emulsions with nano-sized dispersed phases [2]. Acoustic streaming, or propagation of ultrasonic waves through a fluid, has for a long time been known to create a constant vortex movement. Acoustic wave propagation produces two different kinds of streaming – Eckart and Rayleigh streaming [11].

Because of a fluid's viscosity in bulk flow, the energy of waves is attenuated in Eckart streaming, and streaming is formed as a result of this. The tension in Stokes shear-wave layer at a solid barrier generates large-scale streaming refers as outer streaming in Rayleigh streaming [3]. Rayleigh [4] pioneered the science of acoustic streaming and considered a lengthy pipe and looked at a standing wave in it. Westervelt [5], Nyborg [6], Schlichting [7], and Lighthill [8] continued studies later. All of the researchers concentrated on linear acoustics and showed how the wave energy is captivated by fluid-viscosity outcome. Stuart [9] presented streaming Reynolds number

DOI: 10.1201/9781003229865-6

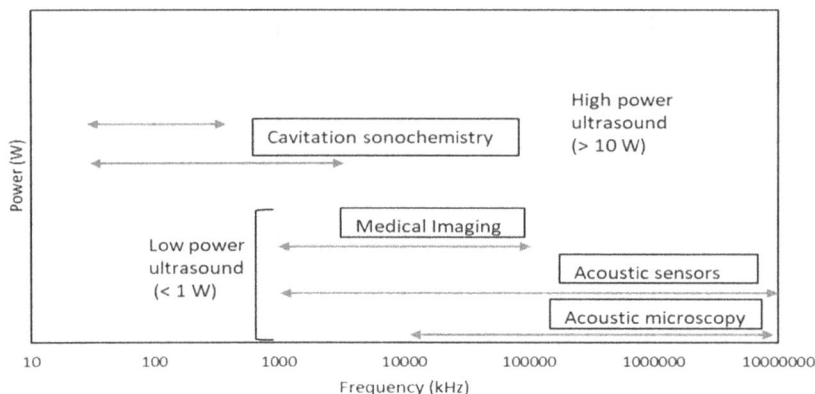

Figure 6.1 Usage of ultrasound depending on power and frequency [1].

(Re) and indicated that acoustic streaming should be studied using nonlinear terms if Re_s is greater than 1. Several researchers investigated audio streaming for heat transfer enhancement in many regimes, including convection [10–23], boiling [14, 15], melting [16], and so on. Delouei et al. [17] used ultrasonic vibration to evaluate the influence of input turbulent flow into a circular pipe under constant wall temperature condition. They discovered that when Re surges, heat transfer enhancement diminishes. The result shows that at Reynolds numbers above 8,500, the Nusselt number (Nu) doesn't increase in existence of ultrasonic vibration.

Many industries have found use for magnetic fluid dynamics in the production of a magnetic field, including heat exchangers, combustors, and mixing chambers. Flow condition and applied magnetic field from the outside channel are both important factors in the mobility of magnetic fluid in a channel. Besides, it is crucial to consider how an imposed magnetic field affects the stability and distributive behavior, as well as flow condition. The mathematical modeling and numerical simulations appear to be a hopeful addition to the ongoing development of above-stated applications. To achieve a certain goal, it's first vital to gain a basic understanding of fluid dynamics while taking magnetic field impacts into account. Steve Papell at NASA created the unique class of fluids known as magnetic fluids in 1963. In the presence of a magnetic field, magnetic fluid flow can solve a broad range of technological difficulties.

Researchers and scientists on the subject of magnetic fluid flow have conducted extensive experimental, numerical, and theoretical research works. Hayat et al. [18] investigated laminar and turbulent fluid flow (2D) while considering both ferro-hydrodynamics (FHD) and magneto-hydrodynamics (MHD). Takeuchi et al. [19] conducted research work on turbulent flow. Malekzadesh et al. [20] studied the effect of magnetic field on flow (steady) while taking into account the MHD phenomenon. Khashan et al. [21]

considered a liquid flowing through a two-dimensional tube under the influence of a magnetic field. Hayat et al. [22] studied peristaltic MHD liquid flow in a pipe/channel. The magnetic field effect on peristaltic movement of liquids in an adjustable cylinder was investigated by Abd-Alla et al. [23]. In the presence of a continuous magnetic field, fluid flow was investigated using computational fluid dynamics (CFD) by many researchers [24–30]. Outside magnetic fields have an uncertain impact on laminar flow, according to Azizian et al. [31]. Dritselis and Knaepen [32] numerically recreated magnetic fluid flow in the presence of a uniform transverse magnetic field. A spatially varying external magnetic field's influence on two-dimensional fluid flow was numerically assessed by Turk et al. [33]. Under a magnetic field, Akram and Nadeem [34] conducted a thorough investigation of peristaltic liquids. Saha and Chakrabarti [35] computationally investigated the effect of Re on liquids (ferromagnetic nature) keeping the MHD phenomenon in mind. Maatki et al. [36] performed numerical research on the effect of magnetic field on fluid flow (3D) using the finite volume method (FVM). A numerical analysis of natural convection was presented by Mebarek-Oudina [37]. Maatki et al. [38] used FVM to quantitatively study the effect of magnetic fields on electromagnetic parameters. The impact of magnetic fields on natural convection was numerically examined by Kolsi et al. [39]. Zaim et al. [40] looked at MHD fluid flow with natural convection in mind. Mebarek-Oudina et al. [41] examined the impact of a transverse magnetic field on the generation of entropy while taking MHDs into account. Using statistics, it was determined how external magnetic fields affected the flow of water-based nanofluids, and the MHDs principle was used to determine how a magnetic field affected the flow of electrically conducting fluids [42]. Swain et al. [43] investigated the effects of changing magnetic fields on a hybrid nanofluid (water based). Saha and Chakrabarti [44] investigated a 2D turbulent flow in a square channel that was subjected to a magnetic field. Flow of viscoelastic fluids under a transverse magnetic field was explored by Abo-Dahab et al. [45].

6.2 MECHANISMS

6.2.1 Ultrasound waves

Vibrational energy, which is a kind of mechanical energy, is created from electrical energy through an ultrasonic system. Then, mechanical energy is transferred to a sonicated medium. Some part of the input energy is wasted as heat, while another part causes cavitation. Chemical, physical, and biological impacts are produced by a fraction of the cavitational energy. Acoustic streaming and cavitation are two of the four impacts of significance in heat transfer augmentation, observed due to ultrasound propagation in a liquid.

6.2.1.1 Acoustic streaming

When a tiny bubble is subjected to ultrasound and is surrounded by a liquid medium, a phenomenon known as acoustic streaming is occurring. Toroidal eddy currents form when an oscillatory action causes the liquid surrounding a bubble to move. As ultrasonic frequency is increased, the magnitude of these eddy currents decreases, resulting in less micro-streaming. Micro-streaming is thus the result of a low-frequency, low-pressure reaction to micro-bubble oscillations.

6.2.1.2 Acoustic cavitation

When powerful ultrasound waves travel through liquid, the local pressure drops below the vapor pressure during the sound wave's rarefaction period, causes bubbles to develop, which then burst when they migrate to a high-pressure location resulting in significant local heating and high pressure. Acoustic cavitation is the production, progress, and collapse phenomenon of bubbles that results in heat transfer enhancement as an outcome as shown in Figure 6.2. The thermal and velocity boundary layers are broken when a bubble burst close to a solid-liquid contact, reducing thermal resistance and generating micro-turbulence. Thus, heat transfer is enhanced.

6.2.2 Magnetic field

The use of magnetic nanofluids (MNFs) impacted by an external magnetic field for heat transfer improvement has several advantages over conventional nanofluids (non-MNFs) [46]. In typical energy conversion and cooling devices, there may be no moving elements required to make a fluid flow. Temperature differential and a nonuniform magnetic field, which can be created using a permanent magnet system, generate MNF current. The direction and kind of fluid flow are determined by system configuration. As a result, thermomagnetic convection is substantially stronger than gravitational convection. External magnetic fields can be used to tune the thermophysical characteristics of MNFs (thermal conductivity and viscosity) [47, 48].

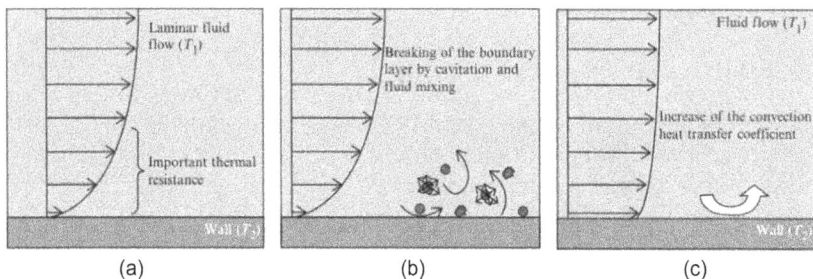

Figure 6.2 Heat transfer mechanism in acoustic cavitation [1]. (a) Typical velocity boundary layer; (b) Cavitation and microagitation; and (c) Modified boundary layer profile.

6.2.2.1 Preparation of MNF

Superparamagnetic nanoparticles are diffused in a nonmagnetic carrier to create MNF. In heat transfer applications, MNF is subjected to magnetic fields, magnetic field gradients, and/or gravity fields, which could cause particle sedimentation in fluid. It is obvious that the latter plays a crucial role in particle sedimentation, changing MNF stability, because size variation of magnetic nanoparticles is directly related to the engagement range of magnetic nanoparticles in applied fields [47]. When the thermal energy of particles exceeds magnetic and gravitational energies, particle stability against particle sedimentation can be assured. Odenbach [47] determined the maximum particle size to be $d < (6k_BT/\mu_0M_0\pi H)^{1/3}$ for MNF in magnetic field presence and $d < (6k_BT/\Delta\rho gh\pi)^{1/3}$ for MNF in gravitational field presence. Furthermore, during synthesis, magnetic nanoparticle aggregation must be prevented definitely. In theory, particle aggregation increases the active diameter of particles, causing sedimentation to destabilize suspension. In this scenario, maximum particle diameter (d) was calculated as $d < (144k_BT/\mu_0M_0^2)$, which corresponds to a maximum interaction energy when two interacting particles collide [47]. Recently, magnetic nanoparticles made of metal and metal oxide with the suitable size distribution have been synthesized [49]. Metallic nanoparticles such as Co, Fe, and Ni can be created from the reduction of metal complexes and metal-salts [49, 50]. Magnetic nanoparticles of metal oxides (Fe_3O_4, γ-Fe_2O_3) are frequently used in MNF because of their chemical stability. Processes like chemical coprecipitation, microemulsion, and phase transfer have all been used to make metal oxide magnetic nanoparticles [49, 51, 52]. Stability of MNFs is not dependent on the existence of the necessary superparamagnetic particles in MNF [53]. A suspension of magnetic nanoparticles in a carrier fluid will not be stable due to the London–van der Waals and magnetic forces present, leading to irreversible particle aggregation and eventually sedimentation. When creating stable MNF, repulsive forces between magnetic nanoparticles must be added to offset London–van der Waals forces and dipole–dipole magnetic interactions. There are two approaches to create a repulsive mechanism between particles: coating them with a polymer surfactant, which creates an entropic repulsion (and/or charging their surfaces), which results in a coulombian repulsion [52, 53]. It is worth noting that the mechanism to be utilized should primarily be determined by the properties of carrier fluids and particles. Ultrasonic equipment (or high-speed homogenizer) is used to disperse material in the presence of a polymer surfactant.

6.2.2.2 Thermal conductivity of MNF

The concept behind employing a suspension for heat transfer was to enhance the thermal conductivity of standard heat transfer fluids by including greater thermal conductivity nanoparticles [54]. As a result, nanofluids generated with metallic or metallic oxide nanoparticles and carbon nanotubes dominated some earlier studies on thermal conductivity of nanofluids [54–58]. Investigations on the thermal conductivity of MNF were impeded by the fact that magnetic materials used in MNF have a relatively low thermal conductivity, especially without an external magnetic field. Recent research has shown, however, that using a solid material (of high thermal conductivity) as a suspension to enhance thermal conductivity of a carrier fluid is not necessarily a good idea [59]. This discovery, together with the prospect of manipulating MNF thermal conductivity by magnetic field, could be probable reasons for increased interest in MNF thermal conductivity recently.

6.3 RESULTS

6.3.1 Use of ultrasound waves

There has been a steady increase in studies on convective heat transfer and other phase change processes. Due to its specific characteristics inside the propagation medium, ultrasonography is increasingly regarded as a capable and efficient tool for improving heat transmission. The propagation of ultrasonic waves causes some hydrodynamic phenomena (Figure 6.3). Most important of which are cavitation and acoustic streaming for heat transfer enhancement [60]. A high-powered ultrasonic propagation in liquid causes cavitation bubbles when intermolecular attraction forces surpass negative pressure [60–64].

6.3.1.1 Phase change heat transfer

Power ultrasonography is particularly useful for improving phase transition processes in which heat and mass are transferred simultaneously. Acoustic

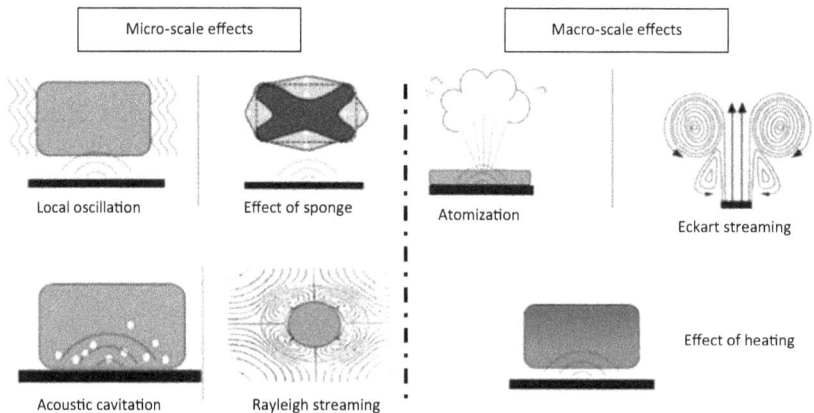

Figure 6.3 Various ultrasound phenomena [45].

cavitation and microstreaming cause a strong vibration inside a liquid, which improves the heat and mass transfer efficiencies of many processes.

Cavitation bubbles operate as nuclei in a freezing process, increasing nucleation rate and contributing to the creation of tiny, uniformly dispersed ice crystals resulting in greater product quality [65, 66]. Boiling is another important form of heat transmission, with several regimes being explored in the presence of ultrasonic vibrations by scientists [67]. Compared to saturated conditions, subcooled boiling under ultrasonic propagation is considerably intensified. Because of increased attenuation from bubbles at greater heat fluxes, ultrasonography has a low efficacy for saturated-boiling heat transfer [68]. Bubble nucleation and growth are affected by acoustic cavitation, as is bubble detachment from the heating surface. As a result, vapor does not rise up but forms a film on the heated surface of the wall, significantly improving boiling heat transmission [69]. The position of the heated surface relative to the ultrasonic transducer affects heat transmission, with antinodes offering the highest results [70]. Increasing ultrasonic intensity and decreasing frequency both improve the boiling process [71]. Ultrasound has a variety of effects on drying, including reduction of the drying time, providing higher-quality dried materials, and increasing energy efficiency. Heating and sponge effects are factors that help ultrasonic waves in drying process improvement. Heating a solid medium can improve mass transfer by increasing the mass diffusivity of moisture. Sponge effect enables liquids to flow from the inside of a solid material to its surface. Given that liquid-releasing forces can exceed surface tension, solid bodies can develop tiny channels, increasing mass exchange with the environment [72]. Furthermore, micro-vibration enhances gas-side heat and mass transmission coefficients by increasing gas turbulence along a solid surface. Synergy effects can increase heat and mass transmission by minimizing internal and external resistance. Ultrasound has been used to regenerate liquid and solid desiccant in dehumidification cycles. Several fascinating studies [73–75] have looked into and verified good impact of ultrasound on the effectiveness of silica-gel regeneration phenomena. Micro-droplets created by atomization increase heat and mass transmission during liquid desiccant regeneration [76, 77]. Ultrasound, in general, has a significant impact on phase transition processes because it causes special effects inside gas, liquid, and solid media. Key forces that influence phase transition processes include acoustic cavitation in liquids, heating and sponge effects in solids, and acoustic streaming in gaseous media.

6.3.1.2 Ultrasound wave heat transfer

Acoustic streaming and cavitation are two main hydrodynamic processes that happen in a fluid medium when ultrasonic waves are irradiated, as mentioned in Section 6.3.1.1. Acoustic cavitation, which occurs when power ultrasound travels through a liquid media, causes many hydrodynamic

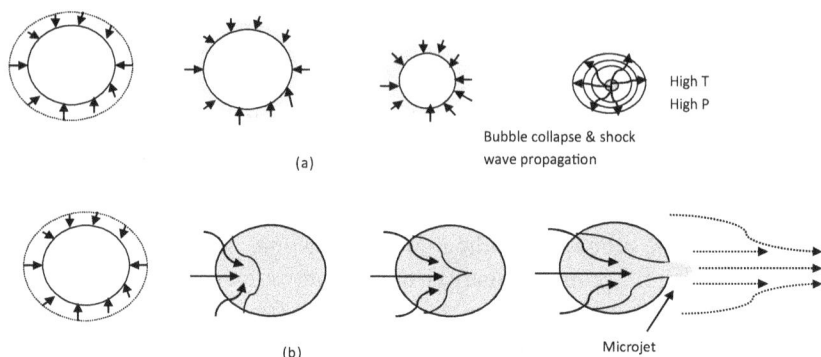

High T
High P

Bubble collapse & shock
wave propagation

(a)

(b)

Microjet

Figure 6.4 (a) Shock wave and (b) Microjet [81].

processes that cause intense bulk convection and accordingly increasing convective coefficient. Microstreaming is caused by limited rhythmic motion of liquid constituents caused by ultrasonic wave propagation named as micro-turbulence [78]. Ultrasound compresses and rarefies fluids. Shock waves (Figure 6.4a) are created inside a liquid when gas bubbles are compressed to their full capacity; as a result, fast pressure rise causes bubbles to abruptly stop and burst with incredible velocity [79]. Microjet (Figure 6.4b) is another turbulence-inducing phenomenon that occurs during power ultrasound propagation. Because of pressure gradient, cavitation bubbles near the interface lose their spherical shape. When a segment of a cavitation bubble is subjected to more pressure, it collapses faster than the remainder of a bubble. Disruption of thermal boundary layer caused by this activity close to the heated surface may result in increased heat transmission. Recirculating flows created as a result of acoustic streaming are principal drivers of heat transfer enhancement at high frequencies. Eckert streaming, which is being investigated more by experts, can boost heat transmission by intensifying turbulence inside a fluid [80].

6.3.1.3 Factors affecting heat transfer improvement using ultrasound

6.3.1.3.1 Frequency

The use of various ultrasonic frequencies to improve natural and forced convective heat transfer has been informed in numerous previous studies. The majority of studies used low-frequency ultrasound (20 to 40 kHz); however, a few concentrated on using high-frequency ultrasound to improve heat transmission. Because cavitation-induced effects diminish with increasing ultrasound frequency, low-frequency ultrasound enhances heat transfer for both free and forced convection [82]. Dehbani et al. [60] examined heat increase effects by ultrasonic waves of 24 kHz and 1.7 MHz frequency.

They compared natural convection with and without ultrasonic propagation of an electrically heated platinum wire. They concluded that, whereas 24 kHz ultrasound produced better heat transfer enhancement, 1.7 MHz ultrasound provided better results, as measured by the ratio of heat absorbed to the dissipated power. Mongkolkitngam et al. [83] conducted another comparative investigation of effects of sonication with different frequencies on natural convective heat transfer, indicating that 40 kHz was most effective. In forced convection heat transfer, however, various investigations have been undertaken utilizing varying frequency. On diverse flow rates, the average augmentation ratio was shown to be inversely proportional to ultrasonic frequency [84]. The effects of individual and combined ultrasonic fields of 20 and 33 kHz on the forced convection of water moving through a tube while being continuously heated were examined by Dhanalakshmi et al. [85]. They found that utilizing 20 kHz ultrasound enhanced heat transfer more than using 33 kHz ultrasound, and that combination of two frequencies, had best efficiency with a slight difference. In order to investigate the impacts of 25 kHz and 2 MHz ultrasonic vibration, Bulliard-Sauret et al. [86] conducted a study on forced conventional heat transfer of flowing water over a rectangular channel. Low frequency was shown to be more effective than high frequency, according to these researchers. Due to acoustic cavitation, the most significant phenomena for increased turbulence, low-frequency ultrasound, in general, has a larger ability to boost heat transfer. Furthermore, using low- and high-frequency ultrasound simultaneously causes a synergy of acoustic cavitation and acoustic streaming, resulting in increased heat transmission. Ultrasonic amplitude percentage, which represents the amount of power applied to a medium during ultrasound propagation and has values of 0% and 100% for no sonication and rated sonication power, respectively, is used to describe this power [87]. The average ultrasound power per unit cross-sectional area perpendicular to the direction of propagation, or intensity, can be used to describe the power of ultrasonic waves [88]. Ultrasound power is stated in the majority of published articles, but ultrasound intensity is expressed in only a few. The strength of ultrasonic waves can considerably disrupt velocity and thermal boundary layers, improving heat transfer [87]. Most studies on heat transfer (free convection) have been done at a constant ultrasound power, but some studies have been done at variable ultrasound powers, and the results show that the convective heat transfer factor improves with increasing ultrasound power [89–91]. The direct proportional relationship between ultrasonic power and abrupt collapse of cavitation bubbles explains this idea. Because the size of a resonant bubble is related to ultrasonic power, effects of bubble implosion become more intense as ultrasound power increases. This causes mixing in the medium and, as a result, an increase in the rate of heat transmission. Kiani et al. [92] observed that ultrasonic intensity improves the cooling rate while testing a stationary copper sphere at intensities ranging from 0 to 4,100 W/m² in ethylene glycol. The highest enhancement factor was attained for the highest intensity. When the intensity is less than 450 W/m², the increasing quantity of intensity is extra beneficial for increasing cooling rate. Alternatively, a tiny increase in intensity resulted in larger changes in the cooling rate. The cooling rate did not significantly increase at higher intensities, though. They attributed these findings to higher intensities of the adverse heating effect. The beneficial effect of ultrasonic propagation gradually diminished as it caused heat to be created at the surface of the sphere owing to acoustic bubble implosion. Musielak and Mierzwa [93] had discovered a linear link between heat transfer increase and ultrasonic power by examining forced convection from air moving on the surface of many solid structures of various forms. Numerous studies on the impact of ultrasound's ability to enhance thermal efficiency on heat exchangers have been conducted. Forced convective heat transfer is the major mechanism.

6.3.1.3.2 Propagation medium

According to the literature, water is the preferred working fluid in forced and free convection heat transfer systems with ultrasonic propagation. It is affordable and accessible. Some medium-effect research studies compared water to other fluids. Zhou et al. [94] studied natural convection from a tube in liquids (acetone, ethanol & water). Due to greater acoustic cavitation bubbles created by acetone's higher vapor pressure, it improved heat transmission the most. Cai et al. [89] utilized sugar and brine water in a natural convective heat transfer system. It was concluded that as sugar water concentration rises, so does heat transfer enhancement factor, while the maximum value of brine is at 3%. For these two solutions, more heat transport study is needed. Some experiments have focused attention on using ultrasonic in a medium containing nanoparticles as a passive heat transfer augmentation approach. Shen et al. [95] studied 28 kHz sonication effects on hot wire convection through a nanofluidic medium. Ultrasound could boost benefits of Al_2O_3 nanoparticles, and it was found that sonication doubled the heat transfer. Azimy et al. [96] reported that using ultrasonic and carbon nanotubes together boosted heat transfer by 200%. Nanofluid concentration increased heat transfer. They put ultrasonic transducers on the channel wall to prevent nanoparticle settlement. FC-72, a dielectric fluid with exceptional chemical and thermal stabilities, also has become of interests to the researchers. Ultrasonic waves of 40 kHz facilitated heat transfer (free convective) between FC-72 and electronic board as a coolant liquid. Few research works have focused on natural convective heat transfer in a gaseous medium and heat transfer augmentation [97–99]. A heated horizontal cylinder's heat exchange with the surrounding air was found to be marginally boosted by 40 kHz sonication, according to Prodanov [100]. Musielak and Mierzwa [93] reported a 250% increase in forced convection in a gaseous medium based on heat transfer between solid samples and flowing air. They also looked at air temperature (40°C and 60°C), and it was concluded that as temperature rises ultrasonic waves provide thermal intensification.

6.3.1.3.3 System geometry

Researchers have looked at a variety of topics, including size and arrangement of fluid containers in forced and natural convection systems, the size and arrangement of channels in natural convection systems, propagation of ultrasound, positioning of ultrasound transducers in relation to test sections, and shape of heating surfaces. A hot wire inside a container of liquid subjected to ultrasonic vibrations is frequent free convection geometry [60, 95, 101–103]. Despite identical ultrasonic properties, geometry affects result (test section geometry) [93]. They investigated how forced convection from air to samples of different forms of stainless steel was affected by ultrasound. The geometry of samples affects the ultrasound capacity to improve them. The most efficient sample is cube shaped. Kiani et al. [92] found an interesting effect of test section-ultrasonic transducer distance. Three 25 kHz ultrasonic transducers placed on the bottom of a fluid container were used to measure the cooling rate of a sphere at various distances. Closer distances to the ultrasonic transducer surface hasten cooling as long as the hot body is not near the fluid's free surface. At the furthest distance from the transducer surface, according to their investigation, the largest heat transfer enhancement factor was found. When the spherical body came closer to the transducer surface, more cavitation clouds were formed, increasing the cooling rate. Most intense cavitation clouds were found around the gas–liquid interface, where streamlines boost cooling rate. Rahimi et al. [103] had investigated the thermal enhancement impact of high-frequency ultrasonic transducers on heat transfer (natural convection) from a horizontal, thin platinum wire to surroundings water. In a cylindrical container, they used five 1.7 MHz ultrasonic transducers, three on the bottom and two on the side walls. Heat transmission was improved more by wall transducers than by bottom transducers. Wall transducers emit ultrasonic waves in the direction of the wire center line. Simultaneous activation of two opposing platinum-wired transducers resulted in lower cooling performance, perhaps due to neutralizing impact of other. One of three ultrasound transducers on the container bottom was aligned with the wire, and it worked better than the other two. Heat transfer augmentation ratios vs. heights of circular rod heaters from liquid tank bottom are correlated with heater heat flux, according to Liu et al. [104]. With increasing heater height and low heat flow, single-phase free convection enhancement ratio was increased. Because experiments were done further from liquid–gas contact than transducer surfaces results are consistent with previous investigations. Chen et al. [105] discovered a different pattern in heat transmission as transducer height increased and interpreted their data differently. In order to study heat transfer (natural convection), they heated a rod (6 mm in diameter) in a thermostat tank with 80 mm of degassed water. Experiments were conducted at three different heights using three 40 kHz, 50 W transducers mounted to the tank bottom (9, 19, 22 mm). Mean heat transfer enhancement ratios were higher than at 19 mm from the transducer surface at 9 and 22 mm. This was a new pattern

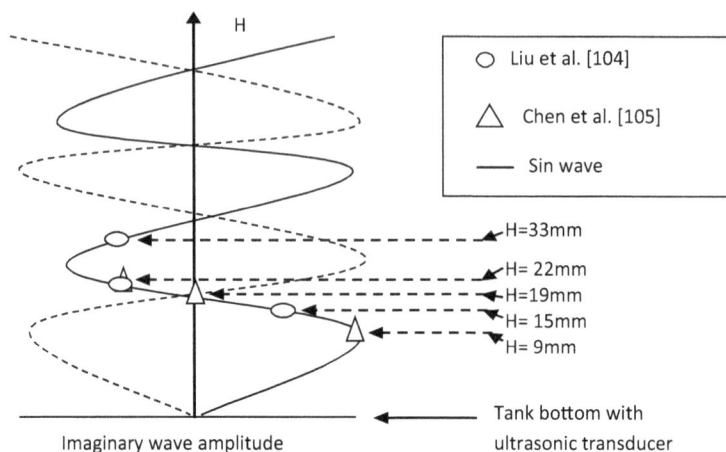

Figure 6.5 Amplitude and test heights of Imaginary ultrasonic wave [105].

from past research since there was no rising tendency with height. Heat transfer enhancement depends on the distance between heater and ultrasonic transducers, as well as wave amplitude created through medium. Whether a surface is in the "node" or "antinode" zone determines the height of the test section from the transducer. "Node" regions have a lower heat transfer enhancement ratio than "anti-node" or "wavy" regions. They showed transducer heights as a sinusoidal function along the direction of propagation on a figure of hypothetical ultrasonic wave amplitude (Figure 6.5). Ultrasonic vibration enhancement was lower in the "node" zone due to less pressure variations, but higher in the "anti-node" or "wavy" region. Mongkolkitngam et al. [109] studied spontaneous convective heat transfer from a vertical cylinder to a triple-frequency ultrasonic transducer (40, 80, and 120 kHz). It was concluded that sonication frequency affects heat transfer ratio based on transducer distance. Wave attenuation caused the enhancement ratio at 40 and 80 kHz to decrease over distance [106]. Because acoustic streaming dominates at 120 kHz, enhancement ratio decreased, but not as much as with low frequency ultrasound. Acoustic streaming promotes global mixing via tank. Heat transfer enhancement was reported to be highest when the heater was directly exposed to ultrasonic wave at 0°.

6.3.1.3.4 Flow rate in forced convective heat transfer

Lee and Choi [107] investigated CO_2-saturated water moving through a straight channel with and without sonication. At higher Reynolds numbers, ultrasound-generated turbulence strength decreases due to lower ultrasonic energy emission per unit mass of the fluid flow. Gaseous cavitation produces stronger turbulence than ultrasound. Gaseous cavitation collapse causes the viscous sublayer to separate, increasing turbulence, but should promote heat transmission. Similar findings were obtained utilizing a small furnace tube with a nonvarying wall heat flux and laminar water flow. Acoustic cavitation and streaming induced by 20 to 33 kHz ultrasonic vibration are less prominent in proportion to fluid bulk turbulence with increased Reynolds number [85]. As the flow rate increases, ultrasonic effectiveness diminishes [90].

6.3.1.4 Heat exchangers involving ultrasound

Heat exchangers that use forced convection have received a lot of attention and studies have been performed to increase their efficiency. Heat exchangers aid transfer of heat between fluids at different temperatures [108] and can be classified by their purpose [109]. Heat exchangers are utilized in HVAC, food processing, chemical processing, and thermal recycling [108]. In recent decades, researchers have studied ultrasound to improve heat exchanger performance. The first study was by Kurbanov and Melkumov [110]. They assumed that ultrasound improves heat exchanger performance by homogenizing velocity vectors across the tube and reducing liquid surface tension at boundaries. Ultrasound interference removes lubricant-based heat resistance.

6.3.2 Use of magnetic field

Applying a magnetic field perpendicular to the direction of flow is a simple way to accomplish this technique. Magnetic characteristics should be present in the working fluid. Natural convection can be reduced with this strategy. Changing applied magnetic field can simply control fluid flow and heat transfer. Numerous scholars have looked into how the magnetic field affects natural convection in vertical annular geometry (VAG). Four basic boundary conditions were examined by Al-Nimr and Alkam [111], who also provided an analytical solution for the influence of magnetic field during NC in a VAG filled with porous material. They came up with formulas for volumetric flow rate, regional Nusselt number, and mixing cup temperature. A numerical study by Sankar et al. [112] examined how a magnetic field affected NC in a fluid with a low Prandtl number. An axial magnetic field decreased heat transmission and flow in shallow cavities. Radial magnetic field, on the other hand, appears to be more effective in taller cavities. When an external magnetic field was imposed, the flow oscillations were found to be subdued. The mean Nusselt number rises as the radii ratio does, while it falls as the Hartmann (Hr) number rises.

Kakarantzas et al. [113] quantitatively investigated the impact of a magnetic field placed on an electrically conducting fluid during NC in VAG. To assess the flow, they employed 3-D DNS for a wide range of Rayleigh and Hr numbers. They noticed turbulent flow in the absence of a magnetic field, which becomes laminar for high Hartmann numbers (Hr > 75) and loses its turbulent behavior in the presence of a strong magnetic field.

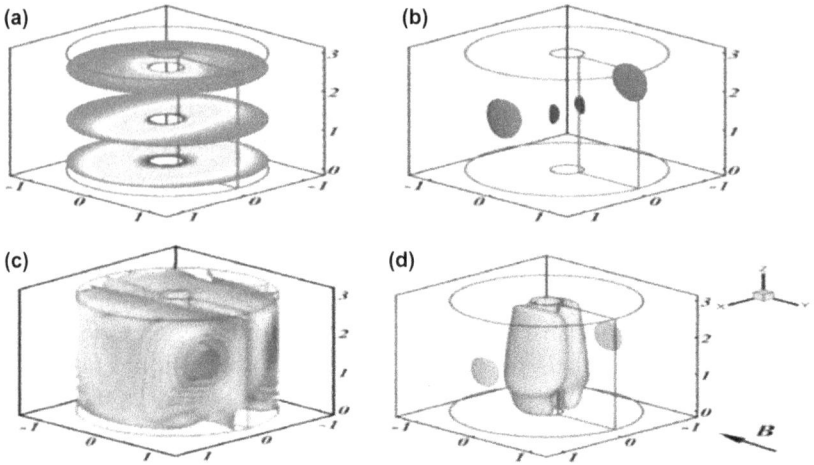

Figure 6.6 (a) Isotherm distribution; (b) velocity distribution; (c) electric potential distribution; and (d) current distribution [113].

They discovered that when a magnetic field is applied, convection heat transfer is reduced, and flow loses its axisymmetry as a result of the formation of Hartmann and Robert layers. For Ra = 105 and Hr = 100, Figure 6.6 depicts the 3D distributions of temperature, vertical velocity, electric potential, and electric current.

Shivashankar [114] investigated the effect of magnetic field on flow (natural convective) in double annuli packed through porous media using numerical simulations. Compared to shallow cavities, they discovered that radial magnetic field was most successful in inhibiting convection in taller cavities. Magnetic field influence on NC (double diffusive) in VAG is investigated by Venkatachalappa et al. [115]. They found that double-diffusive convective flow is only suppressed by the magnetic field at lower buoyancy ratios. The magnetic field, on the other hand, reduces thermal convective flow at larger buoyancy ratios. Jha et al. [116] investigated the impact of a magnetic field during NC in a VAG of an electrically conducting fluid with a microgeometry. In the presence of a radial magnetic field, they were able to get a perfect solution to the governing equations. They investigated the effects of numerous parameters on flow conditions such as flow rate and velocity of fluid, such as Hartmann number, radius ratio, and refraction parameter. The results show that increasing radius ratio improves flow rate and minimizes slip caused by fluid wall interaction, whereas increasing Hr number decreases volume flow rate.

Afrand [117] used molten gallium as a working fluid in a 3D computational simulation of NC in VAG. A perpendicular direction to an annulus mold was used to apply the magnetic field. It has been kept consistent and clear how hot or cold the walls are inside and out. Bottom and top are filled with electrically conducting fluids and are kept adiabatic. Findings show that natural convection is more restricted by magnetic field in narrow annuli than in thick annuli. In the opposite direction of buoyancy force, the electric potential created by a magnetic field, results in a decrease in Lorentz force. To suppress convection, a low Rayleigh number (Ra) was discovered to be advantageous. For improved quality goods, a larger magnetic field should be used to cast high Prandtl number liquid metal. Afrand et al. [118] conducted a computational analysis of the influence of the electric field on the magneto NC in the VAG using potassium as the working fluid for electrical conductivity. It was found that electric and magnetic fields alter the flow axisymmetry and lower the Lorentz forces, respectively.

For the investigation of fluid flow, Marzougui et al. [119] executed a numerical study inside a magnetized channel (porous). Poiseuille–Rayleigh-Benard mixed convection with an aspect ratio of $(A = 5)$ inside a channel has been studied and thermodynamic irreversibility was noted. A consistent transverse magnetic field is also present in experiments. With changes in different parameters such as Hartmann number (Hr), Darcy number (Da), Brinkmann number (Br), and porosity, various features of fluid flow are thoroughly researched. Studies show that the magnetic impact promotes entropy generation, whereas the Darcy number decreases it. In addition, Br variation has a substantial effect on entropy generation. Since 1995, magnetic field impact on NC in VAG has been statistically observed. With rise in Hr, they all noticed a fall in natural convection heat transmission. There have also been significant alterations in flow axisymmetry.

6.4 CONCLUSIONS

In recent years, active approaches have been developed to increase heat transport and thermal system efficiency. Ultrasound irradiation is a highly efficient means to promote heat transmission, especially convective heat transfer. Acoustic cavitation and acoustic streaming increase turbulence and heat transmission depending on ultrasonic wave frequency. This study looked at the effect of ultrasonic power, heat transfer system geometry, propagation medium, and forced convection flow rate in addition to frequency. This also reviews research on the ultrasound impact on heat exchanger performance. Ultrasound increases convective heat transfer and reduces fouling, which improves heat exchanger performance.

Application of MNFs in heat transfer appears to be promising. Many elements of this topic are still difficult to grasp, from the synthesis and characterization of MNF to practical applications through the knowledge of the mechanics behind the reported improvement in heat transmission. Understanding of processes and physical phenomena is required to develop schemes with improved heat transfer when utilizing MNF. Hydrodynamics and heat and mass transfer issues have been addressed. However, some issues such as MNF stability in magnetic and gravitational fields' presence, heat transmission processes, and formulation methods impact on physical characteristics remain to be addressed.

REFERENCES

[1] M. Legay, N. Gondrexon, S. Le Person, P. Boldo, A. Bontemps, Enhancement of heat transfer by ultrasound: Review and recent advances, *Int. J. Chem. Eng.*, Vol. 2011, 1–17, 2011. https://doi.org/10.1155/2011/670108

[2] S.M.M.M. Gheisari, R. Gavagsaz-Ghoachani, M. Malaki, P. Safarpour, M. Zandi, Ultrasonic nano-emulsification – A review, *Ultrason. Sonochem.*, 2018. https://doi.org/10.1016/j.ultsonch.2018.11.005

[3] M. Wiklund, R. Green, M. Ohlin, Acoustofluidics 14: Applications of acoustic streaming in microfluidic devices, *Lab Chip*, Vol. 12, 2438, 2012. https://doi.org/10.1039/c2lc40203c

[4] L. Rayleigh, *Theory of Sound*, 2nd ed., Dover, New York, 1945.

[5] P.J. Westervelt, The theory of steady rotational flow generated by a sound field, *J. Acoust. Soc. Am.*, Vol. 25, 60–67, 1953. https://doi.org/10.1121/1.1907009

[6] W.L. Nyborg, Acoustic streaming due to attenuated plane waves, *J. Acoust. Soc. Am.*, Vol. 25, 68–75, 1953. https://doi.org/10.1121/1.1907010

[7] H. Schlichting, *Boundary Layer Theory*, 7th ed., McGraw-Hill, New York, 1979.

[8] S.J. Lighthill, Acoustic streaming, *J. Sound Vib.*, Vol. 61, 391–418, 1978. https://doi.org/10.1016/0022-460X(78)90388-7

[9] J.T. Stuart, Double boundary layers in oscillatory viscous flow, *J. Fluid Mech.*, Vol. 24, 673–687, 1966. https://doi.org/10.1017/S0022112066000910

[10] S.-V. Orandrou, J.-C. Roy, Y. Bailly, E. Poncet, L. Girardot, D. Ramel, Determination of the heat transfer coefficients for the combined natural and streaming convection on an ultrasonic transducer, *Int. J. Heat Mass Transf.*, Vol. 62, 402–410, 2013. https://doi.org/10.1016/j.ijheatmasstransfer.2013.02.071

[11] Y. Iida, K. Tsutsui, Effects of ultrasonic waves on natural convection, nucleate boiling, and film boiling heat transfer from a wire to a saturated liquid, *Exp. Thermal Fluid Sci.*, Vol. 5, 108–115, 1992. https://doi.org/10.1016/0894-1777(92)90059-E

[12] B. Tajik, A. Abbassi, M. Saffar-Avval, A. Abdullah, H. Mohammad-Abadi, Heat transfer enhancement by acoustic streaming in a closed cylindrical enclosure filled with water, *Int. J. Heat Mass Transf.*, Vol. 60, 230–235, 2013. https://doi.org/10.1016/j.ijheatmasstransfer.2012.12.066

[13] S. Hyun, D.-R. Lee, B.-G. Loh, Investigation of convective heat transfer augmentation using acoustic streaming generated by ultrasonic vibrations, *Int. J. Heat Mass Transf.*, Vol. 48, 703–718, 2005. https://doi.org/10.1016/J.IJHEATMASSTRANSFER.2004.07.048

[14] C. Bartoli, F. Baffigi, Use of ultrasonic waves in sub-cooled boiling, *Appl. Therm. Eng.*, Vol. 47, 95–110, 2012. https://doi.org/10.1016/j.applthermaleng.2012.02.009

[15] J.H. Jeong, Y.C. Kwon, Effects of ultrasonic vibration on subcooled pool boiling critical heat flux, *Heat Mass Transf.*, Vol. 42, 1155–1161, 2006. https://doi.org/10.1007/s00231-005-0079-1

[16] Y.K. Oh, S.H. Park, Y.I. Cho, A study of the effect of ultrasonic vibrations on phasechange heat transfer, *Int. J. Heat Mass Transf.*, Vol. 45, 4631–4641, 2002. https://doi.org/10.1016/S0017-9310(02)00162-X

[17] A. Amiri Delouei, H. Sajjadi, R. Mohebbi, M. Izadi, Experimental study on inlet turbulent flow under ultrasonic vibration: Pressure drop and heat transfer enhancement, *Ultrason. Sonochem.*, Vol. 51, 151–159, 2019. https://doi.org/10.1016/J.ULTSONCH.2018.10.032

[18] T. Hayat, A. Afsar, M. Khan, S. Asghar, Peristaltic transport of a third order fluid under the effect of a magnetic field, *Comput. Math. Appl.*, Vol. 53(7), 1074–1087, 2007.

[19] J. Takeuchi, S.I. Satake, N.B. Morley, T. Kunugi, T. Yokomine, M.A. Abdou, Experimental study of MHD effects on turbulent flow of FlibeSimulent fluid in circular pipe, *Fusion Eng. Des.*, Vol. 83(7), 1082–1086, 2008.

[20] A. Malekzadeh, A. Heydarinasab, M. Jahangiri, Magnetic field effect on laminar heat transfer in a pipe for thermal entry region, *J. Mech. Sci. Technol.*, Vol. 25(4), 877–884, 2011.

[21] S.A. Khashan, E. Elnajjar, Y. Haik, Numerical simulation of the continuous biomagnetic separation in a two-dimensional channel, *Int. J. Multiph. Flow.*, Vol. 37(8), 947–955, 2011.

[22] T. Hayat, S. Noreen, A. Alsaedi, Slip and induced magnetic field effects on peristaltic transport of Johnson-Segalman fluid, *Appl. Math. Mech.*, Vol. 33(8), 1035–1048, 2012.

[23] A.M. Abd-Alla, G.A. Yahya, S.R. Mahmoud, H.S. Alosaimi, Effect of the rotation, magnetic field and initial stress on peristaltic motion of micropolar fluid, *Mec. Dent.*, Vol. 47(6), 1455–1465, 2012.

[24] E. Gedik, H. Kurt, Z. Recebli, C. Balan, Two-dimensional CFD simulation of magnetorheological fluid between two fixed parallel plates applied external magnetic field, *Comput. Fluids*, Vol. 63, 128–134, 2012.

[25] V. Galindo, K. Niemietz, O. Pätzold, G. Gerbeth, Numerical and experimental modelling of VGF-type buoyant flow under the influence of traveling and rotating magnetic fields, *J. Cryst. Growth*, Vol. 360, 30–34, 2012.

[26] B.G. Srivastava, S. Deo, Effect of magnetic field on the viscous fluid flow in a channel filled with porous medium of variable permeability, *Appl. Math Comput.*, Vol. 219(17), 8959–8964, 2013.

[27] H. Aminfar, M. Mohammadpourfard, F. Ghaderi, Two-phase simulation of non-uniform magnetic field effects on biofluid (blood) with magnetic nanoparticles through a collapsible tube, *J. Magn. Magn. Mater.*, Vol. 332, 172–179, 2012.

[28] S. Akram, S. Nadeem, M. Hanif, Numerical and analytical treatment on peristaltic flow of Williamson fluid in the occurrence of induced magnetic field, *J. Magn. Magn. Mater.*, Vol. 346, pp. 142–151, 2013.

[29] P. Bitla, T.V.V. Iyengar, Pulsating flow of an incompressible micropolar fluid between permeable beds with an inclined uniform magnetic field. *Eur. J. Mech. B/Fluids.*, Vol. 48, 174–182, 2014.

[30] M. Muthtamilselvan, D.H. Doh, Magnetic field effect on mixed convection in a lid-driven square cavity filled with nanofluids, *J. Mech. Sci. Technol.*, Vol. 28(1), 137–143, 2014.

[31] R. Azizian, E. Doroodchi, T. McKrell, J. Buongiorno, L.W. Hu, B. Moghtaderi, Effect of magnetic field on laminar convective heat transfer of magnetite nanofluids, *Int. J. Heat Mass Transf.*, Vol. 68, 94–109, 2014.

[32] C.D. Dritselis, B. Knaepen, Mixed convection of a low Prandtl fluid with spatially periodic lower wall heating in the presence of a wall-normal magnetic field, *Int. J. Heat Mass Transf.*, Vol. 74, 35–47, 2014.

[33] O. Turk, C. Bozkaya, M. Tezer-Sezgin, A FEM approach to biomagnetic fluid flow in multiple stenosed channels. *Comput. Fluids*, Vol. 97, 40–51, 2014.

[34] S. Akram, S. Nadeem, Consequence of nanofluid on peristaltic transport of a hyperbolic tangent fluid model in the occurrence of apt (tending) magnetic field, *J. Magn. Magn. Mater.*, Vol. 358, 183–189, 2014.

[35] S. Saha, S. Chakrabarti, MHD modelling and numerical simulation on ferromagnetic fluid flow in a channel, *Int. J. Fluid Mech. Res.*, Vol. 43(1), 79–92, 2016.

[36] C. Maatki, W. Hassen, L. Kolsi, N. AlShammari, M. N. Borjini, H. B. Aissia, 3-D numerical study of hydromagnetic double diffusive natural convection and entropy generation in cubic cavity. *J. Appl. Fluid Mech.*, 2016. DOI: 10.18869/acadpub.jafm.68.235.24820

[37] F. Mebarek-Oudina, Numerical modeling of the hydrodynamic stability in vertical annulus with heat source of different lengths, *Eng. Sci. Technol. Int. J.*, Vol. 20, 1324–1333, 2017.

[38] C. Maatki, L. Kolsi, K. Ghachem, A.S. Alghamdi, M.N. Borjini, H.B. Aissia, Numerical analysis of electromagnetic parameters of thermal dominated double diffusive magneto- convection, *J. Eng. Res.*, Vol. 5(3), 198–215, 2017.

[39] L. Kolsi, S. Algarni, H.A. Mohammed, W. Hassen, E. Lajnef, W. Aich, M.A. Almeshaal, 3D magneto-buoyancy-thermocapillary convection of CNT-water nanofluid in the presence of a magnetic field, *Processes*, Vol. 258(8), 1–16, 2020. DOI: 10.3390/pr8030258

[40] A. Zaima, A. Aissab, F. Mebarek-Oudinac, B. Mahantheshe, G. Lorenzinif, M. Sahnounb, M. El Ganaouig, Galerkin finite element analysis of magneto hydrodynamic natural convection of Cu-water nanoliquid in a baffled U-shaped enclosure, *Propuls. Power Res.*, Vol. 9(4), 383–393, 2020.

[41] F. Mebarek-Oudina, R. Bessaih, B. Mahanthesh, A.J. Chamkha, Z. Raza, Magneto-thermal-convection stability in an inclined cylindrical annulus filled with a molten metal, *Int. J. Numer. Methods Heat Fluid Flow*, 2020. DOI: 10.1108/HFF-05-2020-0321

[42] F. Mebarek-Oudina, F. Redouane, C. Rajashekhar, Convection heat transfer of MgO-Ag water magneto-hybrid nanoliquid flow into a special porous enclosure, *Alger. j. renew energy sustain. dev.*, Vol. 2(1), 84–95, 2020.

[43] K Swain, F Mebarek-Oudina, S.M. Abo-Dahab, Influence of MWCNT/Fe_3O_4 hybrid nanoparticles on an exponentially porous shrinking sheet with chemical reaction and slip boundary conditions, *J. Therm. Anal. Calorim.*, 2020. DOI: https://doi.org/10.1007/s10973-020-10432-4

[44] S. Saha, S. Chakrabarti, Numerical simulation on the magnetic fluid flow through a channel, *Indian J. Eng.*, Vol. 17(47), 117–126, 2020.

[45] S.M. Abo-Dahab, M.A. Abdelhafez, F. Mebarek-Oudina, S.M. Bilal, MHD Casson nanofluid flow over nonlinearly heated porous medium in presence of extending surface effect with suction/injection, *Indian J Phys.*, 2021. https://doi.org/10.1007/s12648-020-01923-z

[46] A.O. Kuzubov, O.I. Ivanova, Magnetic liquids for heat exchange, *Journal dePhysique III France*, Vol. 4, 1–6, 1994.

[47] S. Odenbach *Magnetoviscous effects in ferrofluids*, Springer-Verlang, Berlin, Heidelberg, 2002.

[48] J. Philip, P. Shima, B. Raj. Evidence for enhanced thermal conduction through percolating structures in nanofluids, *Nanotechnology*, Vol. 19, p. 305706, 2008.

[49] V.L. Kolesnichenko, Synthesis of nanoparticulate magnetic materials, in *Magnetic Nanoparticles* (ed. S.P. Gubin), WILEY-VCH VerlagGmbH & Co. KGaA, Weinheim, 2009.

[50] C.-H. Lo, T.-T. Tsung, L.-C. Chen, Ni nano-magnetic fluids prepared by Submerged Arc Nano Synthesis System (SANSS), *JSME Int. J.*, Vol. 48, pp. 750–755, 2005.

[51] S.L. Vékás, D. Bica, M.V. Avdeev, Magnetic nanoparticles and concentrat-edmagnetic nanofluids: Synthesis, properties and some applications, *China Particuology*, Vol. 5, 43–49, 2007.

[52] S.W. Charles, The preparation of magnetic fluids, in *Ferrofluids: Magnetically Controllable*, (ed. S Odenbach), Springer-Verlag, Berlin–Heidelberg, 2002.

[53] S. Odenbach, Recent progress in magnetic fluid research. *J. Phys. Condens. Matter*, Vol. 16, R1135–R1150, 2004.

[54] S.K. Das, S.U.S. Choi, W. Yu, T. Pradeep, *Nanofluids*. Wiley Interscience ed. John Wiley& Sons, Inc, Hoboken. New Jersey, USA, 2009.

[55] H. Chen, S. Witharana, Y. Jin, C. Kim, Y. Ding. Predicting thermal conduc-tivityof liquid suspensions of nanoparticles (nanofluids) based on rheology, *Particulorogy*, Vol. 7, 151–157, 2008.

[56] X.-Q. Wang, A.S. Mujumdar, Heat transfer characteristics of nanofluids: A review, *Int. J. Therm. Sci.*, Vol. 46, 1–19, 2007.

[57] A.K. Singh, Thermal conductivity of nanofluids, *Def. Sci. J.*, Vol. 58: 600–607, 2008.

[58] C.-W. Nan, Z. Shi, Y. Lin, A simple model for thermal conductivity ofcarbon nanotube-based composites, *Chem. Phys. Lett.*, Vol. 375, 666–669, 2003.

[59] T.-K. Hong, H.-S. Yang, C.J. Choi, Study of the enhanced thermal conductivity of Fe nanofluids, *J. Appl. Phys.*, Vol. 97, 064311, 2005.

[60] M. Dehbani, M. Rahimi, M. Abolhasani, A. Maghsoodi, P. Ghaderi Afshar, A.R. Dodmantipi, A.A. Alsairafi, CFD modeling of convection heat transfer using 1.7 MHz and 24 kHz ultrasonic waves: A comparative study, *Heat. Mass. Transf.*, Vol. 50, 1319–1333, 2014.

[61] A. Amiri, P. Sharifian, N. Soltanizadeh, Application of ultrasound treatment for improving the physicochemical, functional and rheological properties of myofibrillar proteins, *Int. J. Biol. Macromol.*, Vol. 111, 139–147, 2018.

[62] M. Ja'fari, S.L. Ebrahimi, M.R. Khosravi-Nikou, Ultrasound-assisted oxida-tive desulfurization and denitrogenation of liquid hydrocarbon fuels: A critical review, *Ultrason. Sonochem.*, Vol. 40, 955–968, 2018.

[63] N. Bhargava, R.S. Mor, K. Kumar, V.S. Sharanagat, Advances in application of ultrasound in food processing: A review, *Ultrason. Sonochem.*, Vol. 70, 105293, 2021. https://doi.org/10.1016/j.ultsonch.2020.105293

[64] Y. Zhang, N. Abatzoglou, Review: Fundamentals, applications and potentials of ultrasound-assisted drying, *Chem. Eng. Res. Des.*, Vol. 154, 21–46, 2020.

[65] L. Qiu, M. Zhang, B. Chitrakar, B. Bhandari, Application of power ultrasound in freezing and thawing processes: Effect on process efficiency and product quality, *Ultrason. Sonochem.*, Vol. 68, 105230, 2020.

[66] S.-Q. Hu, G. Liu, L. Li, Z.-X. Li, Yi Hou, An improvement in the immersion freezing process for frozen dough via ultrasound irradiation, *J. Food Eng.*, Vol. 114(1), 22–28, 2013.

[67] M. Legay, N. Gondrexon, S. Le Person, P. Boldo, A. Bontemps, Enhancement of heat transfer by ultrasound: review and recent advances, *Int. J. Chem. Eng.*, Vol. 2011, 17 pages, 2011.

[68] A. Swarnkar, V.J. Lakhera, Ultrasonic augmentation in pool boiling heat transfer over external surfaces: A review, *Proc. I. MechE.*, Vol. 235, 2099–2111, 2020.

[69] F. Yu, X. Luo, B. He, J. Xiao, W. Wang, J. Zhang, Experimental investigation of flow boiling heat transfer enhancement under ultrasound fields in a minichannel heat sink, *Ultrason. Sonochem.*, Vol. 70, 105342, 2021.

[70] J.S. Sitter, T.J. Snyder, J.N. Chun, Acoustic field interaction with a boiling system under terrestrial gravity and microgravity, *Int. J. Heat. Mass. Transf.*, Vol. 41, 2143–2155, 1998.

[71] B. Li, X. Han, Z. Wan, et al., Influence of ultrasound on heat transfer of copper tubes with different surface characteristics in sub-cooled boiling, *Appl. Therm. Eng.*, Vol. 92, 93–103, 2016.

[72] H.S. Muralidhara, D. Ensminger, A. Putnam, Acoustic dewatering and drying (low and high frequency): State of the art review, *Drying Technol.*, Vol. 3(4), 529–566, 1985.

[73] Y.E. Yao, W. Wang, K. Yang, Mechanism study on the enhancement of silica gel regeneration by power ultrasound with field synergy principle and mass diffusion theory, *Int. J. Heat. Mass. Trans.*, Vol. 90, 769–780, 2015.

[74] K. Yang, Y.E. Yao, S. Liu, B. He, Separation of ultrasonic contributions and energy utilization characteristics of ultrasonic regeneration, *AIChE. J.*, Vol. 60(5), 1843–1853, 2014.

[75] Y.E. Yao, W. Zhang, K. Yang, S. Liu, B. He, Theoretical model on the heat and mass transfer in silica gel packed beds during the regeneration assisted by highintensity ultrasound, *Int. J. Heat. Mass. Trans.*, Vol. 55(23–24), 7133–7143, 2012.

[76] Y. Yao, W. Li, Y. Hu, Modeling and performance investigation on the counterflow ultrasonic atomization liquid desiccant regenerator, *Appl. Therm. Eng.*, Vol. 165, 114573, 2020.

[77] Y. Yao, Z. Zhu, H. Guo, Y. Yu, Modeling and parametric study of the ultrasonic atomization regeneration of desiccant solution, *Int. J. Heat. Mass. Trans.*, Vol. 127, 687–702, 2018.

[78] R. Kuppa, V.S. Moholkar, Physical features of ultrasound-enhanced heterogeneous permanganate oxidation, *Ultrason. Sonochem.*, Vol. 17(1), 123–131, 2010.

[79] R. Pecha, B. Gompf, Microimplosions: Cavitation collapse and shock wave emission on a nanosecond time scale, *Phys. Rev. Lett.*, Vol. 84, 1328–1330, 2000.

[80] O. Bulliard-Sauret, S. Ferrouillat, L. Vignal, A. Memponteil, N. Gondrexon, Heat transfer enhancement using 2 MHz ultrasound, *Ultrason. Sonochem.*, Vol. 39, 262–271, 2017.

[81] B.C.Q. Seah, B.M. Teo, Recent advances in ultrasound-based transdermal drug delivery, *Int. J. Nanomedicine*, Vol. 13, 7749–7763, 2018.

[82] T. Leong, M. Ashokkumar, S. Kentish, The fundamentals of power ultrasound – A review, *Acoustics Australia*, Vol. 39, 54–63, 2011.

[83] T. Mongkolkitngam, M. Fukuta, M. Motozawa, W. Chaiworapuek, Thermal characterization of a heating cylinder under ultrasonic effects, *Int. J. Heat Mass Transf.*, Vol. 175, 121393, 2021.

[84] K. Viriyananon, J. Mingbunjerdsuk, T. Thungthong, W. Chaiworapuek, Characterization of heat transfer and friction loss of water turbulent flow in a narrow rectangular duct under 25–40 kHz ultrasonic waves, *Ultrasonics*, Vol. 114, 106366, 2021.

[85] N.P. Dhanalakshmi, R. Nagarajan, N. Sivagaminathan, B.V.S.S.S. Prasad, Acoustic enhancement of heat transfer in furnace tubes, *Chem. Eng. Process. Process Intensif.*, Vol. 59, 36–42, 2012.

[86] O. Bulliard-Sauret, J. Berindei, S. Ferrouillat, L. Vignal, A. Memponteil, C. Poncet, J.M. Leveque, N. Gondrexon, Heat transfer intensification by low or high frequency ultrasound: Thermal and hydrodynamic phenomenological analysis, *Exp. Therm. Fluid Sci.*, Vol. 104, 258–271, 2019.

[87] M. Palma, C.G. Barroso, Ultrasound-assisted extraction and determination of tartaric and malic acids from grapes and winemaking by-products, *Anal. Chim. Acta*, Vol. 458(1), 119–130, 2002.

[88] D.R. Dance, S. Christofides, A.D.A. Maidment, I.D. McLean, K.H. Ng, *Diagnostic Radiology Physics: A Handbook for Teachers and Students*, IAEA, Vienna, Austria, 2014.

[89] J. Cai, X. Huai, S. Liang, X. Li, Augmentation of natural convective heat transfer by acoustic cavitation, *Front. Energy Power. Eng. China*, Vol. 4(3), 313–318, 2010.

[90] H. Azimy, A.H. MeghdadiIsfahani, M. Farahnakian, A. Karimipour, Experimental investigation of the effectiveness of ultrasounds on increasing heat transfercoefficient of heat exchangers, *Int. Commun. Heat Mass Transf.*, Vol. 127, 105575, 2021.

[91] R.K. Gould, Heat transfer across a solid-liquid interface in the presence of acoustic streaming, *J. Acoust. Soc. Am.*, Vol. 40(1), 219–225, 1966.

[92] H. Kiani, D.-W. Sun, Z. Zhang, The effect of ultrasound irradiation on the convective heat transfer rate during immersion cooling of a stationary sphere, *Ultrason. Sonochem.*, Vol. 19(6), 1238–1245, 2012.

[93] G. Musielak, D. Mierzwa, Enhancement of convection heat transfer in air using ultrasound, *Appl. Sci.*, Vol. 11, 8846, 2021.

[94] D. Zhou, X. Hu, D. Liu, Local convective heat transfer from a horizontal tube in an acoustic cavitation field, *J. Therm. Sci.*, Vol. 13(4), 338–343, 2004.

[95] G. Shen, L. Ma, S. Zhang, S. Zhang, L. An, Effect of ultrasonic waves on heat transfer in Al_2O_3 nanofluid under natural convection and pool boiling, *Int. J. Heat Mass Transf.*, Vol. 138, 516–523, 2019.

[96] H. Azimy, A.H. MeghdadiIsfahani, M. Farahnakian, Investigation of the effect of ultrasonic waves on heat transfer and nanofluid stability of MWCNTs in sono heat exchanger: An experimental study, *Heat. Mass. Transf.*, Vol. 58(3), 467–479, 2022.

[97] B.-G. Loh, S. Hyun, P.I. Ro, C. Kleinstreuer, Acoustic streaming induced by ultrasonic flexural vibrations and associated enhancement of convective heat transfer, *Acoust. Soc. Am.*, Vol. 111(2), 875–883, 2002.

[98] V. Uhlenwinkel, R. Meng, K. Bauckhage, Investigation of heat transfer from circular cylinders in high power 10 kHz and 20 kHz acoustic resonant fields, *Int. J. Therm. Sci.*, Vol. 39(8), 771–779, 2000.

[99] D.R. Lee, B.G. Loh, Smart cooling technology utilizing acoustic streaming, *IEEE. T. Compon. Pack. T.*, Vol. 30(4), 691–699, 2007.

[100] K. Prodanov, Experimental investigation of the effects of acoustic waves on natural convection heat transfer from a horizontal cylinder in air, *A Thesis presented to the Faculty of California Polytechnic State University*, San Luis Obispo, 2021.

[101] Y. Iida, K. Tsutsui, R. Ishii, Y. Yamada, Natural-convection heat transfer in a field of ultrasonic waves and sound pressure, *J. Chem. Eng. Japan.*, Vol. 24(6), 794–796, 1991.

[102] H.-Y. Kim, Y.G. Kim, B.H. Kang, Enhancement of natural convection and pool boiling heat transfer via ultrasonic vibration, *Int. J. Heat. Mass. Transf.*, Vol. 47(12–13), 2831–2840, 2004.

[103] M. Rahimi, M. Dehbani, M. Abolhasani, Experimental study on the effects of acoustic streaming of high frequency ultrasonic waves on convective heat transfer: Effects of transducer position and wave interference, *Int. Commun. Heat Mass Transf.*, Vol. 39(5), 720–725, 2012.

[104] F.C. Liu, S.W. Chen, J.D. Lee, Feasibility study of heat transfer enhancement by ultrasonic vibration under subcooled pool condition, *Heat. Transf. Eng.*, Vol. 39(7–8), 654–662, 2018.

[105] S.-W. Chen, F.-C. Liu, H.-J. Lin, P.-S. Ruan, Y.-T. Su, Y.-C. Weng, J.-R. Wang, J.- D. Lee, W.-K. Lin, Experimental test and empirical correlation development for heat transfer enhancement under ultrasonic vibration, *Appl. Therm. Eng.*, Vol. 143, 639–649, 2018.

[106] S. Vajnhandl, A.M.L. Marechal, Ultrasound in textile dyeing and the decolourization/mineralization of textile dyes, *Dyes. Pigm.*, Vol. 65, 89–101, 2005.

[107] S.Y. Lee, Y.D. Choi, Turbulence enhancement by ultrasonically induced gaseous cavitation in the CO_2 saturated water, *KSME. Int. J.*, Vol. 16(2), 246–254, 2002.

[108] C. Balaji, B. Srinivasan, S. Gedupudi, Chapter 7 - Heat exchangers, *Heat Transfer Engineering, Fundamentals and Techniques*, 199–231, Academic Press, Oxford, UK, 2021.

[109] L. Pekar, Chapter 1 - Introduction to heat exchangers, *Advanced Analytic and Control Techniques for Thermal Systems with Heat Exchangers*, 3–20, Academic Press, Oxford, UK, 2020.

[110] U. Kurbanov, K. Melkumov, Use of ultrasound for intensification of heat transfer process in heat exchangers, in *Proceedings of the International Congress of Refrigeration*, 1–5, (4), Washington, DC, USA, 2003.

[111] M.A. Al-Nimr, M.K. Alkam Magneto hydrodynamics transient free convection in open-ended vertical annuli, *J. Thermophys. Heat Tran.*, Vol. 13, 256–265, 1999. https://doi.org/10.2514/2.6430

[112] M. Sankar, M. Venkatachalappa, I.S. Shivakumara Effect of magnetic field on natural convection in a vertical cylindrical annulus, *Int. J. Eng. Sci.*, Vol. 44, 1556–1570, 2006. https://doi.org/10.1016/j.ijengsci.2006.06.004

[113] S.C. Kakarantzas, A.P. Grecos, N.S. Vlachos, I.E. Sarris, B. Knaepen, D. Carati, Direct numerical simulation of a heat removal configuration for fusion blankets Energy Convers, *Dent. Manage.*, Vol. 48, 2775–2783, 2007. https://doi.org/10.1016/j.enconman.2007.07.024

[114] M. Sankar, Y. Do, Numerical simulation of free convection heat transfer in a vertical annular cavity with discrete heating, *Int. Commun. Heat Mass Tran.*, Vol. 37, 600–606, 2010. https://doi.org/10.1016/j.icheatmasstransfer.2010.02.009

[115] M. Venkatachalappa, Y. Do, M. Sankar, Effect of magnetic field on the heat and mass transfer in a vertical annulus, *Int. J. Eng. Sci.*, Vol. 49, 262–278, 2011. https://doi.org/10.1016/j.ijengsci.2010.12.002

[116] B.K. Jha, M.O. Oni Natural convection flow in a vertical micro-annulus with time-periodic thermal boundary conditions: An exact solution Multidiscip, *Model. Mater. Struct.*, Vol. 14, 1064–1081, 2018. https://doi.org/10.1108/MMMS-09-2017-0111

[117] M. Afrand, D. Toghraie, A. Karimipour, S. Wongwises A numerical study of natural convection in a vertical annulus filled with gallium in the presence of magnetic field, *J. Magn. Magn Mater.*, Vol. 430, 22–28, 2017. https://doi.org/10.1016/j.jmmm.2017.01.016

[118] M. Afrand, Using a magnetic field to reduce natural convection in a vertical cylindrical annulus Int, *J. Therm. Sci.*, Vol. 118, 12–23, 2017. https://doi.org/10.1016/j.ijthermalsci.2017.04.012

[119] S. Marzougui, M. Bouabid, F. Mebarek-Oudina, N. Abu-Hamdeh, M. Magherbi, K. Ramesh, A computational analysis of heat transport irreversibility phenomenon in a magnetized porous channel, *Int. J. Numer. Methods Heat Fluid Flow*, 2020. https://doi.org/10.1108/HFF-07-2020-0418

Chapter 7

Heat transfer augmentation using jet impingement

7.1 INTRODUCTION

Impinging jets can significantly boost heat and mass transmission in industrial applications. Impinging jets may transmit a lot of heat especially at short distance (i.e., nozzle to plate distance) [1, 2]. By directing the fluid flow toward the surface can transfer a significant amount of heat between fluid and surface. Jet impingement heat transfer coefficients can be up to three times greater than those of conventional convection cooling, which uses restricted flow that is parallel to cooled surface. It has numerous engineering applications like cooling of turbine blades [3], drying process of fabric and paper [4], heating furnace [5], glass and metal sheet tempering [6, 7], and processing of food [8], etc. In cooling electronic components, impinging jets are identified to produce high-average and local heat transfer coefficients [9], and the ability to accommodate extraordinarily large heat fluxes is critical for technical advancement in this field [10].

Impinging jet heat transfer is impacted by a variety of factors, and many researchers have studied impinging jets both experimentally and numerically. Several aspects, such as nozzle to plate distance and jet geometry, are studied. Reynolds number (Re) is an important parameter as it has a major impact on heat transfer and flow characteristics. He et al. [11] studied heat transfer properties of single-slot and synthetic jets on a vertical surface in an experimental set up. For both steady and irregular jets, the heat transfer coefficient rises with Re on a vertical surface. In a laminar cross-flow with multiple impinging jets, Guoneng et al. [12] investigated heat transfer from a rectangular plate. They found that raising the Re number improved convective heat exchange. Mohamed et al. [13] examined heat transmission of impinging circular jets. A broad wide range of Re (7,100 to 30,000) was used in the experiments, and the results indicated that as Re of the jets climbed, so did the average Nusselt number (Nu) values. The distance (H) between the nozzle and the target surface, which is often given as a function of jet diameter, is another critical component. Numerical analysis of the turbulent flow structure and heat transport of a pulsed impinging jet revealed that heat transfer is significantly impacted by the separation between the

        DOI: 10.1201/9781003229865-7

pipe edge and the target surface, and that heat transmission peaks at H/D 14.6 [14]. Aldabbagh and Sezai [15] computationally solved the 3D energy equation and Navier–Stokes equations for steady-state circumstances to examine impinging multiple jets flow and heat transfer characteristics. Jet-to-plate spacing was shown to have a significant impact on heat trans-mission. Jet geometry is another factor in the heat transfer of impinging jets. Both physically and numerically, Caliskan et al. [16] studied the impact of jet geometry on the properties of heat transmission for circular and rectan-gular impinging jet arrays. Compared to circular jets, elliptical jets obtained higher Nu with better heat transfer performance. Wen et al. [17] studied impacts of various jet geometry parameters on heat transfer characteristics numerically. Different jet geometries were investigated using the SST turbu-lence model. According to reports, every jet geometry performs well in a variety of applications. The inclination angle was discovered to be a signifi-cant parameter in impinging jets when considering fluid flow and heat trans-fer. Aside from these criteria, there are numerous studies on impinging jets. Kim and Giovannini [18] evaluated turbulence statistics and flow topology of jet impinging on a square area cylinder. A 3D recirculation zone was made visible after detaching from the cylinder, and foci were discovered between the bottom corner and the recirculation's detachment line. Xing and Weigand [19] had examined heat transfer properties of an impinging jet on dimpled and flat surfaces with various cross-flows. It was found that a dimpled surface transfers heat more quickly than a flat surface. The impact of a parallel, secondary, and low Re round jet was examined using 2D laser Doppler anemometry. The data showed that the primary jet entirely absorbed the secondary jet. Skewness and flatness characteristics were found to be useful indicators of small-scale mixing in the study [20]. Wang et al. [21] looked at the impact of Re number on jets expelled from nozzles at 30° and 45° inclination angles. A jet with a 45° inclination angle was found to have the best heat transfer rate.

By examining many parameters impacting heat transfer rate, the current chapter describes hydrodynamic behavior and heat transfer enhancement of surface utilizing various jet impingement techniques. Additionally, to increase heat transfer, multiple jet impingement techniques have been used. In this chapter, first geometrical characteristics of target surface and jet-target surface gap are considered along with several jet excitation methods. Finally, the roles of phase change materials (PCMs) and nanofluids in jet impingement is examined.

7.2 MECHANISM

Understanding the jet impingement process's flow mechanism and heat transfer characteristics is crucial. Nozzle diameter (D) and distance between nozzle and target surface (H) define geometric arrangement. Jet fluid leaves

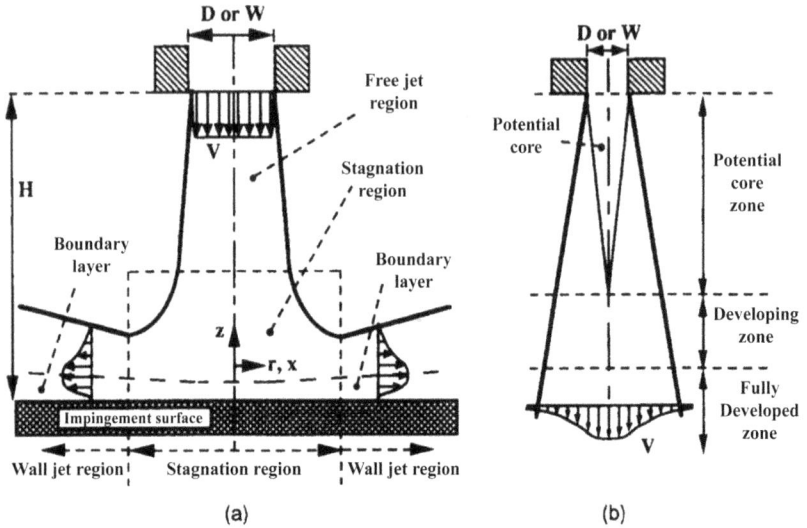

Figure 7.1 (a) Impinging jet regions and (b) free jet regions [5].

the nozzle at a comparatively constant velocity (V) and temperature when there is some ambient turbulence (T). Flow structures of impinging jets can be split into three distinct zones, as illustrated in Figure 7.1. These zones are free jet area, impingement (stagnation) flow area, and wall jet area.

Entrainment of mass, momentum, and energy occur in free jet zone due to shear-driven interaction of outgoing jet and ambient air. The creation of a nonuniform radial velocity profile within the jet, the expansion of the jet with an increase in total mass flow rate, and a change in the temperature of the jet before it impacts the surface are all examples of impacts. A stagnation zone and a jet's radial turning define the impingement zone. Impingement zone boundary layer thickness is approximately constant [22]. A bulk flow in outward radial direction characterizes wall jet zone. The greatest velocity for $0 < H/D < 12$ occurs about one jet diameter from the impingement zone [23]. Heat transfer rate is strongly influenced by the amount of jet turbulence that is subsequently advected into near-wall region. Intense fluid mechanics and thermal interaction have a big impact on heat transfer in stagnation and wall jet zones as well as mean heat transfer above a surface. The developing zone, potential core zone, and developed zone are three zones that make up a free jet region. Figure 7.1(b) depicts these three zones. The potential core's velocity is nonvarying and equal to the exit velocity of the nozzle. Potential core length is determined by the initial velocity profile and the amount of turbulence at the nozzle exit. Potential core zones for axisymmetric jets extend 6–7 diameters from nozzle exit, whereas potential core zones for slot jets extend 4.7–7.7 slot widths [24]. Developing zone is identified by the axial velocity profile produced by significant shear loads near the jet boundary. Turbulence is created as a result of huge shear forces, which encourages entrainment of more fluid. After the development zone, the velocity profile has reached its entire development. A Gaussian velocity distribution fits test observations [25] in this zone, and others have observed that jet broadens linearly and the axial velocity decays linearly in the fully developed zone [26]. An impinging jet is referred to as laminar up to a single-jet Re of approximately 2,500 [27]. Despite the lack of convincing evidence, impinging jets have a transition Re, this figure is commonly employed. Circular free jets can be classified into four distinct regions:

1. Re < 300 is a dispersed laminar jet.
2. A laminar jet with Re more than 300 and less than 1,000.
3. A transition/semi-turbulent jet with Re more than 1,000 and less than 3,000.
4. A turbulent jet with Re greater than 3,000.

A laminar free jet is determined by a variety of criteria including Re_j, separation distance, velocity profile, and whether jet is restricted or not. These parameters have an impact on mixing that occurs at outer boundaries of jet, which turns jet from laminar to turbulent. Radial pressure distribution for a spherical impinging jet has been measured by numerous studies on the impingement surface. This is because radial velocity gradient β is a common form of correlation for Nusselt number. Bernoulli's equation can be used to determine the gradient if impacts of viscous are insignificant and boundary layer velocity is β_r.

$$\beta = \left(\frac{d}{dr} \frac{\sqrt{2\left(p_s - p(r)\right)}}{r} \right) \text{ at } r = 0 \tag{7.1}$$

Consequently, numerous researchers had measured pressure distribution; major heat transfer literature publishes results for β rather than $p(r)$ [28]. The velocity of jet is not having enough time to develop and is largely uniform at modest separation distances ($H/D = 1.2$). Impingement thus takes place within the potential core, and the pressure distribution is identical to that of the inviscid solution. Velocity profile of jet becomes more nonuniform as separation distance grows, and pressure distributions become broader.

7.2.1 Heat transfer characteristics

Heat transfer (Convective) coefficient (h) is defined as (see also Chapter 1),

$$h = \frac{q}{T_w - T_{aw}} \tag{7.2}$$

where q stands for heat flux (convective) and T_{aw} is local adiabatic temperature.

Adiabatic wall temperature (T_{aw}) and h are below presented in terms of Nu (Nusselt number), effectiveness, and recovery factor as,

$$\mathrm{Nu} = \frac{hD}{k} \tag{7.3}$$

and effectiveness,

$$\eta = \frac{T_{aw} - T_r}{T_j^0 - T_\infty} \tag{7.4}$$

or recovery factor,

$$r = \frac{T_{aw} - T_j^0}{\dfrac{V_j^2}{2cp}} \tag{7.5}$$

In published literature, the latter two factors are utilized to correlate and generalize experimental data [29, 30]. For low jet velocities, the temperature of the adiabatic wall equals that of the jet. By averaging the local Nu distribution, mean Nu may be calculated:

$$\overline{\mathrm{Nu}} = \frac{\overline{h}D}{k} = \frac{D}{k}\int_A h\frac{(T_w - T_{aw})dA}{A\Delta T} \tag{7.6}$$

where mean temperature difference $\overline{\Delta T}$ is defined as,

$$\overline{\Delta T} = \overline{(T_w - T_{aw})} = \int_A \frac{(T_w - T_{aw})dA}{A} \tag{7.7}$$

Both axisymmetric and slot jets are covered by these definitions. It is worth noting that average values of $\overline{\mathrm{Nu}}$ and $\overline{\Delta T}$ are likely to vary depending on area (A) across which quantities were averaged. If $T_w - T_{aw}$ is nonvarying over surface and jet is axisymmetric, Eq. (7.6) for the mean Nu reduces to,

$$\overline{\mathrm{Nu}} = \frac{\overline{h}D}{k} = \frac{2}{R^2}\int_0^R \mathrm{Nu}(r)r\,dr \tag{7.8}$$

If heat flux, $q = h\,(T_w - T_{aw})$, is nonvarying over surface πR^2, average Nu becomes

$$\overline{\mathrm{Nu}} = \frac{\overline{h}D}{k} = \frac{qD}{k\,\Delta T_w} \tag{7.9}$$

The focus of the discussion here is on phenomena, knowledge of impacts and data, and understanding gaps, rather than on comparison and suggestion of documented correlations for design purposes. Previous reviews [22, 31–33], which will not be repeated here, have already accomplished this to a great extent.

7.3 INVESTIGATED PARAMETERS

Many researchers are interested in jet impingement because of its vast range of applications and its essential importance. High heat transfer rates are achieved via jet impingement technologies. Target surface geometry, target surface distance, nanofluids use, excitation of jet, PCMs all affect heat transfer rate of an impinging jet. Some of the most important dimensionless geometrical parameters used in jet impingement techniques to enhance heat transfer performance are discussed below. However, several other critical parameters such as nozzle design, jet injection settings, and design of target surface and fluid characteristics will be methodically presented later. The most important geometrical parameters (dimensionless) for optimizing heat transfer performance in jet impingement processes are introduced in the sections that follow.

- Reynolds number: It may be defined as inertial forces divided by viscous forces. It aids in the evaluation of turbulence effects by linking nozzle diameter and jet velocity to fluid viscosity.
- Nusselt number: It is the heat transfer ratio between convection and conduction. It aids in the quantification of thermal augmentation of jet impingement.
- Prandtl number: It may be defined as momentum diffusivity divided by thermal diffusivity and is a thermophysical property of the fluid. Thermal boundary layer effects are reflected by this number.
- Jet and target surface distance: This refers to distance between jet plate and target surface.
- Jet-jet spacing: It speaks of the separation between two neighboring jets. By taking into account jet-to-jet interactions, this is a crucial parameter in enhancing the thermohydraulic performance of the target surface.

7.4 STUDIED GEOMETRIES

Researchers have been stimulated to investigate several fundamental characteristics affecting the heat transfer performance due to advancements in jet impingement system design. Accordingly, literature has numerous modifications in shape of target surface and separation distance of target surface and nozzle to improve rate of heat transfer. Table 7.1 summarizes various target surface forms used for heat transfer improvement by jet impingement.

Table 7.1 Various target surfaces for jet impingement

Ref.	Target geometry	Parameters studied	Results
[34]	Triangle shaped	Re = 900–11,000 Hole shapes used: triangle, circle, and racetrack	The rate of heat transfer is accelerated by the inclusion of triangular roughness.
[35]	Ribs of v shape and dimples of cylindrical shape	Re = 900–11,000 Hole shapes used: triangle, circle, and racetrack	In comparison to ribs alone and dimples alone, a case with both ribs and dimples performed better in terms of heat transfer, thermal performance, and hydraulic performance.
[36]	Dimpled surface	Re = 2,500 Speed = 400 rpm	Dimpled surfaces improved heat transfer in nonrotating circumstances.
[38]	Concave surface	H/d = 3–9 Re = 15,441–37,457	Thermal performance factor increases by 8% to 24%, whereas jet–plate spacing has only a tiny effect.

(Continued)

Table 7.1 (Continued) Various target surfaces for jet impingement

Ref.	Target geometry	Parameters studied	Results
[40]	Circular surface	Re = 10,000 Circular ring height = 0.58, 0.98, 1.42. H/D = 3.0	The average area Nu for the circular ribbed surface increased by 21%.

7.5 RESULTS

Becko [32] has conducted a numerical and experimental study to investigate enhancement in heat transfer for turbine blades (internal cooling) employing dimples and ribs, for Re ranged from 30,000 to 50,000. Internal coolant circulation of a gas turbine blade was stimulated using designs with channel aspect ratios of 2 to 4. Placing dimples between ribs resulted in heat transfer coefficient enhancement and low pressure drop within an acceptable range with respect to rib or dimple configurations. It is worth noting that improved heat transfer via various methods is frequently accompanied by a large rise in flow pressure drop over target plate due to increased interactions of surface and fluid. Consequently, having a system that provides optimum heat transfer enhancement with the least amount of hydraulic loss is always an ideal goal. As a result, simultaneous analysis of heat transmission and friction factor is required to assess the overall performance of a system. Hrycak [33] first introduced the performance enhancement criterion (PEC) as Stanton number divided by friction factor, defined as $(St/St_o)/(f/f_o)^{1/3}$; see also Chapter 2. The recommended method is impractical because PEC < 1 indicates that heat transfer performance is insufficient to satisfy pumping power. When PEC is 1, heat transfer and pumping power are equal, and solution has no impact on operation of the system. PEC > 1 means that heat transfer rate exceeds pumping power; thus, improved system performance is achieved. Furthermore, PEC must be greater than 1 to be a viable heat transfer augmentation method. A PEC was recommended by McInturff et al. [34] as a great tool for keeping track of heat and fluid flow metrics. PEC has also been utilized in a number of heat transfer applications [35–38]. Becko [32] used PEC to evaluate thermal-hydraulic performance of heat transfer during jet impingement. PEC, on other hand, was calculated using Nu and friction factor, as shown in Eqn. (7.11). Reynolds number affects PEC because it affects both friction and Nu variables. Thus, PEC may change for different values of Reynolds number:

$$PEC = \frac{\dfrac{Nu}{Nu_0}}{\left(\dfrac{f}{f_0}\right)^{\frac{1}{3}}} \qquad (7.10)$$

In Eqn. (7.10), Nu denotes augmented Nusselt number, while Nu_0 denotes Nusselt number of a smooth surface. Similarly, f is impinging flow friction factor, while f_0 is reference friction factor for the surface. The use of open-cell metal foams in conjunction with porous media and jet impingement has recently attracted a lot of attention for achieving better cooling rates. Heat sinks (finned metal foam) under impinging jets were studied by Qiu et al. [39] utilizing a combination of numerical and experimental study. At a certain flow rate, the heat transmission of metal foam heat sinks decreases with an increase in foam height, whereas finned metal foam heat sinks experience a rise in heat transfer followed by a considerable reduction. It was also discovered that by increasing effective thickness of fins, bonding substance with a considerable height generates high heat transfer rate, so using finned metal foam is also good option to augment heat transfer. Takeishi [40] investigated finned open-cell aluminum foam thermal performance for an imaging jet application using a numerical model. Results showed that a finned aluminum heat sink had 26% lower thermal resistance than a plate fin–type heat sink at same driving power. Pachpute and Premachandran [41] used a computational model to investigate finned and bare metal foams thermal performance under impingement of air-jet. When impinged jet diameter was equivalent to that of the heated plate, finned metal foam had a higher pressure drop and heat transfer coefficient than bare metal foam. Choi et al. [42] conducted a numerical analysis for thermal and hydraulic performance improvement of heat sinks (finned and bare) when subjected to impinging jet. For same pumping power, finned metal foam heat sinks dissipated three times more heat. By impinging jet arrays on flat and roughened plates, Webb and Eckert [43] numerically monitored heat transmission. To validate the numerical model, a similar experimental setup was created. Both staggered and inline pin-fin layouts might increase heat transfer. Results showed that using an inline roughened pin-fin plate instead of a flat plate resulted in a 34.5% rise in overall heat transfer rate. Yilmaz et al. [44] analyzed micro channel heat sink heat transfer enhancement improvement with jet impingement (MIJ) by inserting dimples into jet surface. MIJs with concave, convex, and mixed dimple architectures were compared to MIJs without dimples in this computational research. Outcomes revealed that MIJs had best cooling performance when it came to convex dimples. MIJs without dimples had the

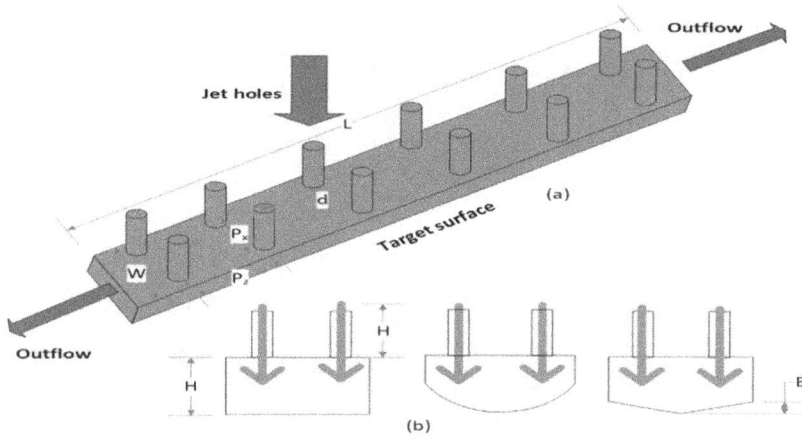

Figure 7.2 Jet impingement model [55].

best cooling performance, followed by MIJs with mixed dimples, and finally MIJs with concave dimples. Additionally, it was found that adding convex dimples to MIJs can lessen flow resistance.

Iasiello et al. [45] investigated cooling performance of jet impingement on non-flat and flat surfaces (V forms and concave) using a computer model (see Figure 7.2). Additional surface configurations of triangular ribs and dimples/protrusions were also integrated into the target surfaces. Compared to flat targets, non-flat target surfaces showed more complex flow patterns. Adopting dimples/protrusions resulted in improvement of both overall and local heat transfer. In flat channels, a higher Reynolds number resulted in a higher average Nu/Nu_0, with sparse arrangement having the highest Nu/Nu_0. Protrusions, on the other hand, were found to have considerable advantages over dimples in terms of heat transfer performance for non-flat targets. Furthermore, protrusions produced a higher f/f_0 than dimples for flat targets. Target surface shapes affect heat flow less than target arrangement and Re. The thermal design of turbine blades and electrical equipment were significantly impacted by the findings.

In an experiment, Yang et al. [46] looked at the heat transmission characteristics of a jet impinging on a heated surface with many protrusions and V grooves. This study employed a Re range of 10,000 to 27,500 (depending on jet diameter). Heat transfer is enhanced by well-spaced multi-protrusions more than by closely spaced V grooves. A numerical study was done by Setareh et al. [47] to look into the flow fields created by a jet impinging over a concave surface. Investigations were done on both swirling and stationary jets. Four spiral and straight grooves were used to inject jets. Rate of heat transfer was boosted by 5% to 10% when a 45° spiral grooves hole was used. Figures 7.3(a) and (b) show contour plots of non-swirling (0°) and

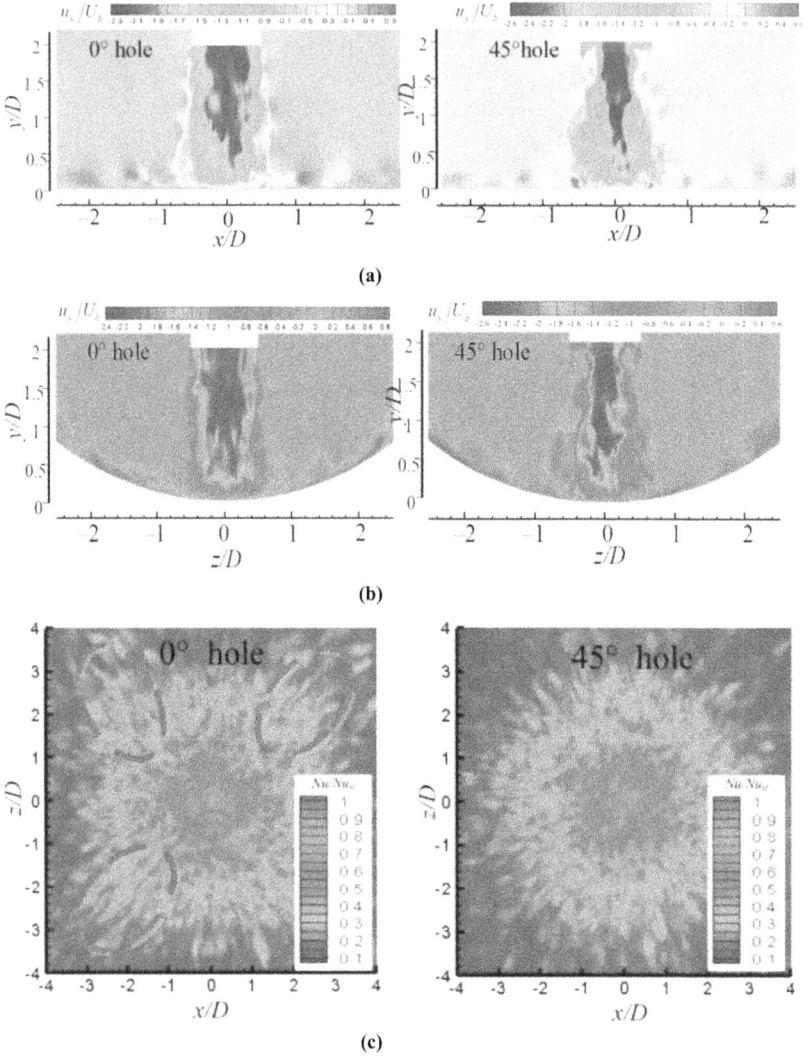

Figure 7.3 (a) XOY plane axial velocity plot; (b) YOZ plane axial velocity plot; (c) Nu/Nu$_{st}$ ratio cloud plots [57].

swirling (45°) jets instantaneous velocity. It is clear that a swirling jet ran into a bigger obstacle than a straight jet. In 45° spinning jets, the majority of the jet was concentrated in the center of the nozzle, which led to a high axial velocity of the potential flow core. Due to the swirl effect, swirling jet also had a higher tangential velocity and a stronger ability to entrain surrounding low-speed airflow. Due to homogeneous heat transmission, the cooling impact on target surface was improved. On the target surface, Figure 7.3(c) depicts Nu ratio (Nu/Nu_{st}), where Nu stands for local Nusselt number and Nu_{st} stands for local Nusselt number (at stagnation point). Difference in heat transmission induced by degradation in the speed of boundary between spiral and center circular holes was eliminated by swirling action of 45° spiral hole grooves. Due to the spiral hole's greater ability to disperse the jet medium, this discrepancy in heat transfer was minimized.

Tepe et al. [58] used jets to impinge on flat plate with ribs in a computational analysis to assess cooling performance. When ribs were placed on surface, the average Nu was lower than no ribs were used. The Nu dispersion was more constant on ribbed surfaces. Tepe et al. [59] had conducted a numerical and experimental study employing jet holes to investigate jet-impinging cooling performance on a rib-roughened surface as shown in Figure 7.4(a). When a rib-roughened target surface is employed, average Nu is increased by 40.32% when expanded jet holes are used, according to the achieved data. Heat transfer enhancement was shown to be more effective with a modest rib height (Hr/Dj) = 0.25. For experimental and computational investigations, contour graphs for local Nu at 0.42 Hr/Dj and 32,500 Re are shown in Figure 7.4(c). Extension of jet holes (due to a reduction in Gj/Dj) increased the local Nu value. It was also shown that the last impingement location did not see a significant increase in heat transfer for any given Gj/Dj number. The middle of the target surfaces had the highest Nu values as a result of the cross-influence, flows which increased in a streamwise direction. However, its effect was observed to disappear beyond $X/Dj > 15$. According to the PEC, the use of expanded jet holes for jet impingement cooling is a practical option for nozzle-target surface spacing of 3.0 or less. At a nozzle-target plate separation of 2.0, the greatest PEC of 1.25 was obtained.

Enhanced surface designs, such as indentations, dimples, and ribs, improve thermal performance of surface by generating turbulence and increasing the rate of convective heat transfer. However, hydraulic performance is just as critical as heat transfer performance. Because of generated disturbance in flow caused by increased surfaces, all studies show an increase in friction factor. The effectiveness of various surfaces has been tested in literature using PECs. It is noticed that dimpled shapes perform better thermally and hydraulically than other surface designs. Another way to achieve substantial turbulent mixing is to reduce jet plate to surface distance, resulting in improved heat transfer. According to experiments, there is a significant amount of turbulence produced as the distance between the target surface and the jet plate/nozzle is decreased, which increases the rate of heat transfer. In light of this, it can be said that jet acceleration and enhanced surface-fluid interactions are essential for producing significant turbulence and, as a result, superior heat transfer performance. These results demonstrate the importance of jet excitations in generating high rates of surface heat transfer. In the next section, jet excitations are considered to increase the heat transfer rate.

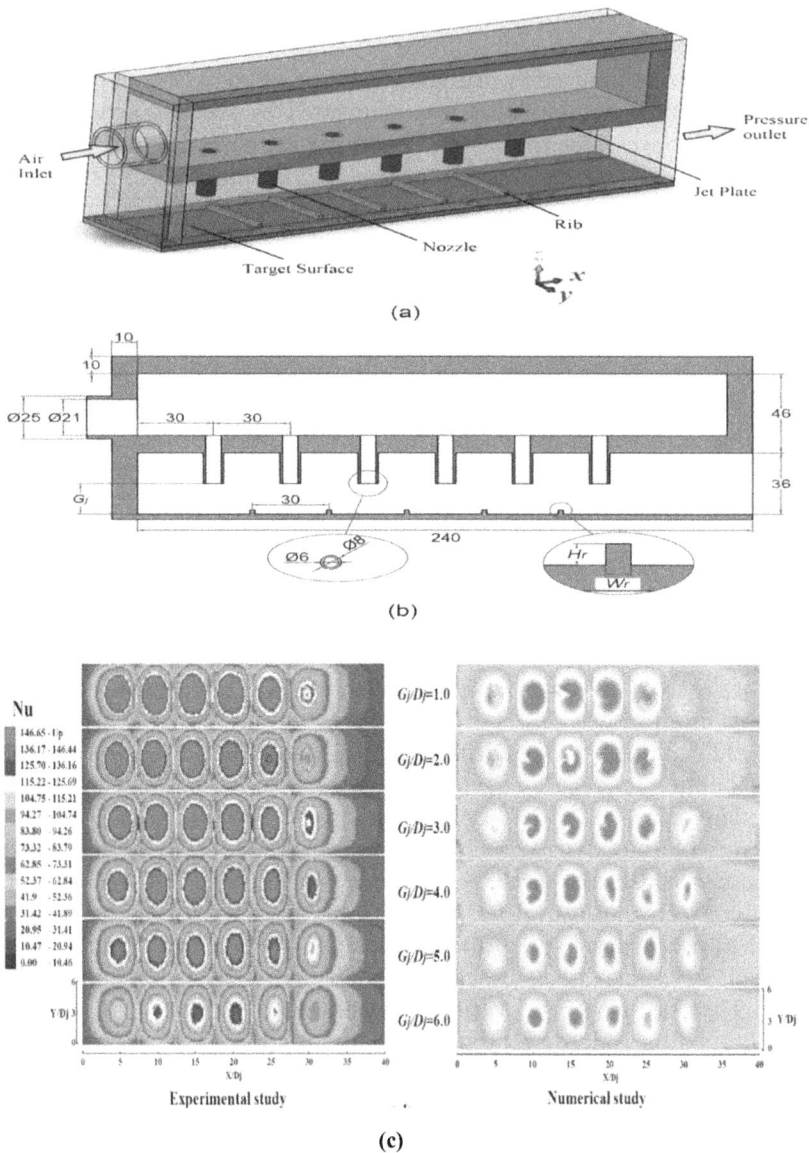

Figure 7.4 (a) 3D view of test section; (b) regional view; (c) influence of Gj/Dj on local Nu [59].

7.6 EXCITED JETS

The use of passive and active excited jets was developed to enhance jet impingement cooling performance. Improved jet turbulence and enhancement in mixing characteristics of jets pose significant implications on thermal system performance. Excited jets can amplify the impact over a greater target region than steady jets [60]. Passive and active jet excitement methods are used to improve heat transfer through jet excitation. Passive jet excitation method does not require any moving parts to intensify heat transmission. Sweeping, swirling, and annular jets are used in a passive approach to improve heat transmission.

Xu et al. [61] studied influence of whirling jets on thermal performance of gas turbine blades by a numerical model. To investigate swirling jet impingement effect on heat transmission, they employed 45 threaded holes rather than circular holes. Jet inclination angle and jet–plate distance affect vortex location and magnitude, but Reynolds number did not have an effect. Wongcharee et al. [62] analyzed impact of a whirling jet on the heat transfer augmentation in an experimental study as shown in Figure 7.5. Tetra-lobed and circular nozzles were also tested for comparison. Data show that increasing Reynolds number and decreasing jet-target separation increases Nu value for all nozzle designs. At a given jet Reynolds number, heat transfer performance of twisted tetra-lobed and normal tetra-lobed nozzles outperformed circular nozzle. Ikhlaq et al. [63] computationally investigated effect of a weak whirling jet on the heat transfer properties of a heated plate. An experiment was carried out with some jet plate separation (H/D = 2, 4, and 6). Results show that weakly swirling jet inflow condition reduces heat transfer at H/D = 2. The swirling and nonswirling jet array impingement flow impact on heat transfer performance was compared using a computational evaluation by Debnath et al. [64]. In their work, two alternative array layouts were used: staggered and inline. Results demonstrated that staggered configurations work better in terms of heat transfer and mixing.

Sapra and Chander [65] evaluated impact of various parameters of design and operation on thermal performance in a twin whirling burner with a tangential entrance. Investigation was done on equivalency ratio, Re, and separation distance. The geometrical design parameters are seen to have a reasonable effect on swirl intensity, resulting in a considerable effect on burners' thermal performance. Both active and passive jet excitations play a

$y/D_h = \infty$ $\qquad y/D_h = 2.0$ $\qquad y/D_h = 3.0$ $\quad y/D_h = 4.0$ $\qquad y/D_h = 5.0$

Figure 7.5 Twisted tetra-lobed nozzle [62].

vibrant role in improvement of thermal performance compared to steady jets, according to documented computational and experimental investigations.

7.7 NANOFLUIDS

Extended recent research on nanofluids has resulted in integration of nano-fluids into jet-impinging heat transmission technology, substituting conventional coolants like water and air. Nanofluids are a cost-effective way to improve heat transfer performance while simultaneously lowering size and weight of cooling system. Using an Al_2O_3-water nanofluid to explore impinging jet thermal performance over various geometries was investigated [66]. The use of nanofluids for a particular particle volume percentage and nozzle-surface distance has been shown to increase heat transmission in experiments. When nanofluids particle volume percentage reached 2.8% and nozzle-surface distance was 5 mm, the heat transfer coefficient is peaked. For jet impingement, a high particle volume percentage, like 6%, did not improve heat transfer. Another experiment employing a jet nanofluid system discovered that the average CPU temperature was 3% and 6% lower than with a liquid jet impingement system and a cooling liquid system, respectively [67]. The goal of research was to find out how well jet nanofluids might transfer heat in the cooling of computer units. For all mass flow rate variations, an impingement cooling system using nanofluids had a higher Nu value than both jet liquid impingement and traditional cooling systems. In an experiment, Zeitoun and Ali [68] impinged an alumina-water nanofluid jet on a horizontally aligned circular surface. Three different nanoparticle volume concentrations (0%, 6.6%, and 10%) were used in the study. According to experiments, increasing nanoparticle concentration at same Re increases heat transfer. By impinging a nanofluid jet on a square mini-fin heat sink, Naphon and Nakharintr [69] evaluated heat transfer performance experimentally. Nanofluid was made of TiO_2 nanoparticles with a 0.2% volume concentration and deionized water as a base fluid. Results reveal that when a nanofluid is present, system transmits heat better than deionized water alone. Zhou et al. [70] investigated submerged impinging jet thermal performance on fin and plate heat sinks using various nanoparticle concentrations of silver–water nanofluids. When a nanofluid was used instead of a base fluid, the heat transfer coefficient was increased by 6.23%, 9.24%, and 17.53% for nanoparticles with weight concentrations of 0.02%, 0.08%, and 0.12%, correspondingly, at same jet speed. Lv et al. [71] found heat transfer increase by impinging an SiO_2 and water-based nanofluid jet on a target surface plane. Heat transfer coefficient of the nanofluid was enhanced by 40% when compared to water considering Re range and volume concentration of nanoparticles, as shown in Figure 7.6.

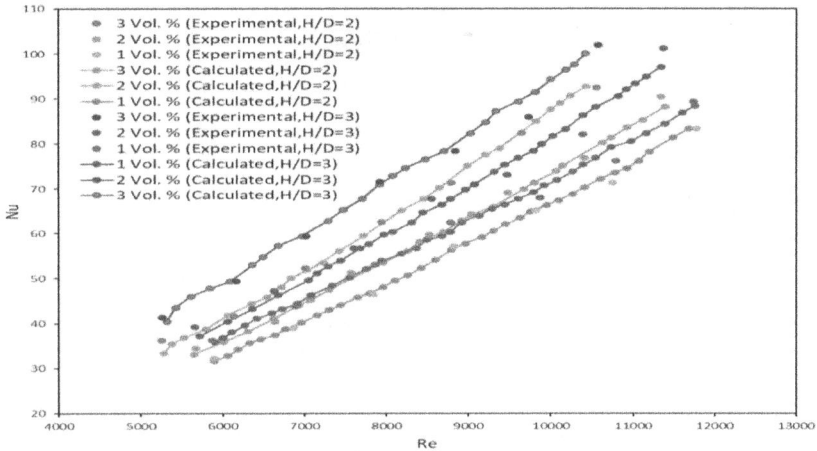

Figure 7.6 Nusselt number variation for SiO_2 nanoparticles in water at $H/D = 2$ and $H/D = 3$ [71].

7.8 PHASE CHANGE MATERIALS

PCMs emit or absorb heat when changing state of phase, e.g., from solid to liquid. In the event, where flow rate and thermal conductivity values are fixed; the heat storage capacity becomes crucial. PCMs with high heat storage capacity, in combination with jet impingement technique, provide a good solution for tackling high heat flux dissipation problems. Parida et al. [72] improved existing cooling process in the high heat dissipating devices by combining high heat transfer jet impingement and high heat storage PCMs. Findings of simulation predictions and experimental data were found to be in good agreement. When heat flux cannot melt the PCM, steady state is reached immediately. Heat source temperature decreased when heat flux was sufficient to partially melt PCM compared to no-PCM situations. To boost device performance, the PCM layer should be designed within the PCM working temperature range; otherwise, it will be useless. Sphere-shaped nano-PCMs can improve system thermal performance. In order to speed up heat transfer, Wu et al. [73] added nano-PCM to water for jet impingement and spray cooling (Figure 7.7). As a phase of paraffin transforms from solid to liquid, nano-PCMs absorb heat. Agglomeration and Paraffin leakage were prevented with the help of encapsulation. Volume percentage of nanoparticles had significant effect on flow parameters (pressure drop) and heat transmission. In comparison to water, the heat transfer coefficients for spray cooling and jet impingement were increased by 70% and 50%, respectively, while utilizing slurry with 28% nanoparticle volume percentage. Nano-PCMs were introduced to air in another investigation to enhance thermal performance of impinging air jet system [74, 75].

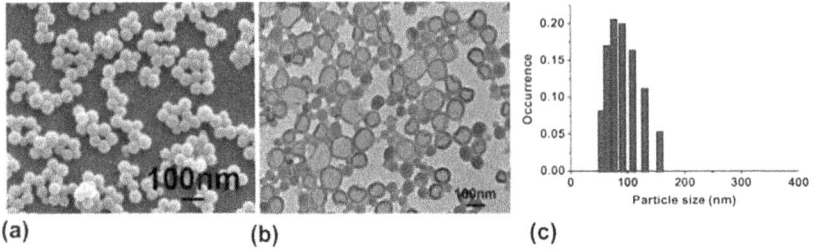

Figure 7.7 Nanoparticles images: (a) utilizing SEM (scanning electron microscopy); (b) using TEM (transmission electron microscopy); (c) encapsulated nanoparticles size distribution [73].

Nanoparticle volume fraction role in heat transfer enhancement and pressure drop was discovered to be substantial. A 2.5% nanoparticle-based air
had 58 times better heat transfer coefficient than pure air.

Researchers' interest has been piqued by recent developments in nanofluids in utilizing nanoparticles in a base fluid to enhance convective heat transfer and as a result, enhanced jet impingement performance. Similarly, in
recent years, utilization of phase transition materials with high latent thermal energy storage capacities in jet impingement systems has received a lot
of interest. Several experimental and numerical studies have shown that
PCMs and nanofluids enhanced cooling of jet impingement. This chapter
analyzed the effects of several factors on cooling performance enhancement,
such as nanofluid flow rate and nanoparticle volume/weight fraction. As
concentration of nanoparticles increases, fluid thermal conductivity is
improved, allowing systems to perform better in terms of cooling. Due to
their high latent heat of fusion, PCMs and NEPCMs (Nanoparticles-
enhanced PCMs) improve heat transfer performance in jet impingement
methods. When a PCM transitions into another phase, it absorbs heat at a
nearly constant temperature. This maintains an effective temperature difference between surface and fluid by a limited increase in the temperature of
the base fluid. Nano-encapsulation can prevent PCM leakage during melting and agglomeration. Consequently, increasing the concentration of nanofluid and nano-encapsulated PCM improves the thermal performance of the
surface up to a certain level.

7.9 CONCLUSIONS

This chapter presented recent works with several jet impingement cooling technologies. It encompasses both experimental and numerical investigations of jet impingement approaches for improving target surface heat
transfer rate. The effects of shape of target surface and jet-target surface

separation on thermal performance were thoroughly discussed. Functions of nanofluids, jet excitations, and PCMs in increasing heat transfer rate are all deeply discussed. In this chapter, the following findings are drawn:

a. It has been found that altering the roughness or shape of a target surface, rather than a flat surface, enhances heat transmission performance. Square, concave, convex, dimple, and triangular are the various shapes/roughness values considered for heat transfer augmentation. The dimple shape works well for fluid flow and heat transfer compared to other designs. Augmentation in heat transfer is achieved in this situation by creating strong flow mixing and extremely turbulent flows. Strong mixing intensities and turbulent flow increase the convective heat transfer.

b. Modification in jet-target surface separation is another parameter described in the analysis of the thermal performance. When jet to target surface distance is narrowed, fluid and turbulence accelerate, resulting in improved heat transfer. Extending jet perforations enhances heat transfer while keeping jet plate to target plate distance as small as possible. Extended jet holes minimize cross-flow velocity, and jet impingement takes place on core region surface, allowing for large heat transfer rates.

c. The chapter also reviewed heat transfer improvement using excited jets (passive and active jets). It has been discovered that these stimulated circular jets outperform steady circular jets in terms of heat transfer enhancement. Enhanced chaotic mixing, fluctuation intensity, flow mixing intensification, and increased turbulence intensity in the stagnation area all contribute to improved heat transfer when using stimulated jets.

d. The scientific community has been pushed to look into the impact of nanofluids and PCMs on heat transfer. Further improvements in heat transfer have been documented after the advent of nanofluids and PCMs. Heat transfer is improved by the nanofluids high thermal conductivity and the huge heat storage capabilities of PCMs. Different types of slurries and nanoparticle volume fractions are provided in PCMs and NEPCMs, and their effects on heat transfer augmentation were analyzed. Still there is a lack of information on utilization of PCMs and NEPCMs.

e. The task of improving heat transfer performance by employing jet impingement cooling technologies is still an important issue. The present chapter contributed to a deeper understanding of jet impingement heat transfer methods, as well as the concept of combining multiple jet impingement cooling ways to provide maximum performance for heat transfer. Nevertheless, more research is needed to study the effect of hybrid nanofluids, PCMs, and NEPCMs on the heat transfer performance.

REFERENCES

[1] V. Narayanan, J. Seyed-Yagoobi, R.H. Page, An experimental study of fluid mechanics and heat transfer in an impinging slot jet flow, *Int. J. Heat Mass Transf.*, Vol. 47, 1827–1845, 2004.

[2] E. Baydar, Y. Ozmen, An experimental and numerical investigation on a confined impinging air jet at high Reynolds numbers, *Appl. Therm. Eng.*, Vol. 25, 409–421, 2005.

[3] B. Han, R.J. Goldstein, Jet-impingement heat transfer in gas turbine systems, *Ann. N. Y. Acad. Sci.*, 934 (1), 147–161, 2001.

[4] S. Polat, Heat and mass transfer in impingement drying, *Dry. Technol.*, Vol. 11(6), 1147–1176, 1993.

[5] R. Viskanta, Heat transfer to impinging isothermal gas and flame jets, *Exp. Therm. Fluid Sci.*, Vol. 6(2), 111–134, 1993.

[6] R. Gardon, J.C. Akfirat, Heat transfer characteristics of impinging two-dimensional air jets, *J. Heat Transf.*, Vol. 88, 101–108, 1966.

[7] R. Gardon, J. Cobonpue, Heat transfer between a flat plate and jets of air impinging on it, *International Developments in Heat Transfer*, in *Proceedings, 2nd International Heat Transfer Conference*, ASME, New York, NY, 1962, pp. 454–460, 1962.

[8] A. Sarkar, et al., Fluid flow and heat transfer in air jet impingement in food processing, *J. Food Sci.*, Vol. 69(4), CRH113–CRH122, 2004.

[9] M. Fabbri, S. Jiang, V.K. Dhir, A comparative study of cooling of high power density electronics using sprays and microjets, *J. Heat Transfer*, Vol. 127(1), 38–48, 2005.

[10] M. Anwarullah, V.V. Rao, K.V. Sharma, Experimental investigation for enhancement of heat transfer from cooling of electronic components by circular air jet impingement, *Heat Mass Transf.*, Vol. 48(9), 1627–1635, 2012.

[11] X. He, J.A. Lustbader, M. Arik, R. Sharma. Heat transfer characteristics of impinging steady and synthetic jets over vertical flat surface, *Int. J. Heat Mass Transf.*, Vol. 80, 825, 2015.

[12] L. Guoneng, X. Zhihua, Z. Youqu, G. Wenwen, D. Cong. Experimental study on convective heat transfer from a rectangular flat plate by multiple impinging jets in laminar cross flows, *Int. J. Therm. Sci.*, ol. 108, 123–131, 2016.

[13] A. Mohamed, T. Mohamed, M. Khairat. Heat transfer due to impinging double free circular jets, *Alex. Eng. J.*, Vol. 54, 281–293, 2015.

[14] M.A. Pakhomov, V.I. Terekhov. Numerical study of fluid flow and heat transfer characteristics in an intermittent turbulent impinging round jet, *Int. J. Therm. Sci.*, Vol. 87, 85–93, 2015.

[15] L.B.Y. Aldabbagh, I. Sezai. Numerical simulation of three-dimensional laminar multiple impinging square jets, *Int. J. Heat Fluid Flow*, Vol. 23, 509–518, 2002.

[16] S. Caliskan, S. Baskaya, T. Calisir. Experimental and numerical investigation of geometry effects on multiple impinging air jets, *Int. J. Heat Mass Transf.*, Vol. 75, 685–703, 2014.

[17] Z.-X. Wen, He Ya-L, X.-W. Cao, C. Yan. Numerical study of impinging jets heat transfer with different nozzle geometries and arrangements for a ground fast cooling simulation device, *Int. J. Heat Mass Transf.*, Vol. 95, 321–335, 2016.

[18] N.-S. Kim, A. Giovannini. Experimental study of turbulent round jet flow impinging on a square cylinder laid on a flat plate, *Int. J. Heat Fluid Flow*, Vol. 28, 1327–1339, 2007.

[19] Y. Xing, B. Weigand. Experimental investigation of impingement heat transfer on a flat and dimpled plate with different crossflow schemes, *Int. J. Heat Mass Transf.*, Vol. 53, 3874–3886, 2010.

[20] A. Vouros, T. Panidis. Influence of a secondary, parallel, low Reynolds number, round jet on a turbulent axisymmetric jet, *Exp. Therm. Fluid Sci.*, Vol. 32, 1455–1467, 2008.

[21] X.K. Wang, J.X. Zheng, D.S. Tan, B. Zhou, S.K. Tan. Study of flow formed by three coplanar impinging pipe jets at inclination angles of 30_ and 45_, *Environ. Fluid Mech.*, Vol. 16, 635–658, 2016.

[22] H. Martin, Heat and mass transfer between impinging gas jets and solid surfaces, *Adv. Heat Transf.*, Vol. 13, 1–60, 1977.

[23] G.N. Abramovich, *The Theory of Turbulent Jets*, MIT Press, Cambridge, MA, 1963.

[24] J.N.B. Livingood, P. Hrycak, Impingement heat transfer from turbulent air stream jets to flat plates-a literature survey, NASA TM X-2778, 1973.

[25] H. Reichardt, *Gesetzmiissigkeiten der freienTurbulenz*, 2nd edn. VDI-Forsch-Heft, Vol. 414. VDI-Verlag, Düsseldorf, p. 30, 1951.

[26] N. Rajaratnam, *Turbulent Jets*, Elsevier, New York, 1976.

[27] S. Polat, B. Huang, A.S. Majumdar, W.J.M. Douglas,Numerical flow and heat transfer under impinging jets: A review, *Ann. Rev. Num. Fluid Mech. Heat Transf.*, Vol. 2, 157–197, 1989.

[28] F. Giralt, C.J. Chia, O. Trass, Characterization of the impingement region in an axisymmetric turbulent jet, *Ind. Chem. Fundam.* Vol. 16, 21–28, 1977.

[29] R.J. Goldstein, K.A. Sobolik, W.S. Seol, Effect of entrainment on the heat transfer to a heated circular air jet impinging on a flat surface, *J. Heat Transf.*, Vol. 112, 608–611, 1990.

[30] J.W. Baughn, A.E. Hechanova, X. Yan, An experimental study of entrainment effects on the heat transfer from a flat surface to a heated circular jet, *J. Heat Transf.*, Vol. 113, 1023–1025, 1991.

[31] S.J. Downs, E.H. James, Jet impingement heat transfer – A literature survey, ASME Paper No. 87-H-35, ASME, New York, 1987.

[32] Y. Becko, *Impingement Cooling—A Review, von Karman Institute for Fluid Dynamics Lecture Series 83*, 1976.

[33] P. Hrycak, Heat transfer from impinging jets: A literature review, AWAL-TR-81- 3054, 1981.

[34] P. McInturff, M. Suzuki, P. Ligrani, C. Nakamata, D.H. Lee Effects of hole shape on impingement jet array heat transfer with small-scale, target surface triangle roughness, *Int. J. Heat Mass Transf.*, Vol. 127, 585–597, 2018.

[35] P. Singh, S. Ekkad Experimental study of heat transfer augmentation in a two-pass channel featuring V-shaped ribs and cylindrical dimples, *Appl. Therm. Eng.*, Vol. 116, 205–216, 2017.

[36] P. Singh, S.V. Ekkad Detailed heat transfer measurements of jet impingement on dimpled target surface under rotation, *J. Therm. Sci. Eng. Appl.*, Vol. 10, 031006, 2018.

[37] R. Vinze, A. Khade, P. Kuntikana, M. Ravitej, B. Suresh, V. Kesavan, S. Prabhu Effect of dimple pitch and depth on jet impingement heat transfer over dimpled surface impinged by multiple jets, *Int. J. Therm. Sci.*, Vol. 145, 105974, 2019.

[38] A. Singh, B.V.S.S.S. Prasad Influence of novel equilaterally staggered jet impingement over a concave surface at fixed pumping power, *Appl. Therm. Eng.*, Vol. 148, 609–619, 2018.

[39] D. Qiu, C. Wang, L. Luo, S. Wang, Z. Zhao, Z. Wang On heat transfer and flow characteristics of jets impinging onto a concave surface with varying jet arrangements, *J. Therm. Anal. Calorim.*, Vol. 141, 57–68, 2019.

[40] K.-I. Takeishi, R. Krewinkel, Y. Oda, Y. Ichikawa Heat transfer enhancement of impingement cooling by adopting circular-ribs or vortex generators in the wall jet region of a round impingement jet, *Int. J. Turbomach. Propuls. Power*, Vol. 5, 17, 2020.

[41] S. Pachpute, B. Premachandran Turbulent multi-jet impingement cooling of a heated circular cylinder, *Int. J. Therm. Sci.*, Vol. 148, 106167, 2019.

[42] E.Y. Choi, Y.D. Choi, W.S. Lee, J.T. Chung, J.S. Kwak Heat transfer augmentation using a rib–dimple compound cooling technique, *Appl. Therm. Eng.*, Vol. 51, 435–441, 2013.

[43] R. Webb, E. Eckert Application of rough surfaces to heat exchanger design, *Int. J. Heat Mass Transf.*, Vol. 15, 1647–1658, 1972.

[44] M. Yilmaz, O. Comakli, S. Yapici, O.N. Sara, M. Yılmaz, S. Yapıcı Performance evaluation criteria for heat exchangers based on first law analysis, *J. Enhanc. Heat Transf.*, Vol. 12, 121–158, 2005.

[45] M. Iasiello, N. Bianco, W.K. Chiu, V. Naso The effects of variable porosity and cell size on the thermal performance of functionally-graded foams, *Int. J. Therm. Sci.*, Vol. 160, 106696, 2020.

[46] S. Yang, Z. Zhao, Y. Zhang, Z. Chen, M. Yang Effects of fin arrangements on thermal hydraulic performance of supercritical nitrogen in printed circuit heat exchanger, *Processes*, Vol. 9, 861, 2021.

[47] M. Setareh, M. Saffar-Avval, A. Abdullah Experimental and numerical study on heat transfer enhancement using ultrasonic vibration in a double-pipe heat exchanger, *Appl. Therm. Eng.*, Vol. 159, 113867, 2019.

[48] R. Sabir, M.M. Khan, N.A. Sheikh, I.U. Ahad, D. Brabazon Assessment of thermo-hydraulic performance of inward dimpled tubes with variation in angular orientations, *Appl. Therm. Eng.*, Vol. 170, 115040, 2020.

[49] S. Feng, J. Kuang, T. Wen, T. Lu, K. Ichimiya An experimental and numerical study finned metal foam heat sinks under imping air jet cooling, *Int. J. Heat Mass Transf.*, Vol. 77, 1063–1074, 2014.

[50] S.S. Feng, J.J. Kuang, T.J. Lu, K. Ichimiya Heat transfer and pressure drop characteristics of finned metal foam heat sinks under uniform imping flow, *J. Electron. Packag.*, Vol. 137, 021014, 2015.

[51] A. Andreozzi, N. Bianco, M. Iasiello, V. Naso Numerical study of metal foam heatsinks under uniform imping flow, *J. Phys. Conf. Ser.*, Vol. 796, 012002, 2017.

[52] N. Bianco, M. Iasiello, G.M. Mauro, L. Pagano Multi-objective optimization of finned metal foam heat sinks: Tradeoff between heat transfer and pressure drop, *Appl. Therm.*, Vol. 182, 116058, 2020.

[53] C. Wan, Y. Rao, P. Chen Numerical predictions of jet impingement heat transfer on square pin-fin roughened plates, *Appl.Therm. Eng.*, Vol. 80, 301–309, 2015.

[54] X. Huang, W. Yang, T. Ming, W. Shen, X. Yu Heat transfer enhancement on a microchannel heat sink with imping jets and dimples, *Int. J. Heat Mass Transf.*, Vol. 112, 113–124, 2017.

[55] Q. Jing, D. Zhang, Y. Xie Numerical investigations of impingement cooling performance on flat and non-flat targets with dimple/protrusion and triangular rib, *Int. J. Heat Mass Transf.*, Vol. 126, 169–190, 2018.

[56] K. Nagesha, K. Srinivasan, T. Sundararajan Enhancement of jet impingement heat transfer using surface roughness elements at different heat inputs, *Exp. Therm. Fluid Sci.*, Vol. 112, 109995, 2019.

[57] L. Xu, X. Zhao, L. Xi, Y. Ma, J. Gao, Y. Li Large-Eddy simulation study of flow and heat transfer in swirling and non-swirling impinging jets on a semi-cylinder concave target, *Appl. Sci.*, Vol. 11, 7167, 2021.

[58] A. Tepe, K. Arslan, Y. Yetisken, U. Uysal Effects of extended jet holes to heat transfer and flow characteristics of the jet impingement cooling, *J. Heat Transf.*, Vol. 141, 082202, 2019.

[59] A.U. Tepe, A. Uysal, Y. Yetisken, K. Arslan Jet impingement cooling on a rib-roughened surface using extended jet holes, *Appl. Therm. Eng.*, Vol. 178, 115601, 2020.

[60] H.M. Maghrabie Heat transfer intensification of jet impingement using exciting jets—A comprehensive review, *Renew. Sustain. Energy Rev.*, Vol. 139, 110684, 2021.

[61] L. Xu, Y. Xiong, L. Xi, J. Gao, Y. Li, Z. Zhao Numerical simulation of swirling impinging jet issuing from a threaded hole under inclined condition, *Entropy*, Vol. 22, 15, 2019.

[62] K. Wongcharee, K. Kunnarak, V. Chuwattanakul, S. Eiamsa-ard. Heat transfer rate of swirling impinging jets issuing from a twisted tetra-lobed nozzle, *Case Stud. Therm. Eng.*, Vol. 22, 100780, 2020.

[63] M. Ikhlaq, Y.M. Al-Abdeli, M. Khiadani Nozzle exit conditions and the heat transfer in non-swirling and weakly swirling turbulent impinging jets, *Heat Mass Transf.*, Vol. 56, 269–290, 2019.

[64] S. Debnath, H.U. Khan, Z.U. Ahmed Turbulent swirling impinging jet arrays: A numerical study on fluid flow and heat transfer, *Therm. Sci. Eng. Prog.*, Vol. 19, 100580, 2020.

[65] G. Sapra, S. Chander Effect of operating and geometrical parameters of tangential entrytype dual swirling flame burner on impingement heat transfer, *Appl. Therm. Eng.*, 115936, 2020.

[66] C.T. Nguyen, N. Galanis, G. Polidori, S. Fohanno, C.V. Popa, A. Le Bechec An experimental study of a confined and submerged impinging jet heat transfer using Al_2O_3-water nanofluid, *Int. J. Therm. Sci.*, Vol. 48, 401–411, 2009

[67] P. Naphon, S. Wongwises Experimental study of jet nanofluids impingement system for cooling computer processing unit, *J. Electron. Cool. Therm. Control*, Vol. 1, 38–44, 2011.

[68] O. Zeitoun, M. Ali Nanofluid impingement jet heat transfer, *Nanoscale Res. Lett.*, Vol. 7, 139, 2012.

[69] P. Naphon, L. Nakharintr Nanofluid jet impingement heat transfer characteristics in therectangular mini-fin heat sink, *J. Eng. Phys. Thermophys.*, Vol. 85, 1432–1440, 2012.

[70] M. Zhou, G. Xia, L. Chai Heat transfer performance of submerged impinging jet using silver nanofluids, *Heat Mass Transf.*, Vol. 51, 221–229, 2014.

[71] J. Lv, C. Hu, M. Bai, K. Zeng, S. Chang, D. Gao Experimental investigation of free single jet impingement using SiO_2-water nanofluid, *Exp. Therm. Fluid Sci.*, Vol. 84, 39–46, 2017.

[72] P.R. Parida, S.V. Ekkad, K. Ngo Novel PCM and jet impingement based cooling scheme for high density transient heat loads, in *Proceedings of the 2010 14th International Heat Transfer Conference*, Washington, DC, USA, 8–13 August, 443–450, 2010.

[73] W. Wu, H. Bostanci, L. Chow, S. Ding, Y. Hong, M. Su, J. Kizito, L. Gschwender, C. Snyder Jet impingement and spray cooling using slurry of nanoencapsulated phase change materials, *Int. J. Heat Mass Transf.*, Vol. 54, 2715–2723, 2011.

[74] W. Wu, H. Bostanci, L.C. Chow, Y. Hong, S.J. Ding, M. Su, J.P. Kizito Jet impingement heat transfer using air-laden nanoparticles with encapsulated phase change materials, *J. Heat Transf.*, Vol. 135, 052202, 2013.

[75] H.R. Seyf, Z. Zhou, H. Ma, Y. Zhang Three dimensional numerical study of heat-transfer enhancement by nano-encapsulated phase change material slurry in microtube heat sinks with tangential impingement, *Int. J. Heat Mass Transf.*, Vol. 56, 561–573, 2013.

Chapter 8

Heat transfer augmentation using nanofluids

8.1 INTRODUCTION

Although many passive and positive methods have been employed to improve the heat transfer efficiency, still the thermal conductivities of gas and liquid heat transfer fluids are very poor compared with metal materials. In order to break the limitation of liquid and gas fluids thermal properties, researchers attempt to add small metal particles into the liquids. It is expected that the suspension process will result in advantages like high thermal conductivities and improved flow characteristics. The particles may settle but if the particle size is reduced to nanometer level, the suspensions have been found to maintain a long-term stability.

The suspensions with dispersed nanoparticles (NP) in the base fluid (BF) are called nanofluids (NFs), and this technique is one of the rapid development areas in materials and thermal science for the decades. The most common types of NPs are metals (Cu, Fe, Zn, Ag, Au, Ti, etc.) [1, 2], metal oxides (Al_2O_3, CuO, TiO_3, MgO, Fe_3O_4, etc.) [3–5], and carbon materials (graphene, graphite, diamond, fullerene, and single- or multiwalled carbon nanotubes) [6–8]. Also, there are some complex NPs, such as metal alloy, metal carbides, and metal nitride NPs (Cu-Zn, Fe-Ni, Ag-Cu, SiC, B4C, ZrC, SiN, TiN, and AlN) [9–11], multimetal oxide NPs ($CuZnFe_4O_4$, $NiFe_2O_4$, and $ZnFe_2O_4$)[12, 13], and hybrid NPs (CNT-Cu, CNT-Au, $Cu-Al_2O_3$, and diamond-nickel) [14, 15], etc. For the BFs, the most common BF is water, heat transfer oils (transformer oil, engine oil, vegetable oil, etc.) [16–18], and organic liquids (ethylene glycol (EG), acetone, and decane) [6], etc. Some other mixture base fluids are also employed, such as water–EG [19], glycerol–EG [20], propylene glycol–water [21], molten salt [22], ionic liquid (IL) [23], etc.

The various kinds of NPs and BFs are considered and synthesized into several kinds of NFs. For this new and huge research field, the researchers have performed a number of works with different aspects. In this chapter, we are mainly concerned about the relevance of augmented heat transfer. The following aspects are summarized: (a) preparation and stability, (b) thermophysical properties, and (c) applications and challenges.

DOI: 10.1201/9781003229865-8

8.2 PREPARATION AND STABILITY

The preparation of nanofluids and to keep them stable is the first step and the basis for the usage in heat transfer processes. More simple operation, less cost and long-time stability are demanded and widely investigated.

8.2.1 Preparation

8.2.1.1 Two-step method

The two-step method is the most employed to prepare various kinds of NFs. So, for instance, Cu/water [24], Al_2O_3/water [25], Au/water, Ag/toluene [26], SiC-TiO_2/oil [27], graphene-Al_2O_3/water [28], and MgO-MWCNT/ethylene glycol [29] have been prepared following this method. The preparation process is shown in Figure 8.1 [30]. First, the NPs are dried to powders and spread into the base fluid. Then, ultrasonic vibration is carried out, and surfactants are added to disperse the NPs and reduce the agglomeration.

This technique is the most economic method to develop the NFs in large quantities. However, there are still some drawbacks for this method, namely the nano-powders have high demand to be kept dry during storage, transportation, and dispersion stages. Most of the surfactants cannot endure temperatures above 333 K [31], which reduce the working window of the NFs. Besides, the sonication or microwave time are both big effect on the thermophysical properties of the suspensions as found by Wang et al. [32]. A short action-time could cause a significant enhancement in thermal conductivity but with high viscosity and poor stability. On the other hand, a long action-time could reduce the viscosity but reduce the thermal conductivity.

8.2.2 One-step method

To reduce the agglomeration of the NPs in the suspensions, a one-step method was developed to synthesize NFs, which combines the NPs production and NFs synthesis together. In this method, the base fluid is placed in a rotating cylinder with an adjustable heating and cooling. As the cylinder rotates, a thin fluid film is created on the inside surface of the cylinder. The particle-source is placed in an evaporator and heated to steam by the heater,

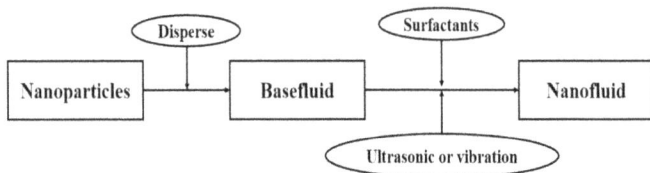

Figure 8.1 Two-step preparation process.

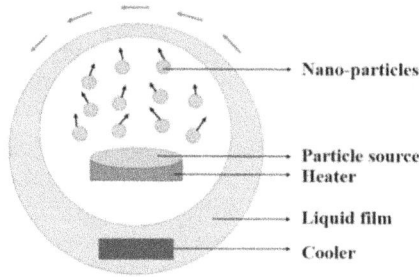

Figure 8.2 One-step preparation process.

and then the evaporated gas is absorbed by the liquid film to synthesize the NF. This preparation process is shown in Figure 8.2.

As the processes of dyeing, storage, transportation, and separation of NPs are canceled, the agglomeration is reduced, and the stability of the suspension is improved. In addition, the simplified process is beneficial to prevent oxidation of metallic particles, so the one-step technique is commonly employed to prepare the NFs containing metal NPs. Zhang et al. [33] synthesized the IL-based nonspherical gold NF by the one-step method. Lee et al. [34] prepared a ZnO/ethylene glycol (EG) NF by a pulsed-wire evaporation one-step method. Munkhbayar et al. [35] dispersed silver NPs into the MWCNT/water NFs using a one-step technique with pulse power evaporation method. Zhu et al. [36] developed a novel one-step method for processing copper NFs by reducing $CuSO_4.5H_2O$ with $NaHPO_2.H_2O$ in EG under a microwave power.

However, the major disadvantages of this technique are that it is difficult to utilize it in large production due to high economy cost of the material vapor deposition method to process the NPs, and it is compatible with only low vapor pressure base fluids.

8.2.2.1 Other novel methods

Apart from the above two main techniques, a few other novel techniques have been developed by various researchers. The phase-transfer method was developed and mainly used for the processing of organosols with noble metals; due to the techniques for direct synthesis of noble metal NPs, non-polar organic liquids were at disadvantage because of the poor solubility of the corresponding metal ion precursors [37]. Feng et al. [38] employed the aqueous organic phase-transfer technique to synthesize gold, silver, and platinum NPs with water. Chen and Wang [39] also used this method to obtain silver and gold NPs. Wang et al. [40] also reported using the phase-transfer based wet chemistry method to prepare an Au/VP-1 NF. Yu et al. [41] employed the phase-transfer method for processing the Fe_3O_4/kerosene

NFs, and by using this method it overcame the previously disadvantage "time-dependence of the thermal conductivity characteristic."

A one-step chemical technique had been recently performed, which is different compared to the traditional one-step method (physical measures). The synthesis process is using reactants A (e.g., Cu^{2+}) and B (e.g., OH^-) in liquid phase to produce the precursor C (e.g., $Cu(OH)_2$). The additives are then separated into the liquid C. Lastly, the solution C transfers into NF D (e.g., CuO/water) under ultrasonic or microwave power[42]. Wei et al. [43, 44] employed the chemical method to synthesize Cu_2O/water and CuS-Cu_2S/$CuSO_4$ NFs. Liu et al. [45] prepared a Cu/water NF by the chemical reduction method. A wet chemical technique was performed to process a CuO/H_2O NF, and the results showed that the microstructures could be synthesized by different the operation parameters [46].

Posttreatment method was developed for the poorly dispersed suspensions that containing agglomerated NPs [47]. Hwang et al. [48] studied different physical posttreatment methods based on two-step method, including stirrer, ultrasonic bath, ultrasonic disruptor, and high-pressure homogenizer to clear the stability of the prepared NFs. Liu et al. [49] used posttreatment technique for the processing of mono-disperse binary $FePt$-Fe_3O_4NPs. The buffer layers produced with ZnO NPs were posttreated through zinc acetate dehydrate in methanol by Oh et al. [50].

8.2.3 Stability

A stability suspension and no agglomeration of the NPs are prerequisites in commercial application of NFs. Because of the high-specific surface area and surface activity, the NPs are easy to approach and agglomerate due to Brownian motion. This could lead to the NPs settlement and clogging, which may block the passages and disadvantage for thermal conductivity improvement. Therefore, the mechanisms of the NPs clogging were investigated, and various kinds of techniques were developed to keep long-term stability of suspensions. The methods can be divided into chemical and physical techniques [51, 52].

8.2.3.1 Chemical techniques

The stability of the suspensions depends on the van der Waals attraction and electrical double-layer repulsive forces, according to the colloidal stability developed by Derjaguin, Verway, Landau, and Overbeek (DVLO) [53, 54]. When the attractive force is more significant than the repulsive force, the particles could agglomeration together. Therefore, maintaining the repulsive force as the dominant force between the NPs is the key point to keep the suspension stability. There are two methods to increase the repulsion of NPs, one repulsion is in steric and another repulsion is due to the charge force, as illustrated in Figure 8.3. [51].

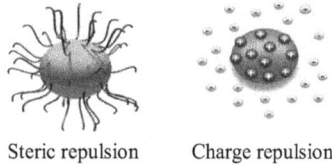

Steric repulsion   Charge repulsion

Figure 8.3 Methods to increase the repulsion [51].

In the above two methods, the addition of surfactants technique and surface modification technique are developed based on the steric repulsion concept. The addition of surfactants technique is dispersion the surfactants into the suspensions, and the dispersants molecular could attach the surface of the NPs and prevent the adhesion between the NPs. The dispersants are long-chain hydrocarbons and are usually composed of a hydrophobic tail portion and a hydrophilic polar head group. They can be defined as four types based on the composition of the head: nonionic surfactants without charge groups in the head, anionic surfactants with negatively charged head groups, cationic surfactants with positively charged head groups, and amphoteric surfactants with zwitterionic head groups [51]. The selection of the surfactants is demanded by the characteristics of the NPs and base fluids. However, some disadvantages are reported in the published literatures: (a) an additional thermal resistance between the NPs and base fluids; (b) cannot withstand high temperature; (c) a relative long stabilization time, within several months [52].

The surface modification method is attaching functional groups on the NPs, and a long-term stability of the suspensions is obtained, which is also called the surfactant-free technique. The surface functionalization of silica NPs was performed to increase the stability by Yang and Liu [55]. Also, Chen et al. [56] treated the multiwalled carbon nanotube (MWNT) surfaces through mechanochemical reaction method. Chen and Xie [57] also employed the wet-mechanochemical reaction to functionalize the single-walled carbon nanotubes (SWNTs) and double-walled carbon nanotubes (DWNTs). To improve the stability of the titania/diglyme NF, surface modification was used on titania NPs by Joni et al. [58]. The common zinc oxide (ZnO) NPs were modified by poly methacrylic acid (PMAA) in an aqueous liquid. The hydroxyl groups of ZnO NPs surface are interacted with carboxyl groups (COO-) of PMAA and process into poly zinc methacrylate complex on the surface of ZnO NPs [59]. A wet chemistry technique coupled with the chemical functionalization and dispersion-centrifugation cycle was employed on the SWNTs. It is cleared that $K_2S_2O_8$ treatment produces hydrophilic groups just like carboxyl and hydroxyl on the surfaces of varying carbons, while this method is little damage on the structure of the NPs [60]. The surface modification technique has many benefits as no dispersants can avoid the loss of thermal conductivity and a slight increase of the viscosity, and no limitation of temperature in applications. However, the process of chemical reactions needs accurate control and is still in laboratory preparation and accordingly hard for large-amount production.

Concerning the two chemical techniques to increase the NFs stability, the charge repulsion concept is achieved by controlling the PH values of the NFs and strengthens the repulsive forces among the NPs. The definition of isoelectric point (IEP) is the PH value of the particular molecule without net electric charge, or hydration forces are approximately zero [61]. The NF cannot be long-time stability as the PH is close to the IEP, because the zeta potential and the repulsive forces are zero at IEP among the NPs, and the NPs trend to cluster and chain together. As the PH value and NP concentration of the NFs are close to the optimized values, the surface charge is improved and lead to a more frequency attack on the surface hydroxyl groups and phenyl sulfonic group by potential-determining ions (H+, OH−, and phenyl sulfonic group) occur, and the stability of colloidal particles become improved. For Al_2O_3 and Cu, the optimal PH values are 8.0 and 9.5, respectively, to obtain good dispersion of the NPs in suspensions, [62]. Goudarzi et al. [63] experimentally studied the effect of different PH values of CuO/water and Al_2O_3/water NFs on the thermal efficiency of solar collectors. The experimental results showed that with the PH value of NFs far from IEP, the thermal efficiency of the collector is improved. For CuO/water NF at PH = 3, the improved thermal efficiency was approximately 52% compared with the CuO/water NF at PH = 10.5. However, for Al_2O_3/water NF at PH = 10.5 the thermal efficiency of the collector was approximately 64.5% larger than the Al_2O_3/water NF at PH = 9.2. According to a study by Zhang et al. [64], the PH value of TiO_2/water NF that is farther away from the determined IEP (PH = 6.5) causes more stability of the NF. This method also possesses its limitation, which is that due to the PH value of NFs it is not suitable for acidic and alkaline media as the wall surfaces could be corroded [65].

8.2.3.2 Physical techniques

The ultrasonic vibration and homogenization are the most used physical methods, which are different from the chemical techniques. The physical techniques do not modify the surface structure and properties of NPs, but only destroy the NPs' agglomerations. It is easy to operate but needs to be carried out regularly. Chen and Wen [66] employed the ultrasonic-aided fabrication method to disperse the spherical and plate-shaped gold NPs into a water base fluid. Amrollahi et al. [67] studied the effect of time of ultrasonic vibration on the thermal conductivity and volumetric heat capacity of carbon/EG NF. Sadri et al. [68] experimentally investigated the thermal conductivity and viscosity of CNT/water NFs, under the different sonication time. They found that the ultrasonication technique is an effective method to control the size and shape of NPs. The size of agglomerated size and number of the particles groups were obviously declined by ultrasonic disruption, and this resulted in an increase of the thermal conductivity. In contrast, the viscosity of MWCNT/water NF was found to be maximal at a sonication time of 7 min and further decreased with an increase of the sonication time.

The ultrasonic method is not always effective as it is acting outside the fluids. Then the high shear homogenizer method acting on the fluids directly is employed to get the stable NFs [69]. The stable silver/water NF was achieved through a high-pressure homogenizer by Filho et al. [70]. Fedele et al. [71] compared the sonication ball milling and high-pressure homogenization methods on stabilization of the SWCNHs/water, TiO$_2$/water, and CuO/water nanofluids and found that the high-pressure homogenization method is more effective. The stirrer, ultrasonic bath, ultrasonic disrupter, and high-pressure homogenization were all used on the CB/water and Ag/water NFs by Hwang et al. [48]. The results also showed that high-pressure homogenization is the most effective technique to destroy the agglomerated NPs groups in base fluids. The factors of high-pressure homogenization also have big influence on the thermal and rheological properties [72–74].

The effect of the ultrasonic and homogenization techniques on the properties of NFs have been explored currently, but more systematic research is needed, and empirical correlations should be further studied.

8.3 THERMOPHYSICAL PROPERTIES

After synthesizing a stable nanofluid, the thermophysical properties (density, thermal conductivity, specific heat, and viscosity) should be well known before it is used as a heat transfer fluid. There are many experimental techniques to determine the properties. However, the thermophysical properties are significantly affected by the following factors: base fluid material, NP material, NP size, NP shape, NP concentration, temperature, additives, clustering, PH value, and preparation technique, as shown in Figure 8.4. Therefore, theoretical models are developed to predict the properties of NFs, based on the testing data. In this chapter, the development of the theoretical models for the thermophysical properties are detailed, and the measurement techniques are briefly introduced.

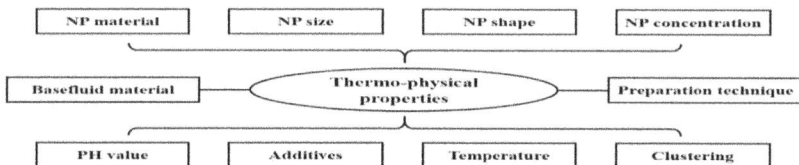

Figure 8.4 Factors affecting on the thermophysical properties of NFs.

8.3.1 Density

The density of NFs is the thermophysical property being most easy to determine. It is defined by the mixing theory as follows:

$$\rho_{nf} = \frac{m_{np} + m_{bf}}{V_{np} + V_{bf}} = \frac{\rho_{np}V_{np} + \rho_{bf}V_{bf}}{V_{np} + V_{bf}} = \varphi\rho_{np} + (1-\varphi)\rho_{bf} \qquad (8.1)$$

where φ is the NP volume concentration. Subscripts nf, np, and bf represent the nanofluid, nanoparticle, and base fluid, respectively. The mass and volume quantities can be found by weighing and measuring cylinder, respectively.

8.3.2 Thermal conductivity

There are many measurement techniques adopted to determine the thermal conductivity of NFs, as shown in Figure 8.5 [75]. Among the techniques, the transient hot-wire method is most popular, and over 90% of the published works have employed this method to measure the thermal conductivity [76].

According to the experimental data, a large number of theoretical models have been developed considering various factors [77]. A summary of theoretical models for NFs thermal conductivity is provided in Table 8.1. The Maxwell model was first proposed to predict the thermal conductivity of low-volume fraction solid-liquid suspensions, and it is mainly employed on spherical particles [78]. Later, several modified Maxwell models were developed [79–82], which considered additional factors. Hamilton and Crosser [79] considered the NP shapes, Wasp [80] was concerned about the high-volume fraction with spherical NPs, Xuan et al. [81] employed the Brownian motion theory and diffusion-limited aggregation model to predict the random motion and aggregation of NPs, and Yu and Choi [82] considered nanolayers between the NPs surface and base fluid.

As the Maxwell model and its modifications cannot well predict the thermal conductivity of all kinds of NFs, the mechanism of the enhanced thermal conductivity of NFs has received attention by scholars. Jang and Choi [83]

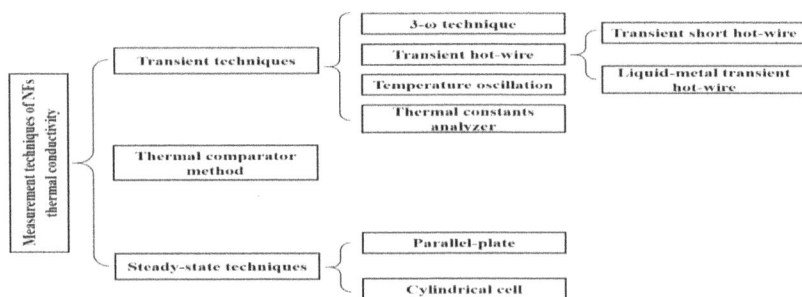

Figure 8.5 Measurement techniques for thermal conductivity of NFs.

Table 8.1 Summary of theoretical models on thermal conductivity of NFs

Researchers	Model expressions	Comments
Maxwell [78]	$$k_{eff} = k_{bf}\frac{k_{np} + 2k_{bf} + 2\varphi(k_{np} - k_{bf})}{k_{np} + 2k_{bf} - \varphi(k_{np} - k_{bf})}$$	Considered the thermal conductivity of NP and base fluid, and the concentrations, but mainly used on spherical particles with low-volume fraction.
Hamilton and Crosser [79]	$$k_{eff} = k_{bf}\frac{k_{np} + (n-1)k_{bf} + (n-1)\varphi(k_{np} - k_{bf})}{k_{np} + (n-1)k_{bf} - \varphi(k_{np} - k_{bf})}$$	NP shape was considered based on the Maxwell model, the n is an empirical shape factor ($n = \psi/3$), ψ is sphericity.
Wasp [80]	$$k_{eff} = k_{bf}\frac{k_{np} + 2k_{bf} - 2\varphi(k_{np} - k_{bf})}{k_{np} + 2k_{bf} + \varphi(k_{np} - k_{bf})}$$	Like the Maxwell model but possesses a high accuracy at high concentrations on spherical NPs.
Xuan et al. [81]	$$k_{eff} = k_{np}\frac{k_{np} + 2k_{bf} + 2\phi(k_{np} - k_{bf})}{k_{np} + 2k_{bf} - \phi(k_{np} - k_{bf})} + \frac{1}{2}\rho_{np}Cp_{np}\varphi\sqrt{\frac{K_B T}{3\pi\mu_{bf}R_{cl}}}$$ where R_{cl} is the apparent radius of the clusters and K_B is Boltzmann constant.	Brownian motion, structure of clustering and fluid temperature are considered.
Yu and Choi [82]	$$k_{eff} = k_{bf}\frac{k_{np} + 2k_{bf} + 2\varphi(k_{np} - k_{bf})(1+\beta)^3}{k_{np} + 2k_{bf} - \varphi(k_{np} - k_{bf})(1+\beta)^3}$$ where $\beta = t/R$ is the ratio of the nanolayer thickness to the original NP radius.	The nanolayer is a significant effect on NF thermal conductivity.
Jang and Choi [83]	$$k_{eff} = k_{bf}(1-\varphi) + 0.01 l k_{np}\varphi + (18\times10^6)\frac{d_f}{d_p}k_{bf}Re_d^2 Pr_{bf}\varphi$$ $$Re_d = C_{RM}d_p/\upsilon, \quad C_{RM} = \frac{K_B T}{3\pi\mu_{bf}d_p l_f}$$ where l_f is diameter of particle mean free path.	The effect of Brownian motion on the thermal conductivity of NFs, and the properties of NP and base fluid, volume fraction, NP size, and temperature are all considered.

(Continued)

Table 8.1 (Continued) Summary of theoretical models on thermal conductivity of NFs

Researchers	Model expressions	Comments
Kumar et al. [84]	$$k_{eff} = k_{bf} + C\,\frac{2K_B T}{\pi v d_p^2\, k_{bf}}\,\frac{\varphi r_b}{(1-\varphi) r_p}\, k_{bf}$$	Considering the particle size, concentration, and temperature effects.
Xue [85]	$$k_{eff} = k_{bf}\,\frac{1 - \varphi + 2\varphi\,\dfrac{k_{np}}{k_{np} - k_{bf}}\,\ln\left(\dfrac{k_{np} + k_{bf}}{2k_{bf}}\right)}{1 - \varphi + 2\varphi\,\dfrac{k_{bf}}{k_{np} - k_{bf}}\,\ln\left(\dfrac{k_{np} + k_{bf}}{2k_{bf}}\right)}$$	Suitable for carbon nanotubes prediction.
Koo and Kleinstreuer [86, 87]	$$k_{eff} = k_{bf}\,\frac{k_{np} + 2k_{bf} + 2\varphi\left(k_{np} - k_{bf}\right)}{k_{np} + 2k_{bf} - \varphi\left(k_{np} - k_{bf}\right)} + 5\times10^4\,\theta\rho_{bf}C_{p bf}\varphi f\left(T,\varphi\right)\sqrt{\frac{K_B T}{\rho_{bf}d_{np}}}$$ $$f\left(T,\varphi\right) = \left(-6.04\varphi + 0.4705\right)T + 1722.3\varphi - 134.63$$ where, θ is the liquid volume fraction which travels with a particle and d_{np} is the diameter of the nanoparticle.	The NPs convection due to the temperature gradient and the Brownian motion are both considered, and the properties of base fluids and NPs are also considered.
Prasher et al. [88]	$$k_{eff} = k_{bf}\,\frac{k_{np}\left(1 + 2\alpha_B\right) + 2k_m + 2\varphi\left[k_{np}\left(1-\alpha_B\right) - k_m\right]}{k_{np}\left(1 + 2\alpha_B\right) + 2k_m - \varphi\left[k_{np}\left(1-\alpha_B\right) - k_m\right]}$$ $$\times\left(1 + A_{eco}\, Re_B^{M_{eco}}\, Pr_{bf}^{0.33}\,\varphi\right)$$ $$k_m = k_{bf}\left(1 + 0.25\,Re_B\,Pr\right),\ Re_B = \frac{1}{v}\sqrt{\frac{18K_B T}{\pi\rho_{np}d_p}},\ \alpha_B = 2R_b k_m/d_{np}$$ where, R_b is the interfacial thermal resistance between the NPs surface and fluids, and Aeco and Meco are two empirical constants.	A coupled model of the Maxwell–Garnett conduction and Brownian movement. Considering the particle size, base fluid, temperature, and thermal interfacial resistance between the particle and fluid.

Leong et al. [89]

$$k_{eff} = \frac{(k_{np} - k_{lr})\varphi k_{lr}(2\beta_1^3 - \beta_2^3 + 1) + (k_{np} + 2k_{lr})\beta_1^3\left[\varphi\beta_2^3(k_{lr} - k_{bf}) + k_{bf}\right]}{\beta_1^3(k_{np} + k_{lr}) - (k_{np} - k_{lr})\varphi(\beta_1^3 + \beta_2^3 - 1)}$$

where $\beta_1 = 1 + t/R$, $\beta_2 = 1 + t/(2R)$, k_{lr} and t are thermal conductivity and thickness of interfacial layers.

Considering the interfacial layer between the NPs surface and fluid.

The volume fraction, particle size, thickness, and thermal conductivity of the interfacial layer are taken into account.

Murshed et al. [90]

$$k_{eff} = k_{st} + k_{dy}$$

$$k_{st} = \left\{ k_{bf}\frac{\varphi\omega(k_{np} - \omega k_{bf})(2\gamma_1^3 - \gamma^3 + 1) + (k_{np} + 2\omega k_{bf})\gamma_1^3\left[\varphi\gamma^3(\omega - 1) + 1\right]}{\gamma_1^3(k_{np} + 2\omega k_{bf}) - (k_{np} - \omega k_{bf})\varphi(\gamma_1^3 + \gamma^3 - 1)}\right.$$

$$\left. + \left\{\varphi^2\gamma^6 k_{bf}\left(3\Lambda^2 + \frac{3\Lambda^2}{4} + \frac{9\Lambda^3}{16}\frac{k_{cp} + 2k_{bf}}{2k_{cp} + 3k_{bf}} + \frac{3\Lambda^2}{2^6} + \ldots\right)\right\}\right\}$$

$$k_{dy} = \frac{1}{2}\rho_{cp}C_{pcp}d_s\left[\sqrt{\frac{3K_B T(1 - 1.5\gamma^3\varphi)}{2\pi\rho_{cp}\gamma^3 r_{np}^3}} + \frac{G_T}{6\pi\eta r_{np}d_s}\right]$$

$$k_{cp} = k_{lr}\frac{2(k_{np} - k_{lr}) + \gamma^3(k_{np} + 2k_{lr})}{(k_{lr} - k_{np}) + \gamma^3(k_{np} + 2k_{lr})}, \; k_{lr} = \omega k_{bf}, \; \Lambda = \frac{k_{cp} - k_{bf}}{k_{cp} + 2k_{bf}}$$

where ω is an empirical parameter depending on the order of fluid molecules at the interface, nature, and surface chemistry of NP.

A coupled model by static and dynamic mechanisms, considering the effects of particle size, nanolayer, Brownian motion, particle surface chemistry, and interaction potential.

(Continued)

Table 8.1 (Continued) Summary of theoretical models on thermal conductivity of NFs

Researchers	Model expressions	Comments
Jiang et al. [91]	$$k_{eff} = k_{bf} \frac{k_{pe} + (n-1)k_{bf} + (n-1)\varphi(k_{pe} - k_{bf})}{k_{pe} + (n-1)k_{bf} - \varphi(k_{pe} - k_{bf})}$$ $$k_{pe} = k_{bf} \frac{2k_{np} + (\beta_i^3 - 1)(k_{np} + k_{lr})}{2k_{lr} + (\beta_i^3 - 1)(k_{np} + k_{lr})}$$ $$k_{lr} = \frac{k_{np}R(1 + t/R - k_{bf}/k_{np})\ln(1 + t/R)}{tk_{bf}\ln[(1 + t/R)k_{np}/k_{bf}]}$$ where $\beta_i = 1 + t/R$ and k_{pe} is the equivalent thermal conductivity of the nanoparticles.	CNT NFs based on the two-dimensional Fourier's law, which considered the interfacial nanolayer and NP shape.
Yang et al. [92]	$$k_{eff} = \frac{(H + 2t)k_{eff\_x} + (R + t)k_{eff\_z}}{H + R + 3T}$$ $$k_{eff\_x} = \frac{A\varphi k_p + (\alpha B + \beta C)\varphi_p k_{lr} + (1 + \alpha + \beta)\varphi_p k_f - k_f}{A\varphi_p + (\alpha B + \beta C)\varphi_p + (1 + \alpha + \beta)\varphi_p - 1}$$ $$k_{eff\_z} = \frac{H + 2t}{H}\varphi_p \left(\frac{(H + 2t)k_{lr}k_p}{2tk_p + Hk_{lr}} + \alpha k_{lr}\right) + k_f - \frac{(H + 2t)(1 + \alpha)k_f\varphi_p}{H}$$	Considering the shape, size, and interfacial layer of nanorod. Different equations which split the conduction in the nanorod with interfacial layer into axial and radial directions, respectively.

Corcione [93]	$$\frac{k_{eff}}{k_{bf}} = 1 + 4.4 Re_p^{0.4} Pr_{bf}^{0.66} \left(\frac{T}{T_{fr}}\right)^{10} \left(\frac{k_{np}}{k_r}\right)^{0.03} \varphi^{0.66}$$ where T_{fr} is the freezing point of the base fluid and Re_p is the nanoparticle Reynolds number.	Based on a large number of experimental data. The ranges of the formula are 10–150 nm (NP diameter), 0.0001–0.071 (volume fraction) and 294–324 K (temperature), respectively.
Afrand et al. [94]	$$\frac{k_{eff}}{k_{bf}} = 0.7575 + 0.3\varphi^{0.323}T^{0.245}$$	Fe_3O_4 magnetic/water NFs. Established by an optimal artificial neural network.
Sidik et al. [95]	$$\frac{k_{eff}}{k_{bf}} = 1.268 \times \left(\frac{T}{80}\right)^{-0.0074} \left(\frac{\varphi}{100}\right)^{0.036}$$	Bio glycol-based Al_2O_3 NFs. Temperatures range from 30°C to 80°C and the volume concentration from 0.1% to 1%.

first proposed the Brownian motion–caused nano-convection as a key nanoscale mechanism effect on the thermal behavior of the NFs. The model considered the thermal-hydraulic properties of NPs and base fluid, volume fraction, NP size, and temperature. The model predictions show good agreement with experimental data containing oxide, metallic, and carbon nanotubes. The effect of particle size, concentration, and temperature on thermal conductivity has been all considered by Kumar et al. [84], and this model predicted well the Au/water NF in the temperature range of 30–60°C. Xue [85] considered the carbon nanotube (CNT) distributions, and performed a new model based on the Maxwell theory. The model results showed good agreement with the experimental data of CNTs/oil and CNTs/decane NFs. The temperature gradient could cause particle convection in the suspensions, and to clear such dynamic phenomena, a thermal conductivity model including temperature, properties of the base fluid, and NP dependence was proposed by Koo and Kleinstreuer [86, 87]. The Brownian motion was considered as well; Prasher et al. [88] developed a coupled model by the Maxwell–Garnett (MG) conduction and Brownian movement, which considered the effects of particle size, base fluid, temperature, and the thermal interfacial resistance between the NPs and fluids. The model was in good agreement with experimental results for water, EG, and oil-based NFs. Leong et al. [89] proposed that the interfacial layer between the NPs surface and liquid has significant effect on the improved the thermal conductivity of nanofluids. The model considers the effects of volume fraction, thickness, particle size, and the interfacial layer features. Murshed et al. [90] proposed a coupled model with static and dynamic mechanisms for thermal conductivity prediction. The effects of Brownian motion, particle size, particle surface chemistry, nanolayer, and interaction potential are taken into account, simultaneously. Jiang et al. [91] deduced a correlation for calculating the thermal conductivity of CNT NFs according to the two-dimensional Fourier's law, which considered the interfacial nanolayer and NP shape. Yang et al. [92] built up different equations which split the conduction or a nanorod with interfacial layer into axial and radial directions, respectively. The allocated proportions of thermal conduction for different directions (axial and radial) in the theoretical model are decided by the ratio of the top and bottom surface areas of the nano rod.

Also, there are some researchers setting up empirical correlations for the specific types of NFs. An empirical correlation was proposed based on a large number of experimental data by Corcione [93]. The model considered the effect of NP diameter, volume fraction, and temperature. The ranges of the correlation are 10–150 nm, 0.0001–0.071, and 294–324 K, respectively. An empirical correlation was proposed to predict the Fe_3O_4 magnetic/water NFs based on an optimal artificial neural network, according to the experimental data by Afrand et al. [94]. An empirical correlation was developed by Sidik et al. [95] to predict the bio glycol-based Al_2O_3 NFs based on the experimental data, within the temperature range from 30°C to 80°C and the volume concentration ranging from 0.1% to 1%.

8.3.3 Specific heat

Specific heat is also a key thermal property for NFs used on heat transfer and thermal storage applications. The differential scanning calorimetry (DSC) following the standard test method ASTM E-1269-05 is a popular method employed in specific heat measurements of NFs. Many commercial instruments are available, such as DSC-1-Stare, manufactured by Mettler Toledo-Switzerland [96]; DSC-200-F3, manufactured by Netzsch-Germany [97, 98]; DSC-7, manufactured by Perkin Elmer [99]; DSC-Model-Q20, manufactured by TA Instruments-USA; and DSC-111 [100], manufactured by Setaram-France [101]. Some researchers developed homemade setups for specific heat measurements of NFs [102–104], as well.

Also, theoretical models for the specific heat of NFs have been proposed based on experimental data, and a summary of the theoretical models are presented in Table 8.2. Pak and Cho [105] first proposed a theoretical model which only considered the volume fraction of NPs. After that, the Pak and Cho model was modified and upgraded by the various researchers. Xuan and Roetzel [106] took account of the effect of density in the prediction of specific heat, and this model was better fitted with experimental data than the Pak and Cho model. Shin and Banerjee [100] developed a new theoretical model for nonaqueous alkali salt-based NFs, which first considered the nanostructures of the cluster of NPs. Also, there are many empirical models obtained by fitting to the experimental results. Vajjha and Das [107] measured three kinds of NPs (Al_2O_3.ZnO and SiO_2) in EG and water-mixed (60:40 by mass) base fluids. A general model was proposed for the specific heat as function of volumetric concentration, temperature, and the specific heat of both the NP and the BF. The average relative error between the correlation results and experimental results was within 2.7%. The cellulose nanocrystal (CNC) separated in an EG–water mixture (40:60 by volume) base fluid was considered by Ramachandran et al. [108]. An empirical mathematical model of the specific heat was developed through the response surface method (RSM) by using central composite design (CCD). The maximum relative error between the correlation predictions and experimental data was 0.72%. A correlation of the specific heat of MWCNTs/heat transfer oil NFs was developed by Pakdaman et al. [109], based on the least square method. For weight concentrations of the order of 0–0.004, and temperatures 313–343 K, the maximum relative error between the experimental results and predicted values was 9.4%. A regression empirical correlation was obtained by Sekhar and Sharma [110], based on 81 experimental investigations of water-based Al_2O_3, CuO, SiO_2, and TiO_2, from the literatures. The temperature varied as 293–323 K, NPs diameter was 15–50 nm, and the volume concentration was 0.01%–4.00%. A hybrid NFs theoretical model was proposed by Moldoveanu and Minea [111], which considered the density and specific heat for each component. Also, a correlation was proposed according to the Sekhar and Sharma model. This correlation is

Table 8.2 Summary of theoretical models on specific heat of NFs

Researchers	Model expressions	Comments
Pak and Cho [105]	$Cp_{nf} = (1-\varphi)Cp_{bf} + \varphi Cp_{np}$	First theoretical model to predict the specific heat of NFs considering the volume fraction of NPs.
Xuan and Roetzel [106]	$Cp_{nf} = \dfrac{\varphi \rho_{np} Cp_{np} + (1-\varphi)\rho_{bf} Cp_{bf}}{\varphi \rho_{np} + (1-\varphi)\rho_{bf}}$	A modified correlation based on Pak and Cho model and considering the effect of nanoparticle and base fluid densities.
Shin and Banerjee [100]	$Cp_{nf} = \dfrac{\varphi_{np}\rho_{np}Cp_{np} + \varphi_{bf}\rho_{bf}Cp_{bf} + \varphi_{ns}\rho_{ns}Cp_{ns}}{\varphi_{np}\rho_{np} + \varphi_{bf}\rho_{bf} + \varphi_{ns}\rho_{ns}}$	A modified correlation based on Pak and Cho model, the ns denotes the nanostructure, but the ρ_{ns}, φ_{ns}, and Cp_{ns} are not easy to obtain by current technology.
Vajjha and Das [107]	$Cp_{nf}/Cp_{bf} = \dfrac{\left[(A \times T) + B \times Cp_{bf}\right]}{C + \varphi}$ NPs — A — B — C Al_2O_3 — 0.0008911 — 0.5179 — 0.4250 SiO_2 — 0.001769 — 1.1937 — 0.8021 ZnO — 0.0004604 — 0.9855 — 0.299	The base fluid is EG/Water mixture fluid (60:40 by mass), and the application ranges of the equation are: 315 K < T < 363 K, 0 < φ < 0.1 for Al_2O_3 and SiO_2 NFs, and 0 < φ < 0.07 for ZnO NF.
Ramachandran et al. [108]	$Cp_{nf}/Cp_{bf} = 0.98154 + 0.001664 \times T$ $+ 0.0057 \times \varphi - 0.000007 \times T^2$ $- 0.0166 - 0.000749 \times \varphi \times T$	CNC/EG-water (40:60 by volume) NFs. Developed through the response surface method by using central composite design.
Pakdaman et al. [109]	$\dfrac{Cp_{bf} - Cp_{nf}}{Cp_{bf}} = (0.0128 \times T + 1.8382)\,\varphi^{0.4779}$	Volume fraction and temperature for MWCNT-heat transfer oil.
Sekhar and Sharma [110]	$Cp_{nf} = 0.8429\left(1+\dfrac{T}{50}\right)^{-0.3037}\left(1+\dfrac{d_p}{50}\right)^{0.4167}$ $\left(1+\dfrac{\varphi}{100}\right)^{2.272}$	Volume fraction, temperature, and particle diameter

(Continued)

Table 8.2 (Continued) Summary of theoretical models on specific heat of NFs

Researchers	Model expressions	Comments
Moldoveanu and Minea [111]	$$Cp_{hybrid} = \frac{\rho_1}{\rho_1 + \rho_2} Cp_1 + \frac{\rho_2}{\rho_1 + \rho_2} Cp_2$$ $$Cp_{nf} = \left(\frac{\rho_{np}}{\rho_{bf}}\right)^{0.2} \left(1 + \frac{d_p}{50}\right)^{0.4167} \left(1 - \frac{\varphi}{100}\right)^{2.272} Cp_{bf}$$	

suitable for ambient temperature and volume concentrations below 5% of three kinds of water-based NFs (Al_2O_3, TiO_2, and SiO_2).

From the above-reviewed literatures, it is evident that the theoretical models of the specific heat of NFs are mainly empirical correlations and are less concerned about the mechanism of the specific heat variations. This is so because the variation tendencies versus volume fraction and temperature are quite different. Roberties et al. [112] reported that the specific heat of Cu/EG NF declined with the increase of NPs volume fraction, and it improved with the increase of temperature. Saeedinia et al. [97] performed that the specific heat of Cu/EO NF declined with increases of NPs volume fraction and temperature. Pakdaman et al. [109] found that the specific heat of MWCNTs/heat transfer oil NFs is remarkably less than that of the BF. In addition, the specific heat is slightly dependent on temperature. Besides, the temperature presented an unpredictable effect on the specific heat of NFs. The volume fraction also presented the opposite trend compared to Mohebbi [113], who found that the specific heat of Si_3N_4/Argon NF is higher than for the base fluid, and increased with the increase of volume fraction. Sonawane et al. [104] found the specific heat of Al_2O_3/aviation turbine fuel NF to first increase and then decline with the increase of volume fraction. The values for 0.3% and 0.5% are higher than that for the base fluid, but the values for 0.1% and 1% are lower than that for the base fluid.

The abovementioned experimental results showed confusing behavior of the specific heat of NFs, and most results were only concerned about the volume fraction and temperature, and thus lack of the size and shape of NPs, etc.

8.3.4 Viscosity

The rheological behavior of NFs is mainly decided by the viscosity. The following test methods are mainly employed to determine the viscosity of NFs: piston-type rheometer, rotational rheometer [114] and the capillary viscometer [115].

Einstein [116] first developed a theoretical model to calculate the viscosity of NFs in 1906. This model was found to only have good precision for NPs volume fraction less than 1%. After the pioneering work of Einstein, extended models were developed by various researchers. Brinkman [117] proposed a modified model that extended the prediction scope of volume fraction below 4%. Bruijin [118] proposed a modified model based on the Brinkman one, which further extended the prediction range of volume fraction to larger concentrations. Guo et al. [119] found a smaller size of NPs could cause a higher viscosity of NFs, and they considered the size of NPs in their correlation. The model predicted well for SiO_2/water NFs. Graham [120] proposed a model considering both the NPs diameter and interparticle distance for spherical NPs. The formula is like the Einstein model. Krieger and Dougherty [121] considered the NFs as non-Newtonian fluids, and developed a correlation which could predict the whole volume fraction range and considered the effect of shear rate. Chen et al. [122] modified the K-D model, and took account of the packing volume fraction of an imaginary sphere containing clusters according to the fractal theory. The non-Newtonian shear thinning behavior of NFs was also considered in the model. Selvakumar and Dhinakaran [123] further modified the Chen et al. model, in which the NPs clustering and interfacial layer features were both taken into account. The correlation is suitable for all NP volume fractions.

The above-reviewed theoretical models of viscosity of NFs only considered the volume fraction and structure of NPs. White and Corfield [124] found that the temperature also has significant effect on the viscosity of NFs, and proposed an empirical model.

Hosseini [125] proposed an empirical model based on dimensionless groups, and considered the viscosity of the base fluid, particle volume fraction, particle size, properties of the surfactant layer, and temperature. The correlation is good agreement with the experimental results for Al_2O_3/water NFs. A correlation for viscosity of Al_2O_3/water NFs, proposed by Abu-Nada [126], depends on the temperature and volume fraction, according to experimental data. The correlation was obtained from a two-dimensional regression, with a maximum error within 5%. Adio et al. [127] also proposed an empirical model based on nondimensional parameters, according to the experimental data for AgO/EG NF, within temperatures 20–70°C, volume fraction 1–5%, and NPs size 1–123 nm. The model considered the essential parameters as NP size, volume fraction, temperature, capping layer thickness, and viscosity of the BF. Godson et al. [128] developed an empirical correlation according to the experimental data of Ag/water NFs by using a linear regression method. The experimental condition included a temperature range of 50–90°C, and the volume fraction ranged from 0.3% to 0.9%. Sekhar and Sharma [110] proposed an empirical correlation according to the experimental data of Al_2O_3/water NFs, for the temperature range 20–70°C, NPs diameter ranging from 13 nm to 100 nm, and the volume fraction ranging from 0.01% to 5.00%. The standard deviation and average deviation between the experimental data and model are 15% and 9%, respectively. The model has considered the effect of temperature, NP diameter, and volume fraction. Table 8.3 summarizes different theoretical models dealing with the viscosity of nanofluids.

Table 8.3 Summary of theoretical models on viscosity of NFs

Researchers	Model expressions	Comments
Einstein [116]	$\mu_{nf} = \mu_{bf}\left(1 + 2.5\varphi\right)$	First theoretical model, valid for volume fractions less than 1%.
Brinkman [117]	$\mu_{nf} = \mu_{bf}/\left(1-\varphi\right)^{2.5}$	An upgraded Einstein model, with the prediction range extended to volume fractions below 4%.
Bruijin [118]	$\mu_{nf} = \mu_{bf}/\left(1 - 2.5\varphi + 1.552\varphi^2\right)$	An upgraded Einstein model, which can be used for larger volume fractions.
Guo et al. [119]	$\mu_{nf} = \mu_{bf}\left(1 + 2.5\varphi + 6.5\varphi^2\right)\left(1 + 350\varphi/d\right)$ where d is the NP diameter.	An empirical correlation considering the effect of particle size with good prediction of SiO_2/water NFs.
Graham [120]	$\mu_{nf} = \mu_{bf}\left(1 + 2.5\varphi + \dfrac{4.5}{\left(l/d\right)\left(2 + l/d\right)\left(1 + l/d\right)^2}\right)$ where l is the minimum separation distance between two spheres.	A theoretical model for spherical NPs, considering the NP diameter and the particles distance
Krieger and Dougherty [121]	$\mu_{nf} = \mu_{bf}\left(1 - \left(\varphi/\varphi_m\right)\right)^{[\eta]\varphi_m}$ where φ_m is the maximum particle volume concentration and $[\eta]$ is the intrinsic viscosity.	Considering the NFs as non-Newtonian fluids and considering the effect of shear rate on viscosity.
Chen et al. [122]	$\mu_{nf} = \mu_{bf}\left(1 - \left(\varphi_{cl}/\varphi_m\right)\right)^{[\eta]\varphi_m}$ where φ_{cl} is the volume fraction of clusters and related to experimental data.	The non-Newtonian shear thinning behavior of NFs is considered in this model

(Continued)

Table 8.3 (Continued) Summary of theoretical models on viscosity of NFs

Researchers	Model expressions	Comments
Selvakumar and Dhinakaran[123]	$\mu_{nf} = \mu_{bf}\left(1 - (\varphi_{ecl}/\varphi_m)\right)^{[\eta]\varphi_m}$ $$\varphi_{ecl} = \varphi_{cl}(1+\beta)^3$$ where φ_{ecl} is the effective volume fraction of the cluster groups and β is the ratio of interfacial layer distance to the average cluster radius.	Considering clustering of NPs and interfacial layer features. The correlation is suitable for all NP volume fractions.
White and Corfield [124]	$$\ln\frac{\mu_{nf}}{\mu_{bf}} \approx a + b\left(\frac{T_0}{T}\right) + c\left(\frac{T_0}{T}\right)^2$$ where a, b, and c are empirical factors and T_0 is the reference temperature.	The model considers temperature as the main variable affecting the viscosity.
Hosseini [125]	$$\frac{\mu_{nf}}{\mu_{bf}} = \exp\left[m + \alpha\left(\frac{T}{T_0}\right) + \beta(\varphi) + \gamma\left(\frac{d}{1+r}\right)\right]$$ where m is a factor decided by the properties of the NFs, while α, β, and γ are factors decided by experimental data.	Considering the temperature, particle volume fraction, particle size, viscosity of the BF, and properties of the surfactant layer.
Abu-Nada [126]	$$\mu_{nf} = -0.155 - \frac{19.582}{T} + 0.794\varphi + \frac{2094.47}{T^2} - 8.11\frac{\varphi}{T}$$ $$- \frac{27463.863}{T^3} + 0.127\varphi^3 + 1.6044\frac{\varphi^2}{T} + 2.1754\frac{\varphi}{T^2}$$	Experimental data of Al$_2$O$_3$/water NFs. Considering the temperature and volume fraction, with a maximum error within 5%.

Adio et al. [127]

$$\frac{\mu_{nf}}{\mu_{bf}} = 1 + a_0\varphi + a_1\left(\frac{T}{T_0}\right)\varphi + a_2\left(\frac{d_p}{h}\right)\varphi + a_3\left(\frac{T}{T_0}\right)\varphi + a_4\left[\left(\frac{d_p}{h}\right)\right]^2\varphi$$

$$+ a_5\left[\left(\frac{T}{T_0}\right)\varphi\right]^2 + a_6\varphi^2 + a_7\left(\frac{T}{T_0}\right)^2\varphi^{1/3}$$

where $a_0 = 7.0764$, $a_1 = -0.1246$, $a_2 = -0.0346$, $a_3 = -0.0024$, $a_4 = -1.2357$, $a_5 = 53.6946$, and $a_6 = 0.0436$; T_0 is the reference temperature equal to 20°C; d_p is the particle diameter; and h is the thickness of the capping layer equal to 1 nm.

Experimental data of MgO/EG NFs:
Volume fractions less than 5%;
Temperature range 20–70°C;
Particle size range 21–125 nm.
The model also considers the capping layer thickness, and viscosity of BF.

Godson et al. [128]

$$\mu_{nf}/\mu_{bf} = 1.005 + 0.497\varphi - 0.1149\varphi^2$$

Experimental data of Ag/water NFs:Temperature range 50–90°C;
Volume fraction range 0.3%–0.9%.

Sekhar and Sharma [110]

$$\mu_{nf} = 0.935\left(1+\frac{T}{70}\right)^{0.5602}\left(1+\frac{d_p}{80}\right)^{-0.05915}\left(1+\frac{\varphi}{100}\right)^{10.51}$$

Experimental data of Al$_2$O$_3$/water NFs:
Temperature range 20–70°C;
NPs diameter range 13–100 nm;
Volume fraction range 0.01%–5.00%.

8.4 APPLICATIONS AND CHALLENGES

8.4.1 Applications

The enhanced thermal conductivity and specific heat of NFs are widely recognized, and sometimes NFs are regarded as next-generation heat transfer and energy storage fluids. Possible applications can be found in heat exchangers [129–132], heat pipes [133], nuclear reactors [134], fuel cells, electronics cooling [135–137], refrigerators and air conditioners [138], engine cooling and vehicle thermal management [139], solar energy collectors and storage [140], building energy savings [141], etc.

Tzeng et al. [139] investigated CuO and Al_2O_3/engine oil NFs to improve the cooling efficiency of a four-wheel-drive transmission system. Tsai et al. [135] employed the gold/water NF as the working fluid in a heat pipe, which was used to cool the CPU or desktop PCs. Also, there are many studies considering various NFs in various heat exchangers, such as, CuO, Fe_3O_4, SiC, SiO_2, and Al_2O_3-water NFs in double-pipe heat exchanger [129, 130], graphene nanoplatelets-EG-water NFs (Gnp-EGW) miniature plate heat exchanger [131], Al_2O_3-water and MWCNT-water chevron plate heat exchanger [132], Al_2O_3-water microchannel heat exchanger [142], Al_2O_3-kerosene NFs in vertical tube heat exchanger at supercritical pressures [143], etc. They all have been confirmed that NFs as heat transfer fluids have a positive effect on heat transfer efficiency, but the pressure drop is also increased significantly.

In addition, many researchers found some specific NFs having higher thermal conductivity and specific heat than the BFs, which is beneficial in reducing the thermal storage materials. Shin and Banerjee [144] proposed SiO2/chloride salt eutectic ($BaCl_2$, NaCl, $CaCl_2$, and LiCl) high-temperature NFs to be used in solar thermal-energy storage with the specific heat enhanced by 14.5%. Wu et al. [145] studied the Cu/paraffin NF used in residential building for effective use of solar energy. The results showed that the phase change heat transfer performance was improved by the added NPs. For 1% mass fraction NF, the melting periodic can be reduced by 13.1%. Liu et al. [146] developed a phase change material by suspending TiO_2 NPs in saturated $BaCl_2$ aqueous solution. The cool storage experiments showed excellent phase change performance and remarkably high thermal conductivity of the NFs, which indicate a potential for substitution of conventional PCMs in cool storage applications.

The abovementioned water, oil, and EG-based NFs are most satisfied normal temperature (<100°C), and lost stability at high temperatures. Here the molten salt-based NFs (MSBNFs) and IL-based NFs (ILBNFs) are developed to satisfy some specific engineering applications, which demand the working mediums at medium (100–400°C) and high temperatures (>400°C)[61]. The silica NPs separated into inorganic salt (Li_2CO_3–K_2CO_3 62:38 mol) was first developed by Shin and Banerjee [144]. They performed that the Cp can be augmented by 25%. IL is a specific type of molten salt with the melting temperature less than 100°C [147]. The ILs received wide attentions because of the low vapor pressure, high thermal stability, and adjustable thermal properties, through changing the suitable cation and anion. Due to the wide working window of the IL from normal temperature, it is promising instead of molten salt used as commercial heat transfer fluid. Addition of the MWCNT NPs into the ILs was first developed by Castro et al. in 2009 [148], and proved that it can improve the λ. According to the above features, some follow-up studies on MSBNFs and ILBNFs were performed [23, 149–154], and most used on solar energy storage and heat transfer working fluids.

Besides using NFs as heat transfer fluids or thermal-energy storage materials, other applications are possible, for example, for friction reduction, magnetic sealing, and optical application and as detergents, and biomedical applications. In this chapter, the major concern is on opportunities for heat transfer augmentation of the NFs, so other applications are not highlighted.

8.4.2 Challenges

Many problems and challenges of NFs need to be resolved before their commercial application. The main challenges are as follows: (a) the preparation process is complicated and the cost might be high; (b) the NPs can easily aggregate and deposit on surfaces, leading to bad stability of NFs; (c) the influencing factors on the thermal physical properties are many, resulting in big differences among measured data; (d) unclear mechanisms of the changes in thermal physical properties by adding NPs, and different kinds of NFs present different results; (e) no theoretical model can accurately predict the thermal physical properties for all NFs and the operating window (volume fraction, temperature, etc.). Additional systematic experiments and mature theoretical models are needed to overcome these challenges and to promote the commercialization.

REFERENCES

[1] D. Li, W. Xie, W. Fang, Preparation and properties of copper-oil-based nanofluids, *Nanoscale Res. Lett.*, Vol. 6, 373–373, 2011.
[2] V. Kumar, N. Pandya, B. Pandya, A. Joshi, Synthesis of metal-based nanofluids and their thermo-hydraulic performance in compact heat exchanger with multi-louvered fins working under laminar conditions, *J. Therm. Anal. Calorim.*, Vol. 135, 2221–2235, 2019.
[3] A. Asadi, F. Pourfattah, Heat transfer performance of two oil-based nanofluids containing ZnO and MgO nanoparticles; a comparative experimental investigation, *Powder Technol.*, Vol. 343, 296–308, 2019.
[4] P. Keblinski, J.A. Eastman, D.G. Cahill, Nanofluids for thermal transport, *Mater. Today*, Vol. 8, 36–44, 2005.

[5] K.S. Suganthi, K.S. Rajan, Metal oxide nanofluids: Review of formulation, thermo-physical properties, mechanisms, and heat transfer performance, *Renew. Sustain. Energy Rev.*, Vol. 76, 226–255, 2017.

[6] H. Xie, H. Lee, W. Youn, M. Choi, Nanofluids containing multiwalled carbon nanotubes and their enhanced thermal conductivities, Vol. 94, 4967–4971, 2003.

[7] T. Tyler, O. Shenderova, G. Cunningham, J. Walsh, J. Drobnik, G. McGuire, Thermal transport properties of diamond-based nanofluids and nanocomposites, *Diamond Relat. Mater.*, Vol. 15, 2078–2081, 2006.

[8] M. Piratheepan, T.N. Anderson, An experimental investigation of turbulent forced convection heat transfer by a multi-walled carbon-nanotube nanofluid, *Int. Comm. Heat cMass Transf.*, Vol. 57, 286–290, 2014.

[9] K. Takenaka, N. Saidoh, N. Nishiyama, A. Inoue, Fabrication and nano-imprintabilities of Zr-, Pd- and Cu-based glassy alloy thin films, *Nanotechnology*, Vol. 22, 105302, 2011.

[10] K.N. Dinh, Q. Liang, C.-F. Du, J. Zhao, A.I.Y. Tok, H. Mao, Q. Yan, Nanostructured metallic transition metal carbides, nitrides, phosphides, and borides for energy storage and conversion, *Nano Today*, Vol. 25, 99–121, 2019.

[11] X. Li, S. Yun, C. Zhang, W. Fang, X. Huang, T. Du, Application of nano-scale transition metal carbides as accelerants in anaerobic digestion, *Int. J. Hydrogen Energy*, Vol. 43, 1926–1936, 2018.

[12] S. Stankic, S. Suman, F. Haque, J. Vidic, Pure and multi metal oxide nanoparticles: synthesis, antibacterial and cytotoxic properties, *J. Nanobiotechnol.*, Vol. 14, 73, 2016.

[13] S. Balamurugan, A.R. Balu, V. Narasimman, G. Selvan, K. Usharani, J. Srivind, M. Suganya, N. Manjula, C. Rajashree, V.S. Nagarethinam, Multi metal oxide CdO–Al$_2$O$_3$–NiO nanocomposite—Synthesis, photocatalytic and magnetic properties, *Mater. Res. Express*, Vol. 6, 015022, 2018.

[14] J. Sarkar, P. Ghosh, A. Adil, A review on hybrid nanofluids: Recent research, development and applications, *Renew. Sustain. Energy Rev.*, Vol. 43, 164–177, 2015.

[15] G. Huminic, A. Huminic, Hybrid nanofluids for heat transfer applications – A state-of-the-art review, *Int. J. Heat Mass Transf.*, Vol. 125, 82–103, 2018.

[16] Q. Wang, M. Rafiq, Y. Lv, C. Li, K. Yi, Preparation of three types of transformer oil-based nanofluids and comparative study on the effect of nanoparticle concentrations on insulating property of transformer oil, *J. Nanotechnol.*, Vol. 2016, 5802753, 2016.

[17] V. Eswaraiah, V. Sankaranarayanan, S. Ramaprabhu, Graphene-based engine oil nanofluids for tribological applications, *ACS Appl. Mater. Interfaces*, Vol. 3, 4221–4227, 2011.

[18] J. Li, Z. Zhang, P. Zou, S. Grzybowski, M. Zahn, Preparation of a vegetable oil-based nanofluid and investigation of its breakdown and dielectric properties, *IEEE Electr. Insul. Mag.*, Vol. 28, 43–50, 2012.

[19] M. Afrand, E. Abedini, H. Teimouri, How the dispersion of magnesium oxide nanoparticles effects on the viscosity of water-ethylene glycol mixture: Experimental evaluation and correlation development, *Physica E*, Vol. 87, 273–280, 2017.

[20] S. Akilu, K.V. Sharma, T.B. Aklilu, M.S.M. Azman, P.T. Bhaskoro, Temperature dependent properties of silicon carbide nanofluid in binary mixtures of glycerol-ethylene glycol, *Procedia Eng.*, Vol. 148, 774–778, 2016.

[21] S. Manikandan, K.S. Rajan, New hybrid nanofluid containing encapsulated paraffin wax and sand nanoparticles in propylene glycol-water mixture: Potential heat transfer fluid for energy management, *Energ. Conver. Manag.*, Vol. 137, 74–85, 2017.

[22] B. Muñoz-Sánchez, J. Nieto-Maestre, I. Iparraguirre-Torres, A. García-Romero, J.M. Sala-Lizarraga, Molten salt-based nanofluids as efficient heat transfer and storage materials at high temperatures. An overview of the literature, *Renew. Sust. Energ. Rev.*, Vol. 82, 3924–3945, 2018.

[23] N.J. Bridges, A.E. Visser, E.B. Fox, Potential of nanoparticle-enhanced ionic liquids (NEILs) as advanced heat-transfer fluids, *Energy Fuel*, Vol. 25, 4862–4864, 2011.

[24] Y. Xuan, Q. Li, Heat transfer enhancement of nanofluids, *Int. J. Heat Fluid Flow*, Vol. 21, 58–64, 2000.

[25] T.R. Barrett, S. Robinson, K. Flinders, A. Sergis, Y. Hardalupas, Investigating the use of nanofluids to improve high heat flux cooling systems, *Fusion Eng. Des.*, Vol. 88, 2594–2597, 2013.

[26] H.E. Patel, S.K. Das, T. Sundararajan, A.S. Nair, B. George, T. Pradeep, Thermal conductivities of naked and monolayer protected metal nanoparticle based nanofluids: Manifestation of anomalous enhancement and chemical effects, Vol. 83, 2931–2933, 2003.

[27] B. Wei, C. Zou, X. Yuan, X. Li, Thermo-physical property evaluation of dia-thermic oil based hybrid nanofluids for heat transfer applications, *Int. J. Heat Mass Transf.*, Vol. 107, 281–287, 2017.

[28] N. Ahammed, L.G. Asirvatham, S. Wongwises, Entropy generation analysis of graphene–alumina hybrid nanofluid in multiport minichannel heat exchanger coupled with thermoelectric cooler, *Int. J. Heat Mass Transf.*, Vol. 103, 1084–1097, 2016.

[29] O. Soltani, M. Akbari, Effects of temperature and particles concentration on the dynamic viscosity of MgO-MWCNT/ethylene glycol hybrid nanofluid: Experimental study, *Physica E*, Vol. 84, 564–570, 2016.

[30] T. Le Ba, O. Mahian, S. Wongwises, I.M. Szilágyi, Review on the recent prog-ress in the preparation and stability of graphene-based nanofluids, *J. Therm. Anal. Calorim.*, Vol. 142, 1145–1172, 2020.

[31] H.Ş. Aybar, M. Sharifpur, M.R. Azizian, M. Mehrabi, J.P. Meyer, A review of thermal conductivity models for nanofluids, *Heat Transf. Eng.*, Vol. 36, 1085–1110, 2015.

[32] J.J. Wang, R.T. Zheng, J.W. Gao, G. Chen, Heat conduction mechanisms in nanofluids and suspensions, *Nano Today*, Vol. 7, 124–136, 2012.

[33] H. Zhang, H. Cui, S. Yao, K. Zhang, H. Tao, H. Meng, Ionic liquid-stabilized non-spherical gold nanofluids synthesized using a one-step method, *Nanoscale Res. Lett.*, Vol. 7, 583, 2012).

[34] G.J. Lee, C.K. Kim, M.K. Lee, C.K. Rhee, S. Kim, C. Kim, Thermal conduc-tivity enhancement of ZnO nanofluid using a one-step physical method, *Thermochimica Acta*, Vol. 542, 24–27, 2012.

[35] B. Munkhbayar, M.R. Tanshen, J. Jeoun, H. Chung, H. Jeong, Surfactant-free dispersion of silver nanoparticles into MWCNT-aqueous nanofluids prepared by one-step technique and their thermal characteristics, *Ceram. Int.*, Vol. 39, 6415–6425, 2013.

[36] H.T. Zhu, Y.S. Lin, Y.S. Yin, A novel one-step chemical method for preparation of copper nanofluids, *J. Colloid Interface Sci.*, Vol. 277, 100–103, 2004.

[37] J. Yang, J.Y. Lee, J.Y. Ying, Phase transfer and its applications in nanotechnology, *Chem. Soc. Rev.*, Vol. 40, 1672–1696, 2011.

[38] X. Feng, H. Ma, S. Huang, W. Pan, X. Zhang, F. Tian, C. Gao, Y. Cheng, J. Luo, Aqueous–organic phase-transfer of highly stable gold, silver, and platinum nanoparticles and new route for fabrication of gold nanofilms at the oil/water interface and on solid supports, *J. Phys. Chem. B.*, Vol. 110, 12311–12317, 2006.

[39] Y. Chen, X. Wang, Novel phase-transfer preparation of monodisperse silver and gold nanoparticles at room temperature, *Mater. Lett.*, Vol. 62, 2215–2218, 2008.

[40] C. Wang, J. Yang, Y. Ding, Phase transfer based synthesis and thermophysical properties of Au/Therminol VP-1 nanofluids, *Prog. Nat. Sci.: Mater. Int.*, Vol. 23, 338–342, 2013.

[41] W. Yu, H. Xie, L. Chen, Y. Li, Enhancement of thermal conductivity of kerosene-based Fe3O4 nanofluids prepared via phase-transfer method, *Colloids Surf. A Physicochem. Eng. Asp.*, Vol. 355, 109–113, 2010.

[42] L. Wang, J. Fan, Nanofluids research: Key issues, *Nanoscale Res. Lett.*, Vol. 5, 1241–1252, 2010.

[43] X. Wei, T. Kong, H. Zhu, L. Wang, CuS/Cu$_2$S nanofluids: Synthesis and thermal conductivity, *Int. J. Heat Mass Transf.*, Vol. 53, 1841–1843, 2010.

[44] X. Wei, H. Zhu, T. Kong, L. Wang, Synthesis and thermal conductivity of Cu$_2$O nanofluids, *Int. J. Heat Mass Transf.*, Vol. 52, 4371–4374, 2009.

[45] M.-S. Liu, M.C.-C. Lin, C.Y. Tsai, C.-C. Wang, Enhancement of thermal conductivity with Cu for nanofluids using chemical reduction method, *Int. J. Heat Mass Transf.*, Vol. 49, 3028–3033, 2006.

[46] H. Zhu, D. Han, Z. Meng, D. Wu, C. Zhang, Preparation and thermal conductivity of CuO nanofluid via a wet chemical method, *Nanoscale Res. Lett.*, Vol. 6, 181, 2011.

[47] Z. Zhang, J. Cai, F. Chen, H. Li, W. Zhang, W. Qi, Progress in enhancement of CO$_2$ absorption by nanofluids: A mini review of mechanisms and current status, *Renew. Energy*, Vol. 118, 527–535, 2018.

[48] Y. Hwang, J.-K. Lee, J.-K. Lee, Y.-M. Jeong, S.-I. Cheong, Y.-C. Ahn, S.H. Kim, Production and dispersion stability of nanoparticles in nanofluids, *Powder Technol.*, Vol. 186, 145–153, 2008.

[49] Z. Liu, C. Wu, L. Niu, G. Yang, K. Wang, W. Pei, Q. Wang, Post-treatment method for the synthesis of monodisperse binary FePt-Fe$_3$O$_4$ nanoparticles, *Nanoscale Res. Lett.*, Vol. 12, 540, 2017.

[50] S. Oh, I. Jang, S.-G. Oh, S.S. Im, Effect of ZnO nanoparticle morphology and post-treatment with zinc acetate on buffer layer in inverted organic photovoltaic cells, *Sol. Energy*, Vol. 114, 32–38, 2015.

[51] W. Yu, H. Xie, A review on nanofluids: Preparation, stability mechanisms, and applications, *J. Nanomater.*, Vol. 2012, 1–17, 2012.

[52] Babita, S.K. Sharma, S.M. Gupta, Preparation and evaluation of stable nanofluids for heat transfer application: A review, *Exp. Therm. Fluid Sci.*, Vol. 79, 202–212, 2016.

[53] T. Missana, A. Adell, On the applicability of DLVO theory to the prediction of clay colloids stability, *J. Colloid Interface Sci.*, Vol. 230, 150–156, 2000.

[54] I. Popa, G. Gillies, G. Papastavrou, M. Borkovec, Attractive and repulsive electrostatic forces between positively charged latex particles in the presence of anionic linear polyelectrolytes, *J. Phys. Chem. B.*, Vol. 114, 3170–3177, 2010.

[55] X. Yang, Z.-H. Liu, A kind of nanofluid consisting of surface-functionalized nanoparticles, *Nanoscale Res. Lett.*, Vol. 5, 1324–1328, 2010.

[56] L. Chen, H. Xie, Y. Li, W. Yu, Nanofluids containing carbon nanotubes treated by mechanochemical reaction, *Thermochimica Acta*, Vol. 477, 21–24, 2008.

[57] L. Chen, H. Xie, Surfactant-free nanofluids containing double- and single-walled carbon nanotubes functionalized by a wet-mechanochemical reaction, *Thermochimica Acta*, Vol. 497, 67–71, 2010.

[58] I.M. Joni, A. Purwanto, F. Iskandar, K. Okuyama, Dispersion stability enhancement of titania nanoparticles in organic solvent using a bead mill process, *Ind. Eng. Chem. Res.*, Vol. 48, 6916–6922, 2009.

[59] E. Tang, G. Cheng, X. Ma, X. Pang, Q. Zhao, Surface modification of zinc oxide nanoparticle by PMAA and its dispersion in aqueous system, *Appl. Surf. Sci.*, Vol. 252, 5227–5232, 2006.

[60] H. Jia, Y. Lian, M.O. Ishitsuka, T. Nakahodo, Y. Maeda, T. Tsuchiya, T. Wakahara, T. Akasaka, Centrifugal purification of chemically modified single-walled carbon nanotubes, *Sci. Technol. Adv. Mate.*, Vol. 6, 571–581, 2005.

[61] W. Wang, Z. Wu, B. Li, B. Sundén, A review on molten-salt-based and ionic-liquid-based nanofluids for medium-to-high temperature heat transfer, *J. Therm. Anal. Calorim.*, Vol. 136, 1037–1051, 2018.

[62] J. Huang, X. Wang, Q. Long, X. Wen, Y. Zhou, L. Li, Influence of pH on the stability characteristics of nanofluids, in: *2009 Symposium on Photonics and Optoelectronics*, 1–4, 2009.

[63] K. Goudarzi, F. Nejati, E. Shojaeizadeh, S.K. Asadi Yousef-abad, Experimental study on the effect of pH variation of nanofluids on the thermal efficiency of a solar collector with helical tube, *Exp. Therm. Fluid Sci.*, Vol. 60, 20–27, 2015.

[64] H. Zhang, S. Qing, Y. Zhai, X. Zhang, A. Zhang, The changes induced by pH in TiO_2/water nanofluids: Stability, thermophysical properties and thermal performance, *Powder Technol.*, Vol. 377, 748–759, 2021.

[65] D. Wen, Y. Ding, Experimental investigation into the pool boiling heat transfer of aqueous based γ-alumina nanofluids, *J. Nanopart. Res.*, Vol. 7, 265–274, 2005.

[66] H.-J. Chen, D. Wen, Ultrasonic-aided fabrication of gold nanofluids, *Nanoscale Res. Lett.*, Vol. 6, 198–198, 2011.

[67] A. Amrollahi, A.A. Hamidi, A.M. Rashidi, The effects of temperature, volume fraction and vibration time on the thermo-physical properties of a carbon nanotube suspension (carbon nanofluid), *Nanotechnology*, Vol. 19, 315701, 2008.

[68] R. Sadri, G. Ahmadi, H. Togun, M. Dahari, S.N. Kazi, E. Sadeghinezhad, N. Zubir, An experimental study on thermal conductivity and viscosity of nanofluids containing carbon nanotubes, *Nanoscale Res. Lett.*, Vol. 9, 151–151, 2014.

[69] Z. Berk, Chapter 7 – Mixing, in *Food Process Engineering and Technology* (ed. Z. Berk), 175–194, Academic Press, San Diego, 2009.

[70] E.P. Bandara Filho, O.S.H. Mendoza, C.L.L. Beicker, A. Menezes, D. Wen, Experimental investigation of a silver nanoparticle-based direct absorption solar thermal system, *Energ. Conver. Manage.*, Vol. 84, 261–267, 2014.

[71] L. Fedele, L. Colla, S. Bobbo, S. Barison, F. Agresti, Experimental stability analysis of different water-based nanofluids, *Nanoscale Res. Lett.*, Vol. 6, 300, 2011.

[72] C.H. Bi, Z.M. Yan, P.L. Wang, A. Alkhatib, J.Y. Zhu, H.C. Zou, D.Y. Sun, X.D. Zhu, F. Gao, W.T. Shi, Z.G. Huang, Effect of high pressure homogenization treatment on the rheological properties of citrus peel fiber/corn oil emulsion, *J. Sci. Food Agric.*, Vol. 100, 3658–3665, 2020.

[73] Y. Asakuma, S. Miyauchi, T. Yamamoto, H. Aoki, T. Miura, Homogenization method for effective thermal conductivity of metal hydride bed, *Int. J. Hydrogen Energy*, Vol. 29, 209–216, 2004.

[74] X. Zhu, B. Lundberg, Y. Cheng, L. Shan, J. Xing, P. Peng, P. Chen, X. Huang, D. Li, R. Ruan, Effect of high-pressure homogenization on the flow properties of citrus peel fibers, Vol. 41, e12659, 2018.

[75] M.H. Esfe, M. Afrand, An updated review on the nanofluids characteristics, *J. Therm. Anal. Calorim.*, Vol. 138, 4091–4101, 2019.

[76] G. Paul, M. Chopkar, I. Manna, P.K. Das, Techniques for measuring the thermal conductivity of nanofluids: A review, *Renew. Sustain. Energy Rev.*, Vol. 14, 1913–1924, 2010.

[77] B.A. Bhanvase, D.P. Barai, S.H. Sonawane, N. Kumar, S.S. Sonawane, Chapter 40 - Intensified heat transfer rate with the use of nanofluids, in *Handbook of Nanomaterials for Industrial Applications* (ed. C. Mustansar Hussain), 739–750, Elsevier, New Jersey, USA, 2018.

[78] J.C. Maxwell, *A Treatise on Electricity and Magnetism*, Clarendon Press, Oxford, 1873.

[79] R.L. Hamilton, O.J.I. Crosser, Thermal conductivity of heterogeneous two-component systems, *Ind. Eng. Chem. Fundam.*, Vol. 1, 187–191, 1962.

[80] E.J. Wasp, J.P. Kenny, R.L.J.S.B.M.H. Gandhi, Solid--liquid flow: Slurry pipeline transportation. [Pumps, valves, mechanical equipment, economics, Vol. 1, 6343851, 1977.

[81] Y. Xuan, Q. Li, W. Hu, Aggregation structure and thermal conductivity of nanofluids, *AIChE J.*, Vol. 49, 1038–1043, 2003.

[82] W. Yu, S. Choi, The role of interfacial layers in the enhanced thermal conductivity of nanofluids: A renovated Maxwell model, *J. Nanopart. Res.*, Vol. 5, 167–171, 2003.

[83] S. Pil Jang, S.U. Choi, Effects of various parameters on nanofluid thermal conductivity, *ASME J. Heat Mass Transf.*, Vol. 129, 617–623, 2007.

[84] D.H. Kumar, H.E. Patel, V.R. Kumar, T. Sundararajan, T. Pradeep, S.K. Das, Model for heat conduction in nanofluids, *Physi. Rev. Lett.*, Vol. 93, 144301, 2004.

[85] Q.Z. Xue, Model for thermal conductivity of carbon nanotube-based composites, *Physi. B: Conden. Matt.*, Vol. 368, 302–307, 2005.

[86] J. Koo, C. Kleinstreuer, A new thermal conductivity model for nanofluids, *Appli. Therm. Engine.*, Vol. 6, 577–588, 2004.

[87] J. Koo, C. Kleinstreuer, Impact analysis of nanoparticle motion mechanisms on the thermal conductivity of nanofluids, *Int. Commu. Heat Mass Trans.*, Vol. 32, 1111–1118, 2005.

[88] R. Prasher, P. Bhattacharya, P.E. Phelan, Brownian-motion-based convective-conductive model for the effective thermal conductivity of nanofluids, *ASME J. Heat Mass Trans.*, Vol. 128, 588–595, 2006.

[89] K. Leong, C. Yang, S.M.S. Murshed, A model for the thermal conductivity of nanofluids–The effect of interfacial layer, *J. Nanopar. Resear.*, Vol. 8, 245–254, 2006.

[90] S. Murshed, K. Leong, C. Yang, A combined model for the effective thermal conductivity of nanofluids, *Appl. Therm. Engine.*, Vol. 29, 2477–2483, 2009.

[91] H. Jiang, Q. Xu, C. Huang, L. Shi, The role of interfacial nanolayer in the enhanced thermal conductivity of carbon nanotube-based nanofluids, *J. Nanopar. Resear.*, Vol. 118, 197–205, 2015.

[92] L. Yang, X. Xu, W. Jiang, K. Du, A new thermal conductivity model for nanorod-based nanofluids, *Appl. Therm. Eng.*, Vol. 114, 287–299, 2017.

[93] M. Corcione, Empirical correlating equations for predicting the effective thermal conductivity and dynamic viscosity of nanofluids, *Ener. Conver. Manag.*, Vol. 52, 789–793, 2011.

[94] M. Afrand, D. Toghraie, N. Sina, Experimental study on thermal conductivity of water-based Fe_3O_4 nanofluid: development of a new correlation and modeled by artificial neural network, *Int. Commu. Heat Mass Trans.*, Vol. 75, 262–269, 2016.

[95] A.M. Khdher, N.A.C. Sidik, W.A.W. Hamzah, R. Mamat, An experimental determination of thermal conductivity and electrical conductivity of bio glycol based Al_2O_3 nanofluids and development of new correlation, *Int. Commu. Heat Mass Trans.*, Vol. 73 75–83, 2016.

[96] A.K. Starace, J.C. Gomez, J. Wang, S. Pradhan, G.C. Glatzmaier, Nanofluid heat capacities, Vol. 110, 124323, 2011.

[97] M. Saeedinia, M.A. Akhavan-Behabadi, M. Nasr, Experimental study on heat transfer and pressure drop of nanofluid flow in a horizontal coiled wire inserted tube under constant heat flux, *Exp. Therm. Fluid Sci.*, Vol. 36, 158–168, 2012.

[98] M. Saeedinia, M.A. Akhavan-Behabadi, P. Razi, Thermal and rheological characteristics of CuO–Base oil nanofluid flow inside a circular tube, *Int. Comm. Heat Mass Transf.*, Vol. 39, 152–159, 2012.

[99] M.X. Ho, C. Pan, Optimal concentration of alumina nanoparticles in molten Hitec salt to maximize its specific heat capacity, *Int. J. Heat Mass Transf.*, Vol. 70, 174–184, 2014.

[100] D. Shin, D. Banerjee, Specific heat of nanofluids synthesized by dispersing alumina nanoparticles in alkali salt eutectic, *Int. J. Heat Mass Transf.*, Vol. 74, 210–214, 2014.

[101] C.A. Nieto de Castro, S.M.S. Murshed, M.J.V. Lourenço, F.J.V. Santos, M.L.M. Lopes, J.M.P. França, Enhanced thermal conductivity and specific heat capacity of carbon nanotubes ionanofluids, *Int. J. Therm. Sci.*, Vol. 62, 34–39, 2012.

[102] R.S. Vajjha, D.K. Das, Specific heat measurement of three nanofluids and development of new correlations, *J. Heat Transfer*, Vol. 131, 071601, 2009.

[103] S.M.S. Murshed, Determination of effective specific heat of nanofluids, *J. Exp. Nanosci.*, Vol. 6, 539–546, 2011.

[104] S. Sonawane, K. Patankar, A. Fogla, B. Puranik, U. Bhandarkar, S. Sunil Kumar, An experimental investigation of thermo-physical properties and heat transfer performance of Al_2O_3-aviation turbine fuel nanofluids, *Appl. Therm. Eng.*, Vol. 31, 2841–2849, 2011.

[105] B.C. Pak, Y.I. Cho, Hydrodynamic and heat transfer study of dispersed fluids with submicron metallic oxide particles, *Exp. Heat Transf. Int. J.*, Vol. 11, 151–170, 1998.

[106] Y. Xuan, W. Roetzel, Conceptions for heat transfer correlation of nanofluids, *Int. J. Heat Mass Transf.*, Vol. 43, 3701–3707, 2000.

[107] R.S. Vajjha, D.K. Das, Specific heat measurement of three nanofluids and development of new correlations, *J. Heat Transfer*, Vol. 131, 071601, 2009.

[108] K.K. Kaaliarasan Ramachandran, Devarajan Ramasamy, Mahendran Samykano, Lingenthiran Samylingam, Faris Tarlochan, Gholamhassan Najafi, Evaluation of specific heat capacity and density for cellulose nanocrystal-based nanofluid, Vol. 51, 169–186, 2018.

[109] M.F. Pakdaman, M. Akhavan-Behabadi, P. Razi, An experimental investigation on thermo-physical properties and overall performance of MWCNT/heat transfer oil nanofluid flow inside vertical helically coiled tubes, *Exp. Therm. Fluid Sci.*, Vol. 40, 103–111, 2012.

[110] Y.R. Sekhar, K. Sharma, Study of viscosity and specific heat capacity characteristics of water-based Al_2O_3 nanofluids at low particle concentrations, *J. Exp. Nanosci.*, Vol. 10, 86–102, 2015.

[111] G.M. Moldoveanu, A.A. Minea, Specific heat experimental tests of simple and hybrid oxide-water nanofluids: Proposing new correlation, *J. Mol. Liq.*, Vol. 279, 299–305, 2019.

[112] E. De Robertis, E.H.H. Cosme, R.S. Neves, A.Y. Kuznetsov, A.P.C. Campos, S.M. Landi, C.A. Achete, Application of the modulated temperature differential scanning calorimetry technique for the determination of the specific heat of copper nanofluids, *Appl. Therm. Eng.*, Vol. 41, 10–17, 2012.

[113] A. Mohebbi, Prediction of specific heat and thermal conductivity of nanofluids by a combined equilibrium and non-equilibrium molecular dynamics simulation, *J. Mol. Liq.*, Vol. 175, 51–58, 2012.

[114] S. Hamze, D. Cabaleiro, T. Maré, B. Vigolo, P. Estellé, Shear flow behavior and dynamic viscosity of few-layer graphene nanofluids based on propylene glycol-water mixture, *J. Mol. Liq.*, Vol. 316, 113875, 2020.

[115] M. Gupta, V. Singh, R. Kumar, Z. Said, A review on thermophysical properties of nanofluids and heat transfer applications, *Renew. Sustain. Energy Rev.*, Vol. 74, 638–670, 2017.

[116] A. Einstein, A new determination of molecular dimensions, *Ann. Phys.*, Vol. 19, 289–306, 1906.

[117] H.C. Brinkman, The viscosity of concentrated suspensions and solutions, *J. Chem. Phys.*, Vol. 20, 571–571, 1952.

[118] H. de Bruijn, The viscosity of suspensions of spherical particles. (The fundamental η-c and φ relations), Vol. 61, 863–874, 1942.

[119] G. Shun-song, L. Zhong-yang, W. Tao, Z. Jia-fei, C. Ke-fa, Viscosity of Monodisperse Silica Nanofluids, *Bull. Chin. Ceram. Soc.*, Vol. 5, 012, 2006.

[120] A.L. Graham, On the viscosity of suspensions of solid spheres, *Appl. Sci. Res.*, Vol. 37, 275–286, 1981.

[121] I.M. Krieger, T.J. Dougherty, A mechanism for non-Newtonian flow in suspensions of rigid spheres, *Trans. Soc. Rheol.*, Vol. 3, 137–152, 1959.

[122] H. Chen, Y. Ding, C. Tan, Rheological behaviour of nanofluids, *New J. Phys.*, Vol. 9, 367, 2007.

[123] R.D. Selvakumar, S. Dhinakaran, Effective viscosity of nanofluids—A modified Krieger–Dougherty model based on particle size distribution (PSD) analysis, *J. Mol. Liq.*, Vol. 225, 20–27, 2017.

[124] F.M. White, I. Corfield, *Viscous Fluid Flow*, McGraw-Hill, New York, USA, 2006.

[125] S.M. Hosseini, A. Moghadassi, D. Henneke, A new dimensionless group model for determining the viscosity of nanofluids, *J. Therm. Analy. Calori.*, Vol. 100, 873, 2010.

[126] E. Abu-Nada, Effects of variable viscosity and thermal conductivity of Al_2O_3–water nanofluid on heat transfer enhancement in natural convection, *Int. J. Heat Fluid Flow*, Vol. 30, 679–690, 2009.

[127] S.A. Adio, M. Mehrabi, M. Sharifpur, J.P. Meyer, Experimental investigation and model development for effective viscosity of MgO–ethylene glycol nanofluids by using dimensional analysis, FCM-ANFIS and GA-PNN techniques, *Int. Comm. Heat cMass Transf.*, Vol. 72, 71–83, 2016.

[128] L. Godson, B. Raja, D.M. Lal, S. Wongwises, Experimental investigation on the thermal conductivity and viscosity of silver-deionized water nanofluid, *Exp. Heat Transf.*, Vol. 23, 317–332, 2010.

[129] D. Zheng, J. Du, W. Wang, J.J. Klemeš, J. Wang, B. Sundén, Analysis of thermal efficiency of a corrugated double-tube heat exchanger with nanofluids, *Energy*, Vol. 256, 124522, 2022.

[130] Z. Wu, L. Wang, B. Sundén, Pressure drop and convective heat transfer of water and nanofluids in a double-pipe helical heat exchanger, *Appl. Therm. Eng.*, Vol. 60, 266–274, 2013.

[131] Z. Wang, Z. Wu, F. Han, L. Wadsö, B. Sundén, Experimental comparative evaluation of a graphene nanofluid coolant in miniature plate heat exchanger, *Int. J. Therm. Sci.*, Vol. 130, 148–156, 2018.

[132] D. Huang, Z. Wu, B. Sunden, Pressure drop and convective heat transfer of Al_2O_3/water and MWCNT/water nanofluids in a chevron plate heat exchanger, *Int. J. Heat Mass Transf.*, Vol. 89, 620–626, 2015.

[133] N.K. Gupta, A.K. Tiwari, S.K. Ghosh, Heat transfer mechanisms in heat pipes using nanofluids – A review, *Exp. Therm. Fluid Sci.*, Vol. 90, 84–100, 2018.

[134] J. Buongiorno, L.W. Hu, 8. Innovative Technologies: Two-Phase Heat Transfer in Water-Based Nanofluids for Nuclear Applications Final Report, in Massachusetts Institute of Technology Cambridge, MA 02139-4307, 2009, pp. Medium: ED; Size: 347 p.

[135] C.Y. Tsai, H.T. Chien, P.P. Ding, B. Chan, T.Y. Luh, P.H. Chen, Effect of structural character of gold nanoparticles in nanofluid on heat pipe thermal performance, *Mater. Lett.*, Vol. 58, 1461–1465, 2004.

[136] J. Wang, Z. Zhai, D. Zheng, L. Yang, B.J.H.T.E. Sundén, Investigation of heat transfer characteristics of Al_2O_3-water nanofluids in an electric heater, Vol. 42, 1765–1774, 2020.

[137] Z. Chen, D. Zheng, J. Wang, L. Chen, B. Sundén, Experimental investigation on heat transfer characteristics of various nanofluids in an indoor electric heater, *Renew. Energy*, Vol. 147, 1011–1018, 2020.

[138] W. Jiang, G. Ding, H. Peng, Measurement and model on thermal conductivities of carbon nanotube nanorefrigerants, *Int. J. Therm. Sci.*, Vol. 48, 1108–1115, 2009.

[139] S.C. Tzeng, C.W. Lin, K.D. Huang, Heat transfer enhancement of nanofluids in rotary blade coupling of four-wheel-drive vehicles, *Acta Mech.*, Vol. 179, 11–23, 2005.

[140] Q. Xiong, S. Altnji, T. Tayebi, M. Izadi, A. Hajjar, B. Sundén, L.K.B. Li, A comprehensive review on the application of hybrid nanofluids in solar energy collectors, *Sustain. Energy Technol. Assess.*, Vol. 47, 101341, 2021.

[141] D.P. Kulkarni, D.K. Das, R.S. Vajjha, Application of nanofluids in heating buildings and reducing pollution, *Appl. Energy*, Vol. 86, 2566–2573, 2009.

[142] J. Wang, K. Yu, M. Ye, E. Wang, W. Wang, B. Sundén, Effects of pin fins and vortex generators on thermal performance in a microchannel with Al2O3 nanofluids, *Energy*, Vol. 239, 122606, 2022.

[143] D. Huang, X.-Y. Wu, Z. Wu, W. Li, H.-T. Zhu, B. Sunden, Experimental study on heat transfer of nanofluids in a vertical tube at supercritical pressures, *Int. Comm. Heat Mass Transf.*, Vol. 63, 54–61, 2015.

[144] D. Shin, D. Banerjee, Enhancement of specific heat capacity of high-temperature silica-nanofluids synthesized in alkali chloride salt eutectics for solar thermal-energy storage applications, *Int. J. Heat Mass Transf.*, Vol. 54, 1064–1070, 2011.

[145] S. Wu, H. Wang, S. Xiao, D. Zhu, Numerical simulation on thermal energy storage behavior of Cu/paraffin nanofluids PCMs, *Procedia Eng.*, Vol. 31, 240–244,2012.

[146] Y.D. Liu, Y.G. Zhou, M.W. Tong, X.S. Zhou, Experimental study of thermal conductivity and phase change performance of nanofluids PCMs, *Microfluid Nanofluidics*, Vol. 7, 579, 2009.

[147] J.O. Valderrama, R.A. Campusano, Melting properties of molten salts and ionic liquids. *Chemical Homology, Correlation, and Prediction, Comptes Rendus Chimie*, Vol. 19, 654–664, 2016.

[148] C. Nieto de Castro, M. Lourenço, A. Ribeiro, E. Langa, S. Vieira, P. Goodrich, C. Hardacre, Thermal properties of ionic liquids and ionanofluids of imidazolium and pyrrolidinium liquids, *J. Chem. Eng. Data*, Vol. 55, 653–661, 2009.

[149] P. Andreu-Cabedo, R. Mondragon, L. Hernandez, R. Martinez-Cuenca, L. Cabedo, J.E. Julia, Increment of specific heat capacity of solar salt with SiO2 nanoparticles, *Nanoscale Res. Lett.*, Vol. 9, 582, 2014.

[150] M. Chieruzzi, G.F. Cerritelli, A. Miliozzi, J.M. Kenny, L. Torre, Heat capacity of nanofluids for solar energy storage produced by dispersing oxide nanoparticles in nitrate salt mixture directly at high temperature, *Sol. Energy Mater Sol. Cells.*, Vol. 167, 60–69, 2017.

[151] M. Chieruzzi, G.F. Cerritelli, A. Miliozzi, J.M. Kenny, Effect of nanoparticles on heat capacity of nanofluids based on molten salts as PCM for thermal energy storage, *Nanoscale Res. Lett.*, Vol. 8, 448,2013.

[152] E.B. Fox, A.E. Visser, N.J. Bridges, J.W. Amoroso, Thermophysical properties of nanoparticle-enhanced ionic liquids (NEILs) heat-transfer fluids, *Energy Fuel*, Vol. 27, 3385–3393, 2013.

[153] W. Chen, C. Zou, X. Li, An investigation into the thermophysical and optical properties of SiC/ionic liquid nanofluid for direct absorption solar collector, *Sol. Energy Mater Sol. Cells.*, Vol. 163, 157–163, 2017.

[154] F. Wang, L. Han, Z. Zhang, X. Fang, J. Shi, W. Ma, Surfactant-free ionic liquid-based nanofluids with remarkable thermal conductivity enhancement at very low loading of graphene, *Nanoscale Res. Lett.*, Vol. 7, 314, 2012.

Chapter 9

Performance evaluation methods for different heat transfer techniques

9.1 INTRODUCTION

Passive enhanced heat transfer technologies attempt to improve thermal efficiency but the flow resistance is deterioration, simultaneously. An effective performance evaluation of these techniques is needed to assess the increase in the ratios of the heat transfer, pressure drop, and a comprehensive index. Many evaluation systems are developed and used in the published literature, and different kinds of evaluation criteria have their own purpose and physical meaning.

The convective heat transfer coefficient (h) and pressure drop (ΔP) are the basic indices commonly used for single-phase heat transfer [1], and there are many derived indices based on them, that is, Nusselt number (Nu), j factor, and friction factor (f). These relative evaluation parameters are introduced in many heat transfer textbooks [1–3], and collectively they are here referred to as the common method. The basic physical meaning of h and ΔP lead to some limitations. These two indices represent the heat transfer between the wall surface and fluid flow, and pressure differences between two cross-sections, respectively. For complex vortex flow fields, these evaluation methods fail to describe the thermal and dynamic transfer caused by the big vortices [4].

To increase the understanding of augmented heat transfer structures, the evaluation of the entropy generation is employed to calculate the energy conversion efficiency for both heat transfer and flow dissipation integration. This led to entransy, a new thermodynamic concept, based on an analogy between heat conduction and charge transport. The multi-objective optimization method has been introduced in the heat transfer research field during the last decade. This approach with multiple conflicting targets has allowed enhancement of the heat transfer and improvement of heat exchanger design.

The many evaluation techniques on single-phase convective heat transfer introduced above have been prevalent in the literature from 1990 to 2020

DOI: 10.1201/9781003229865-9

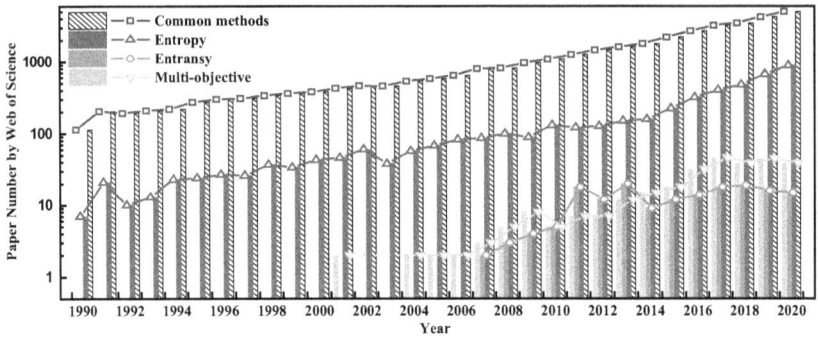

Figure 9.1 Statistics on published work for different evaluation techniques on single-phase convective heat transfer (1990–2020).

with the statistics from the web of science on publications shown in Figure 9.1. The common methods are the most prevalent, an order of magnitude greater than the entropy generation method, and two orders of magnitude greater than the entransy method and multi-objective method. It is foreseeable that the common methods still will be the most used evaluation techniques, and the entropy generation method presents a good increasing tendency close to the common methods, over decades. The growth rate of entransy and multi-objective methods tends to be gentle, which may need a breakthrough on existing research limitations.

9.2 PERFORMANCE ASSESSMENT BASED ON THE FIRST LAW OF THERMODYNAMICS

The most widely used evaluation techniques are based on the first law of thermodynamics: convective heat transfer coefficient (h), and pressure drop (ΔP), Eqs. (9.1) and (9.2), [1], respectively.

$$h = \frac{Q}{A \cdot \Delta T} = \frac{q}{\Delta T} \tag{9.1}$$

$$\Delta P = P_{in} - P_{out} \tag{9.2}$$

where Q is total heat flow (W), A is heat transfer area (m²), ΔT is the temperature difference between the wall and fluid (K), q is average heat flux (W/m²), P_{in} is inlet pressure, and P_{out} is outlet pressure (Pa).

From h and ΔP equations, the dimensionless Nusselt number (Nu), Darcy and Fanning friction factors (f_{Darcy} and $f_{Fanning}$) are developed. The Nu indicates the intensity of convective heat transfer as the ratio of the convective heat transfer and the thermal conductivity of the fluid layer; see Eq. (9.3). The physical meaning of the friction factor is the ratio of viscous force to inertial force; see Eq. (9.4).

$$Nu = \frac{h \cdot De}{\lambda} \tag{9.3}$$

$$f_{Darcy} = 4f_{Fanning} = \frac{2\tau}{\rho \cdot u^2} = \frac{2\Delta P \cdot De}{\rho \cdot u^2 \cdot L} \tag{9.4}$$

where De is the hydraulic diameter (m), τ is local shear stress (kg/m/s^2), ρ is fluid density (kg/m^3), u is velocity (m/s), and L is length of heat transfer channel (m).

The Stanton number (St) and Colburn j factor are dimensionless indices that represent the heat transfer performance. The St is a ratio of convective heat transfer and the thermal capacity of fluid. The Stanton number is proposed according to the geometric similarity of the momentum and thermal boundary layers. It is commonly employed to assess the ratio between the shear stress (due to flow viscous drag) and the total heat transfer on the wall (due to thermal diffusivity); see Eq. (9.5). To eliminate the assumption of Pr = 1, Colburn [5] suggested that the St number should be multiplied by Pr$^{2/3}$ for pipe or tube flow and then it is called the Colburn j factor, see Eq. (9.6).

$$St = \frac{Nu}{Re \cdot Pr} = \frac{h}{\rho \cdot u \cdot C_p} \tag{9.5}$$

$$j = St \cdot Pr^{2/3} = \frac{Nu}{Re \cdot Pr^{1/3}} \tag{9.6}$$

The abovementioned indices are commonly used to assess the convective thermal and dynamic characteristics. The increase in the performance due to technical enhancements are considered as a ratio of the performance of a modified enhanced surface compared with that of the corresponding smooth surface, where "e" and "s" indicate the enhanced surface and s smooth surface, respectively. Generally, Nu_e/Nu_s, j_e/j_s, and St_e/St_s are used to represent the enhanced thermal performance; $\Delta P_e/\Delta P_s$ and f_e/f_s are often used to represent the change in flow resistance [6]. The Nu_e/Nu_s and f_e/f_s always increase simultaneously. A performance evaluation criterion (PEC) [6] was developed to comprehensively assess the increase in the thermal and dynamic performance, as shown in Eqs. (9.7) and (9.8). If the PEC is above 1, it implies an augment in the heat transfer performance greater than the increase in the flow resistance. This suggests that the enhanced structure provides a benefit while a value below 1 means the enhanced structure presents a disadvantage

of the two PEC formulas, the Eq. (9.8) has been most widely used in recent decades [7–9].

$$PEC = \frac{Nu_e/Nu_s}{f_e/f_s} = \frac{j_e/j_s}{f_e/f_s} \tag{9.7}$$

$$PEC = \frac{Nu_e/Nu_s}{\left(f_e/f_s\right)^{1/3}} = \frac{j_e/j_s}{\left(f_e/f_s\right)^{1/3}} \tag{9.8}$$

More details of the above indices can be found in heat transfer textbooks. Fan et al. [10] developed a new method using these indices in a different way. The Nu_e/Nu_s and f_e/f_s, respectively, are used as two coordinates giving a space that can be divided in to four regions using three characteristic lines, as shown in Figure 9.2. Region 1 is represented as a little heat transfer augment but severely flow resistance deterioration, where the increase of heat transfer rate is less than the increase of pumping power. Region 2 describes the heat transfer augment under the fixed pumping power, where the enhanced surface results in a larger heat transfer coefficient compared with the reference value under the fixed pumping power consumption. In Region 3, augmented heat transfer performance is reached under the same flow resistance, where the enhanced surface reach to a larger heat transfer coefficient compared with the reference value under the same flow resistance. Region 4 is the most advantage region, where the heat transfer enhancement rate is obviously larger than the increase ratio of the friction factor under the fixed

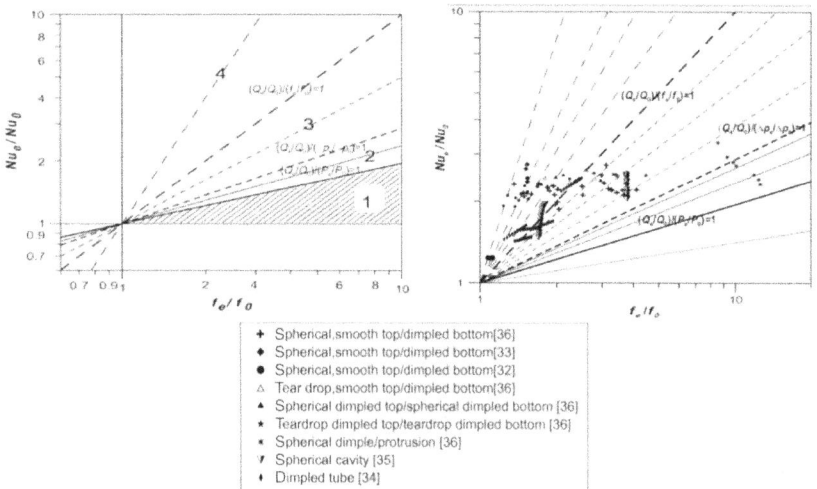

+   Spherical,smooth top/dimpled bottom[36]
♦   Spherical,smooth top/dimpled bottom[33]
●   Spherical,smooth top/dimpled bottom[32]
△   Tear drop,smooth top/dimpled bottom[36]
▲   Spherical dimpled top/spherical dimpled bottom [36]
★   Teardrop dimpled top/teardrop dimpled bottom [36]
✕   Spherical dimple/protrusion [36]
✦   Spherical cavity [35]
♦   Dimpled tube [34]

Figure 9.2 A performance assessment partition between the heat transfer augment and energy-saving [10].

flow rate. The three characteristic lines are $(Q_e/Q_s)/(P_e/P_s) = 1$, $(Q_e/Q_s)/(\Delta P_e/\Delta P_s) = 1$, and $(Q_e/Q_s)/(f_e/f_s) = 1$ forming regions 1 to 4.

Thermal-hydraulic performances of different dimpled surfaces are summarized in Figure 9.2 from references [11–15]. The data is predominately placed in regions 3 and 4, which represent that most dimpled structure surfaces can obviously increase the heat transfer with a preferable saving energy performance. The surfaces with spherical dimples and protrusion on opposite surfaces [15] present poorer performance than other dimpled surfaces, implying this enhanced structure is the least energy-saving.

The above evaluation criteria assess all the performance measures from a whole heat transfer element but fail to reflect the active enhanced heat transfer mechanisms. Guo et al. [16] developed a new evaluation method that calculates the intersection angle (θ) between the velocity vector $(\vec{U})$ and the temperature gradient (∇T) by convective term; see Eq. (9.9).

$$u\frac{\partial T}{\partial x} + v\frac{\partial T}{\partial y} = \vec{U}\cdot\nabla T = |\vec{U}||\nabla T|\cos\theta \qquad (9.9)$$

This method was named the field synergy principle (FSP), with θ as the local field synergy angle. The closer θ is to 0° or 180°, the larger is the absolute value of $\cos\theta$ indicating a larger heat flux, and better enhancement of heat transfer. If the $\cos\theta$ is a positive value, it means the fluid is heated; otherwise, the fluid is cooled. An optimal design of a heat transfer structure will have a small intersection angle (θ) between the velocity vector and the temperature gradient for both heating and cooling processes. To express the synergy of the system, the field synergy number (Fc) is calculated as in Eq. (9.10).

$$Fc = \int_0^1 \left(\bar{U}\cdot\nabla\bar{T}\right)d\bar{y} = \frac{Nu}{Re\cdot Pr} \qquad (9.10)$$

A large Fc implies a better heat transfer enhancement ability [17]. The FSP is commonly used as an effective index to explore enhanced heat transfer mechanism and optimization of the structure. The secondary flow and heat transfer augment mechanism in double-layered microchannels with cavities and ribs were analyzed using the FSP [18]. The intensity of secondary flow (Se), Nusselt number (Nu), field synergy angle (β), and field synergy number (Fc) are shown in Figure 9.3. The results show that the Se of channel C is much more significant than for the other two channels. The Nu for channel C is also the highest, but only a little higher than channel B. The β in channel C is smaller than that of the other two channels, indicating that the decrease of βc causes a better synergy relationship between temperature and velocity distributions. The β is unaffected by increasing Re. The absolute values of Fc and Nu of channel C are slightly higher than those of channel B, but

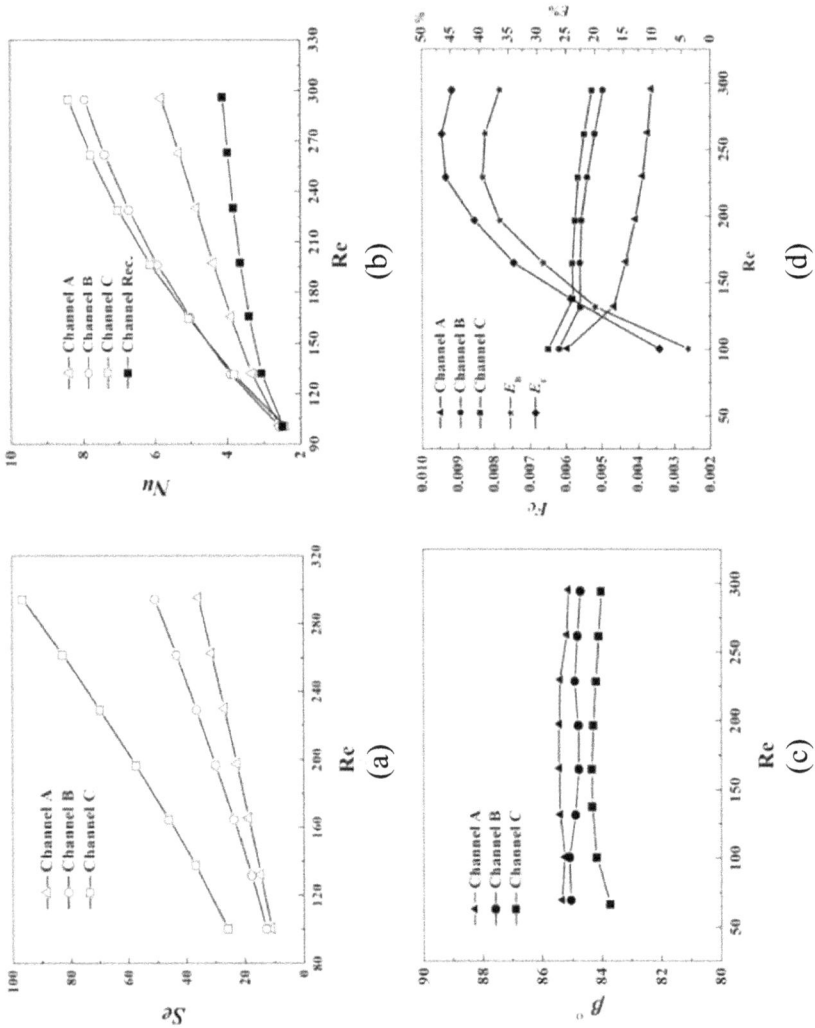

Figure 9.3 Performance for three kinds of double-layered microchannels at laminar flow (Re = 50–300) [18]. (a) Intensity of secondary flow (Se); (b) Nusselt number (Nu); (c) Average field synergy angle (β); and (d) Field synergy number (Fc).

significantly higher than those of channel A. The Fc gradually declines with the increase of Re, opposite to the relationship for Nu. A relative Index of the Fc is also defined in this work, as $Ei \% = (Fc_i - Fc_A)/Fc_A * 100\%$.

The FSP has been used to investigate many complex flow and heat transfer augment mechanisms, for example, laminar flow and heat transfer in helical rectangular tubes [19], curved square channels [20], wavy finned elliptic tubes [21], compressible fluid flow in variable-cross-section pipeline [22], conjugate heat transfer in electronic cooling [23], forced convection in porous medium [24], flow and heat transfer in porous wick of capillary pumped loop evaporator [25], etc.

9.3 PERFORMANCE ASSESSMENT BASED ON SECOND LAW OF THERMODYNAMICS

Systems can be evaluated by the second law of thermodynamics and the energy conversion efficiency of the heat exchanger design. This allows the reduction of energy dissipation in the process of thermal convection, including thermal irreversibility and viscosity irreversibility for single convective heat transfer. The entropy and exergy are both thermodynamic criteria widely used to evaluate thermal systems [26–28].

McClintock [29] first reported the results of heat exchanger design by minimum irreversibility, in 1951, leading to the entropy generation minimization (EGM) method. It received gaining attention and worth deeply exploring in convection heat transfer field [30]. The convective heat transfer method was fundamentally studied and a local entropy production theoretical model was developed by Bejan in 1979 [31]. The Bejan entropy generation model has been employed when designing various kinds of heat exchangers and enhanced heat transfer surfaces.

Guo et al. [32] proposed a new thermodynamic criteria called entransy, in 1998, to represent the heat transfer characteristic of a single-unit or a system in a period according to the definition of heat conduction and charge transport. The entransy dissipation rate was employed to compute the irreversibility of heat transfer. The relationship between the extremum value of the entransy dissipation rate and the maximum convective heat transfer index was defined as the entransy dissipation extremum (EDE) principle using to optimize the convective heat transfer processes [33]. This principle coupled with the other optimization methods has been employed to optimize various kinds of convective heat transfer and produce some new type of enhancement methods [34].

The above two thermodynamics-based methods are most commonly employed in heat exchanger design approaches [35], and are introduced below in detail.

9.3.1 Entropy production (generation) theory

A two-dimensional model for local entropy generation for incompressible flow without internal heat production has been developed as shown in Eqs. (9.11)–(9.15). The total irreversibility includes two independent parts: (a) conduction in the presence of nonzero temperature gradient called heat transfer entropy generation and (b) statistical for viscous dissipation of mechanical power in the fluid flow called friction entropy generation.

$$\dot{S}_{tot}''' = \dot{S}_{ht}''' + \dot{S}_{fr}''' \tag{9.11}$$

$$\dot{S}_{ht}''' = \frac{\lambda}{T_f^2}\left[\left(\frac{\partial T}{\partial x}\right)^2 + \left(\frac{\partial T}{\partial y}\right)^2\right] \tag{9.12}$$

$$\dot{S}_{fr}''' = \frac{\mu}{T_f}\left\{2\left[\left(\frac{\partial u}{\partial x}\right)^2 + \left(\frac{\partial v}{\partial y}\right)^2\right] + \left(\frac{\partial u}{\partial y} + \frac{\partial v}{\partial x}\right)^2\right\} \tag{9.13}$$

$$S_{tot} = \int \dot{S}_{tot}''' \, dV, \quad S_{ht} = \int \dot{S}_{ht}''' \, dV, \quad S_{fr} = \int \dot{S}_{fr}''' \, dV \tag{9.14}$$

$$Be_{local} = \dot{S}_{ht}'''/\dot{S}_{tot}''', \quad Be_{ave} = S_{ht}/S_{tot} \tag{9.15}$$

where λ represents thermal conductivity, μ represents dynamic viscosity, and T_f represents the mean temperature of the fluid. The S and S' represent the total and local entropy generation rate. The foot mark tot, ht, fr, local, and ave, respectively, represent total, heat transfer, friction, local, and average. The Bejan number (Be) is calculated by the ratio of the heat transfer and total entropy generation rates. The Be > 0.5 represents that the heat transfer irreversibility is superior; otherwise the flow friction irreversibility is superior.

An entropy generation number (N) is proposed as a ratio of entropy generation rate of corrugated tube (ct) to that of smooth tube (st), as shown in Eq. (9.16) (N_{s-ht}, N_{s-fr}, and N_{s-tot}). If $N_s < 1$, it represents that their reversibility of the corrugated tube is reduced, and implies that the utilizable energy is improved compared with the smooth tube.

$$N_{s-ht} = \frac{S_{ht-ct}}{S_{ht-st}}, \quad N_{s-fr} = \frac{S_{fr-ct}}{S_{fr-st}}, \quad N_{s-tot} = \frac{S_{tot-ct}}{S_{tot-st}} \tag{9.16}$$

An alternate semi-empirical entropy generation model has been proposed as shown in Eq. (9.17) [31]. The first expression computes the heat transfer entropy generation rate, and the second expression computes the flow viscosity (friction) entropy generation rate:

$$S_{g,tot} = \underbrace{\frac{q'^2}{\pi\lambda T_f^2 Nu}}_{\text{heat transfer}} + \underbrace{\frac{8\dot{m}^3 f}{\pi^2 \rho^2 T_f D^5}}_{\text{flow viscosity}} \tag{9.17}$$

where q' represents the average heat flux per length, m represents the mass flow rate, D is hydraulic diameter, and Nu and f are, respectively, Nusselt number and Darcy friction factor calculated by Gnielinski (9.18) and Filonenko (9.19) empirical correlations [1]:

$$Nu = \frac{(f/8)(Re-1000)Pr}{1.07 + 12.7(f/8)^{1/2}(Pr^{2/3}-1)} \tag{9.18}$$

$$f = (1.82 \log Re - 1.64)^{-2} \tag{9.19}$$

A comparison between the local entropy generation model and semi-empirical entropy generation model for turbulent convective heat transfer in a smooth tube was shown by Wang et al., and reproduced in Figure 9.4 [4]. The results present good agreement for the two entropy generation expressions, and the errors are within ±5%.

Analysis of the entropy generation in the turbulent convective heat transfer mechanism for outward transverse and helically corrugated tubes (TCT

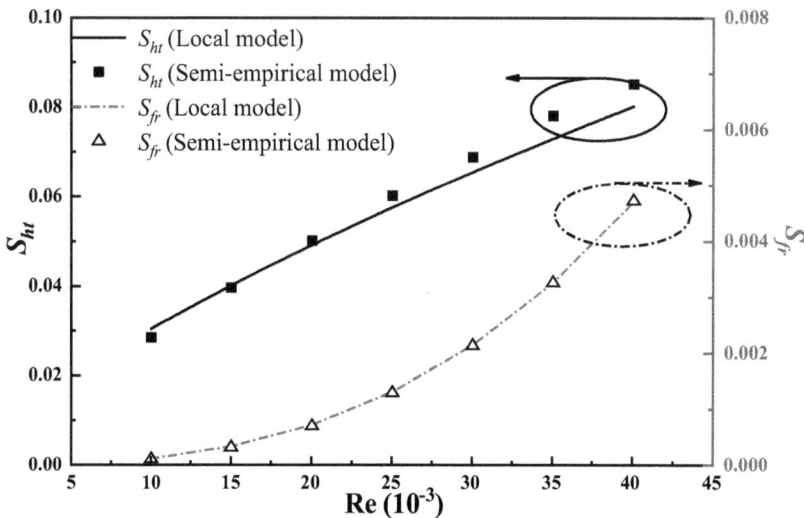

Figure 9.4 Validation of numerical results with Bejan's formula. (Reproduced from the study by Wang et al. [4].)

and HCT) with different corrugation height (hl/D = 0.05, 0.10, and 0.15) was conducted using the local model [4]. The distributions of local heat transfer entropy generation for the four cases at Re = 18,800 are shown in Figure 9.5. In the figures, AP, AP′, SP, and RP mean the beginning of adverse pressure gradient region point, ending of adverse pressure gradient region point, secondary flow separation point, and reattachment point, respectively. The heat transfer entropy generation rate develops at the boundary layer region where the temperature gradient is obvious. Near the separation point (SP), the heat transfer irreversibility decreases due to the fluid that is almost still standing in the main flow direction. The heat transfer irreversibility distribution in the secondary flow region is greater than in the core flow region because the mass transfer by the circulating flow has improved the thermal irreversibility. The heat transfer entropy generation decreases with increasing Hl/D. In the HCT case, there is a slight reduction compared with the TCT because of the rotational flow. This suggests that the spiral flow inhibits the heat transfer entropy generation rate.

Figure 9.6 presents the local friction entropy generation distributions. The viscous dissipation entropy generation is obviously in the snuggle region near the detached vortex and in the boundary layers. The variation in friction irreversibility with the structure characteristics is like that for the heat transfer irreversibility. The significant heat transfer and friction entropy irreversibility are predominantly caused by the detached flow and weakened by the spirally flow.

The distribution of local Be (Be_{local}) is presented in Figure 9.7. It is illustrated that the heat transfer dominant zone ($Be_{local} > 0.5$) is in the sublayer

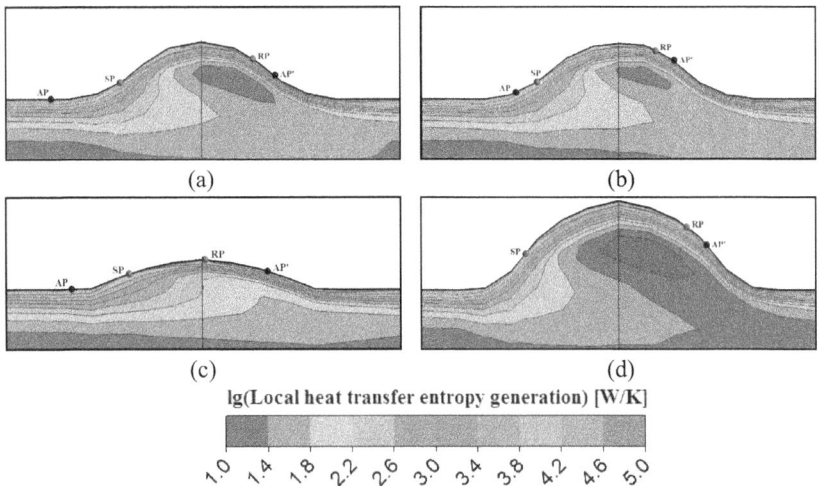

Figure 9.5 Distributions of the local heat transfer entropy generation for TCT and HCTs at Re = 19,000 [4]. (a) TCT with Hl/D=0.10; (b) HCT with Hl/D=0.10; (c) HCT with Hl/D=0.05; and (d) HCT with Hl/D=0.15.

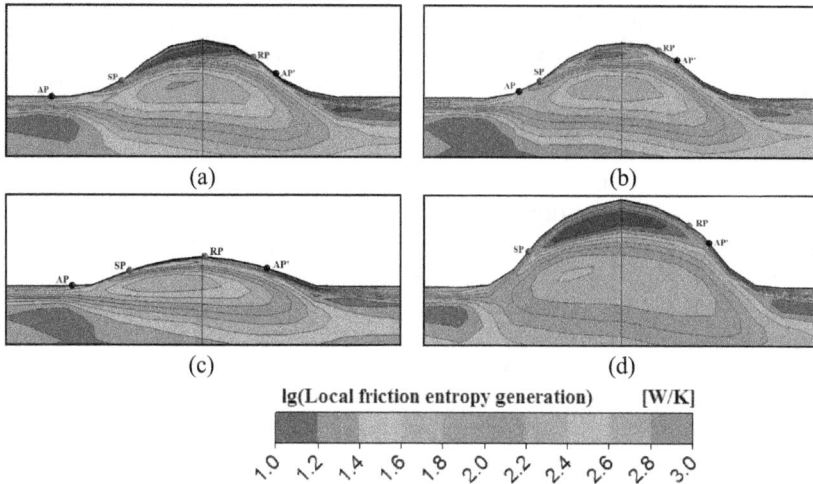

Figure 9.6 Distributions of the local friction entropy generation rate for TCT and HCTs at Re = 19,000 [4]. (a) TCT with Hl/D=0.10; (b) HCT with Hl/D=0.10; (c) HCT with Hl/D=0.05; and (d) HCT with Hl/D=0.15.

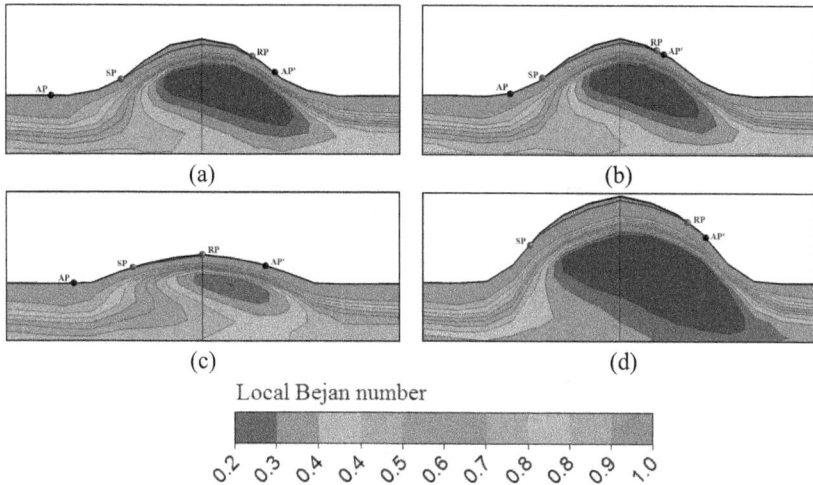

Figure 9.7 Distributions of the local Be for TCT and HCTs at Re = 19,000 [4]. (a) TCT with Hl/D=0.10; (b) HCT with Hl/D=0.10; (c) HCT with Hl/D=0.05; and (d) HCT with Hl/D=0.15.

and buffer layer. The ineffective zone (Be_{local} < 0.2) is in the severely turbulence pulsation zone. In the HCT case, with Hl/D = 0.15, the ineffective zone dominates adversely affecting the heat transfer and energy utilization. There is no obvious difference between the TCT and HCT cases with Hl/D = 0.10, except that the ineffective zone for HCT is smaller than for TCT.

These results show that the local entropy generation describes the global irreversibility of the heat transfer and viscous dissipation that is beneficial for obtaining an enhanced heat transfer mechanism for the complex vortex flow fields. The variations in N_{s-ht}, N_{s-fr}, N_{s-tot}, and Be_{ave} with Re for different corrugation height and pitch ratios are presented in Figure 9.8. The entropy generation rate increases with increased roughness. The heat transfer and friction entropy generation of helical corrugated tube is worse than the transverse corrugated tube due to the secondary flow. The N_{s-tot} is rapidly promoted (Re < 12,000), then it reaches a fixed value (12,000 < Re < 31,000), and finally increases again (Re > 31,000) except for HCT-3, as shown in Figure 9.8©. The N_{s-tot} in the HCT-2 case is below 1 at very small Re indicating that the efficiency of the energy utilization surpasses that of smooth tube with poor heat transfer index at small Re [36]. Thus, it is better to utilize these systems in a range of Re where the N_{s-tot} is constant.

The Be_{ave} presents an exponential decline with increases in Re. At Re < 12,000, the S_{fr} is very small as compared with the S_{ht}, and it is practically negligible. With an increase in the Re, the S_{fr} exponentially promotes and

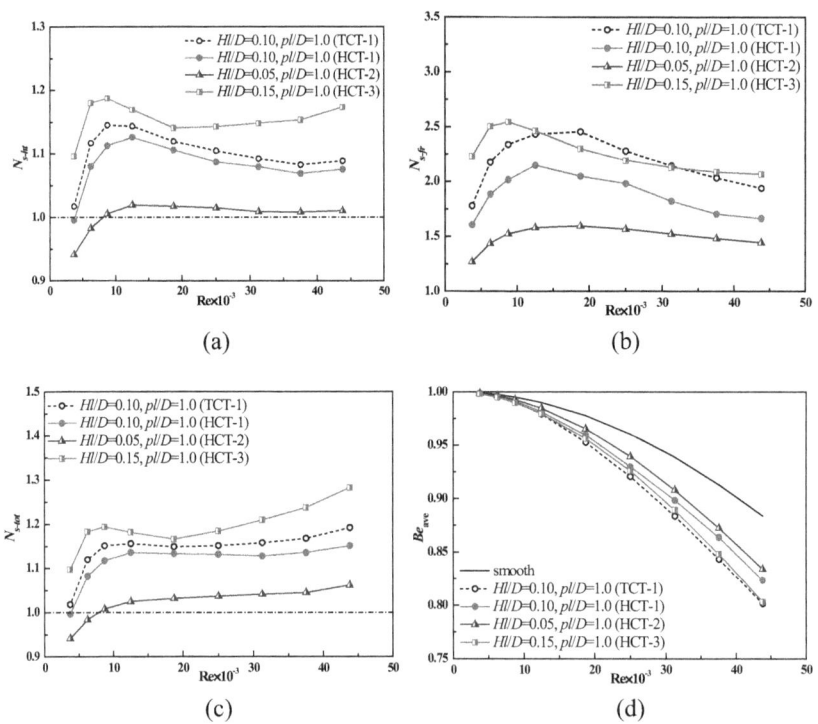

Figure 9.8 Average of N_{s-ht}, N_{s-fr}, N_{s-tot}, and Be_{ave} relative to Re for corrugated tubes. (a) Average of N_{s-ht} versus Re; (b) Average of N_{s-fr} versus Re; (c) Average of N_{s-tot} versus Re; and (d) Average of Beave versus Re.

increasingly contributes to the total entropy generation. The decrease of Be_{ave} is obviously fast for larger and densely corrugations, with a smaller decrease in the rate for the smooth tube. A comprehensive assessment of $N_{s\text{-tot}}$ and Be_{ave} presents that the optimal Re value is below 31,000 and below 18,800 for HCT-3. The results agree with previous studies, showing that HCT-1 is comprehensively better than the TCT-1. It is because the generated rotational flow has little effect on heat transfer, but inhibits the pressure obviously [37].

The entropy generation evaluation methods are used in a number of enhanced heat transfer studies [31]: the heat transfer irreversibility resistance minimization is used for the optimal design a two-steam heat exchanger [38, 39]; entropy generation analysis of an intermediate heat exchanger in a nuclear energy system [40]; entropy optimization of the magnetic hydrodynamics of a Williamson nanofluid on a vertical plate surface with nonlinear thermal radiation [41]; entropy generation investigating the heat transfer of nanoparticles employed in an innovative turbulator [42]; entropy generation study on a nanofluid in a cavity with magnetic field and thermal radiation [43]; entropy generation analysis of the thermal and dynamics in backward-facing steps with different expansion ratios [44]; investigation of convection heat transfer in shell and helically coiled tube heat exchangers by minimizing entropy generation theory [45]; investigation of nanofluid entropy generation rate in a heat exchanger with helical twisted tapes [46]; and investigation of entropy generation rate of supercritical water flow in a hexagon rod bundle [36]. The amount of published research shows that the entropy generation evaluation method is widely employed on various enhanced heat transfer technologies for analysis. It is beneficial for understanding the global heat transfer and flow viscous process and evaluate them under the unified metric. However, it still is not frequently used on industrial heat exchanger design due to the unclear relationship with the traditional evaluation system (h and ΔP), and opposite tendency of heat transfer entropy production under the different thermal boundary conditions (fixed temperature and heat flux) [35].

9.3.2 Entransy theory

The heat transport potential capacity developed by Guo et al. [32] gives an analogy to electrical conduction, fluid flow, and heat conduction, called entransy, with the differential form expressed as follows:

$$dG = Mc_v T dT \tag{9.20}$$

where $Mc_v T$ represents the thermal energy stored, c_v represents the constant volume-specific heat, and M represents the mass flow rate. This expression is only suitable for fixed volume system. The integral form of entransy as follows, The entransy is considered as a thermal capacitor that stores thermal energy, and is parallel to an electrical capacitor:

$$G = \frac{1}{2} M c_v T^2, \quad g = \frac{1}{2} \rho c_v T^2 \tag{9.21}$$

An EDE principle was derived for incompressible convective heat transfer processes. This shows that the EDE leads to the best heat transfer characteristic with the maximum convective heat transfer index. In an analog to the electrical resistance, the entransy dissipation-based thermal resistance (R_g) for a thermal convection is calculated as the ratio of the entransy dissipation rate (Φ_g) to the square of the heat transfer rate (Q) [47]:

$$R_g = \frac{\Phi_g}{Q^2} \tag{9.22}$$

Alternatively, R_g can be calculated as the ratio of the square of the average temperature difference (ΔT) to the entransy dissipation rate (Φ_g):

$$R_g = \frac{(\Delta T)^2}{\Phi_g} \tag{9.23}$$

These two expressions, Eqs. (9.22) and (9.23), show that minimizing the R_g is represented to minimize the Φ_g for a specific heat flux or a specific temperature difference. The EDE principle (EDEP) for thermal convection is equivalent to the minimum entransy dissipation-based thermal resistance principle (minimum R_g) results in a maximum heat transfer index for these two types of boundary conditions. The EDEP can be employed on three kinds of heat transfer methods (heat conduction, heat convection, and thermal radiation) [48–50], and is also suitable for heat exchanger optimization [51].

Five examples were compared by investigating the Nu, FSP, and EDEP [52]: turbulent flow in H-type finned tubes (Example 1); laminar flow in finned tubes with and without vortex generators (Example 2); laminar flow in five-row finned tubes (Example 3); turbulent flow in dimpled tubes (Example 4); and turbulent air flow in composite porous structure (Example 5). The entransy dissipation maximum optimal principle is used for the fixed wall temperature boundary, and the entransy dissipation minimum optimal principle is used for fixed heat flux boundary, as shown in Figure 9.9.

Two examples concerning the optimization of thermal convection in a 2D cavity (Example 1) and in a straight circular tube (Example 2) are studied comparatively in reference [35]. The optimization results obtained by the two methods present obvious differences, as seen in related works [53, 54]. The two kinds of evaluation method using thermodynamics cause some controversies not being detailed here [55–57]. The entropy generation method is still dominantly used on convective heat transfer analysis at present, as Figure 9.1 showed.

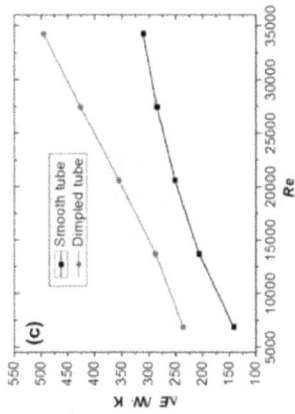

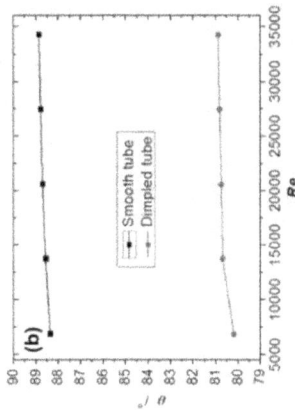

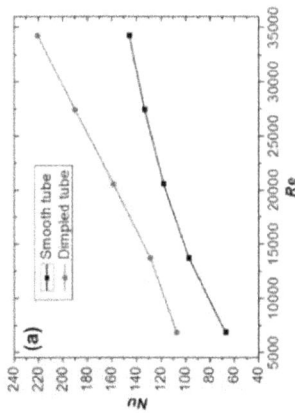

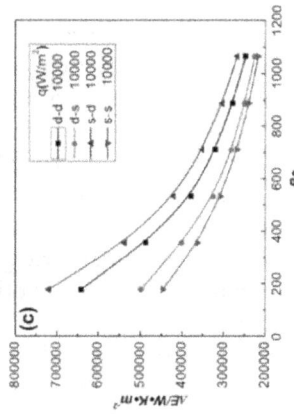

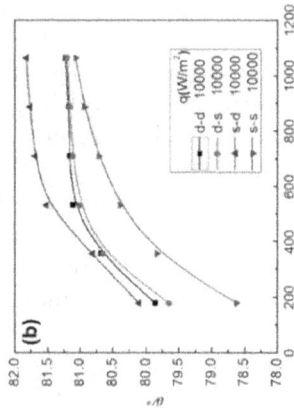

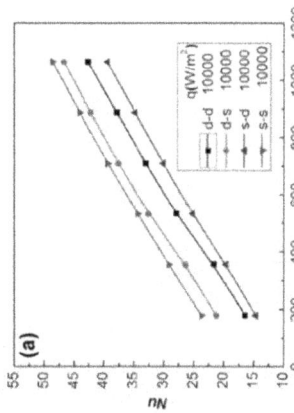

Figure 9.9 Comparisons of Nu, FSP, and EDEP for Examples 4 and 5 from reference [52]. (a) Comparisons of Nu, FSP and EDEP for Example 4 with constant temperature; and (b) Comparisons of Nu, FSP and EDEP for Example 5 with constant heat flux boundary.

9.4 MULTI-OBJECTIVE OPTIMIZATION AND EVALUATION

A multi-objective optimal and evaluation method was recently introduced to heat exchanger design from the operational research field. This method is employed to solve the problem of multiple conflicting targets converting them to a single target evaluation. It has been used for various kinds of heat exchanger (HE) designs: shell-and-tube HE [58–60]; plate fine HE [61]; spiral wound HE [62]; helical baffle HEs [63]; helical coils HE [64]; printed circuit HE [65, 66]; tube HE with porous media [67]; plane-shaped HE in an automotive exhaust thermoelectric generator [68]; HE in an irreversible Brayton cycle system [69]; and many others.

The process of multi-objective optimization technique includes two steps: regression model establishment and multi-objective solution optimization. The first part can be performed by the response surface method (RSM) [70], which includes: design of experiment (DOE), regression model fitting, regression, and variance analyses. The second part, the multi-objective problem-solving, can be performed by many methods: genetic algorithm (GA) [71], simulated annealing algorithm [72], ant colony algorithm [73], artificial neural network (ANN) [74, 75], combing ANN and GA methods [76], combining ANN and metaheuristic algorithms [77], etc. The nondominated sorting genetic algorithm II (NSGA-II) is commonly employed for heat exchanger optimal design. Figure 9.10 presents a multi-objective optimization flowchart of a helically corrugated tube HE [78].

This optimization sets three targets and three factors. The three targets are Nu, the Poiseuille number (f_{Re}), and the overall heat transfer performance (PEC).

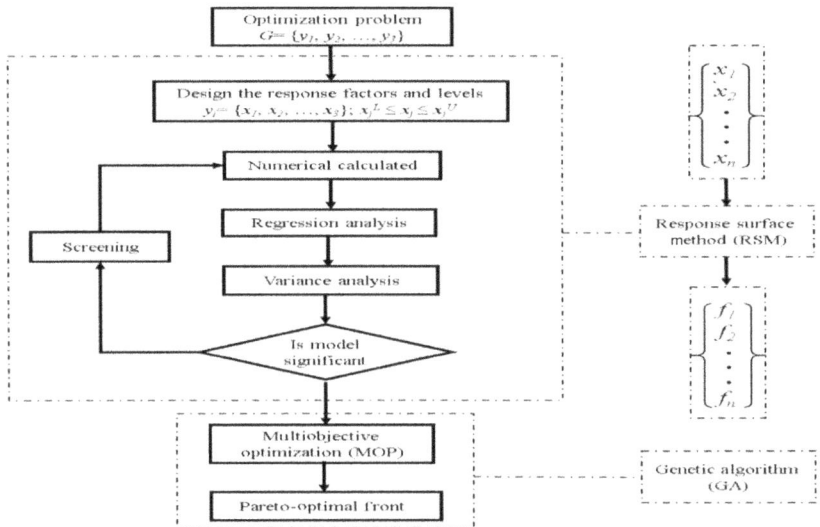

Figure 9.10 Flowchart for multi-objective optimization of corrugated tube HE.

The three factors are the Re, dimensionless corrugation pitch (pl/D), and dimensionless corrugation height (Hl/D). The target is to maximize Nu and PEC while minimizing f_{Re}. The factors of Re, pl/D, and Hl/D vary from 6,260–43,820, 0.5–2.5, and 0.05–0.15, respectively:

$$\begin{cases} \text{Maximize: } \text{Nu} = f_1(\text{Re}, Hl/D, pl/D) \\ \text{Minimize: } f_{Re} = f_2(\text{Re}, Hl/D, pl/D) \\ \text{Maximize: } PEC = f_3(\text{Re}, Hl/D, pl/D) \\ \text{Subject to: } \begin{smallmatrix} 6260 \le \text{Re} \le 43820 \\ 0.5 \le pl/D \le 2.5 \\ 0.05 \le Hl/D \le 0.15 \end{smallmatrix} \end{cases}$$

A central composite design (S-CCD) was employed for experimental design instead of full factor design (FFD), and this significantly reduces the number of tests [79]. For three levels of FFD, the number of tests needed is 3^k, where k is the number of factors. This means that the number of tests is 9, 27, 81, and 243, corresponding to 2, 3, 4, and 5 factors, respectively. The CCD method can reduce the number of tests significantly. For 3, 4, and 5 factors using CCD, only 15, 25, and 27 tests are required, respectively. The two-factor CCD and three-factor CCD are shown in Figure 9.11. Three factors with five levels of CCD were employed for the experimental design of the helically corrugated tube, requiring 29 experiments, compared with 243 for full factor experimental designs [80]. The five levels of factors, CCD experimental design, and corresponding numerical solutions are shown in Tables 9.1 and 9.2, respectively.

The second-order polynomials expression is employed to fit the response surface regression model, as in Eq. (9.24).

$$Y_i = b_0 + b_1(\text{Re}) + b_2(pl/D) + b_3(Hl/D) + b_{11}(\text{Re})^2 + b_{22}(pl/D)^2 + b_{33}(Hl/D)^2$$
$$b_{33}(Hl/D)^2 + b_{12}(\text{Re})(pl/D) + b_{13}(\text{Re})(Hl/D) + b_{23}(pl/D)(Hl/D)$$

$$(9.24)$$

The parameters for the regression model are decided depending on the analysis of variance (ANOVA). The nonsignificant and interaction terms can

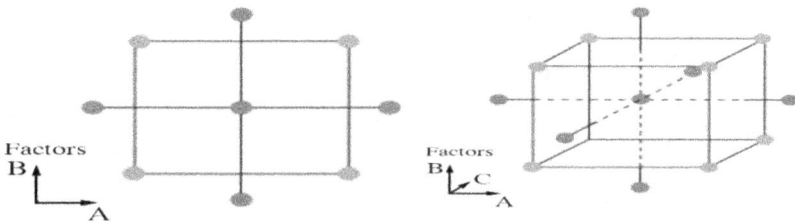

Figure 9.11 Surface central composite design for two and three factor with three levels.

Table 9.1 Experimental design by S-CCD

		Level				
No.	Factors	$-\alpha$	-1	0	$+1$	$+\alpha$
I	Re	6,260	12,520	25,040	37,560	43,820
2	pl/D	0.5	1.0	1.5	2.0	2.5
3	Hl/D	0.05	0.075	0.1	0.125	0.15

Table 9.2 RSM design and corresponding values

No.	Parameters			Response values		
	Re	pl/D	Hl/D	Nu	f_{Re}	PEC
I	25,040	1.5	0.1	103.172	1347.266	1.117
2	37,560	2	0.125	125.578	1711.300	1.010
3	12,520	I	0.075	65.963	902.710	1.217
4	12,520	2	0.075	53.426	681.602	1.083
5	37,560	I	0.075	133.075	2072.850	0.988
6	37,560	2	0.075	120.752	1297.447	1.048
7	37,560	I	0.125	139.893	2768.296	0.943
8	12,520	I	0.125	71.586	1211.971	1.197
9	12,520	2	0.125	62.786	853.929	1.180
10	25,040	2	0.1	96.6 12	1198.158	1.088
11	25,040	I	0.1	108.107	1757.657	1.072
12	25,040	1.5	0.075	97.482	1172.285	1.106
13	12,520	1.5	0.1	64.555	824.712	1.227
14	25,040	1.5	0.125	105.936	1546.215	1.096
15	37,560	1.5	0.1	131837	1776.134	1.031
16	43,820	2.5	0.15	138.618	1813.228	0.991
17	6260	0.5	0.05	43.530	720.914	1.260
18	6260	2.5	0.05	152.820	384.050	1.011
19	43,820	0.5	0.05	28.334	3079.206	0.916
20	43,820	2.5	0.05	123.645	1404.688	0.962
21	43,820	0.5	0.15	168.006	4985.042	0.857
22	6260	0.5	0.15	49.595	1507.588	1.122
23	6260	2.5	0.15	37.769	524.519	1.215
24	25,040	2.5	0.1	90.920	1064.651	1.065
25	25,040	0.5	0.1	114.521	3050.184	0.945
26	25,040	1.5	0.05	92.361	941.015	1.127
27	6260	1.5	0.1	39.276	498.106	1.285
28	25,040	1.5	0.15	105.956	1706.806	1.061
29	43,820	1.5	0.1	144.733	1971.138	1.006

Table 9.3 Analysis of variance for Nu

Source	DF	Seq SS	Adj SS	Adj MS	F	P	AF	AP
Regression	9	41,755	41,755	4,639	1,538.64	0.000	1,516.14	0.000
Linear	3	41,211	2,213	737	244.64	0.000	287.34	0.000
(Re)	1	39,205	2,007	2,007	665.76	0.000	624.32	0.000
(pl/D)	1	1,521	43	43	14.26	0.001	20.02	0.000
(Hl/D)	1	484	17	17	5.76	0.027	8.14	0.000
Square	3	399	399	133	44.11	0.000	91.91	0.010
(Re)^2	1	385	254	254	84.30	0.000	91.91	0.000
(pl/D)^2	1	4	10	10	3.42	0.080	—	—
(Hl/D)^2	1	8	8	8	2.80	0.110	—	—
Interaction	3	144	144	48	15.98	0.000	11.03	0.001
(Re)×(pl/D)	1	119	119	119	39.75	0.000	22.04	0.000
(Re)×(Hl/D)	1	22	22	22	7.54	0.013	1.10	0.050
(pl/D)×(Hl/D)	1	2	2	2	0.65	0.429	—	—
Residual error	19	57	57	3				
Total	28	41,812.4						

R^2 (adequate) = 99.86%, AR^2 (adequate) = 99.79%

be eliminated from the regression model. The ANOVA results of Nu are presented in Table 9.3. More details about the ANOVA of f_{Re} and PEC are available in literature [78].

The R^2 represents the confidence degree of the regression model, which compares with the numerical data, and it is calculated as R^2 = (Regression Seq SS)/(Total-Seq SS). Seq SS calculated as a sum of (true value – calculated value)^2. The F and P values are employed to represent the significance degree of the regression model terms. A larger F or a smaller P indicates a large significance degree, and a term is considered as significant when $P < 0.05$. The R^2 value of the regression model is computed first, and then the insignificant terms ($P > 0.05$) are eliminated, and a modified regression model is gained.

For Nu, three terms of the correlation are removed due to the P value above 0.05. Then the adjusted R^2 (AR^2) value of Nu is 99.79%, which indicates a good agreement between the regression model and the original value. And the significance values for each terms are as follows: (Re) > (Re)^2 > (Re) × (pl/D) > (pl/D) > (Hl/D) > (Re) × (Hl/D). It indicates that Re is the most significant factor of the Nu, while the corrugation pitch and height are also a big influence on Nu variation.

The factors of the regression correlations for Nu, f_{Re}, and PEC are listed in Table 9.4. These can be employed to calculate the Nu, f_{Re}, and PEC values for the HCTs; these correlations are suitable for the geometrical parameters and flow conditions, as presented in Table 9.2.

The sensitivity degrees of the three indexes are calculated in Eq. (9.25). In the results as shown in Figure 9.12, the horizontal axis represents the

Table 9.4 The coefficients of the regression response surface models.

	Nu	f_{Re}	PEC
b_0	17.802	548.550	1.350
b_1	0.004	0.057	-1.707E-05
b_2	-5.644	-1112.46	0.205
b_3	71.967	10206.5	-1.781
b_{11}	-2.570E-08	—	1.331E-10
b_{22}	—	665.867	-0.096
b_{33}	—	—	—
b_{12}	-1.996E-04	-0.031	1.751E-06
b_{13}	0.002	0.335	—
b_{23}	—	-8086.44	0.785

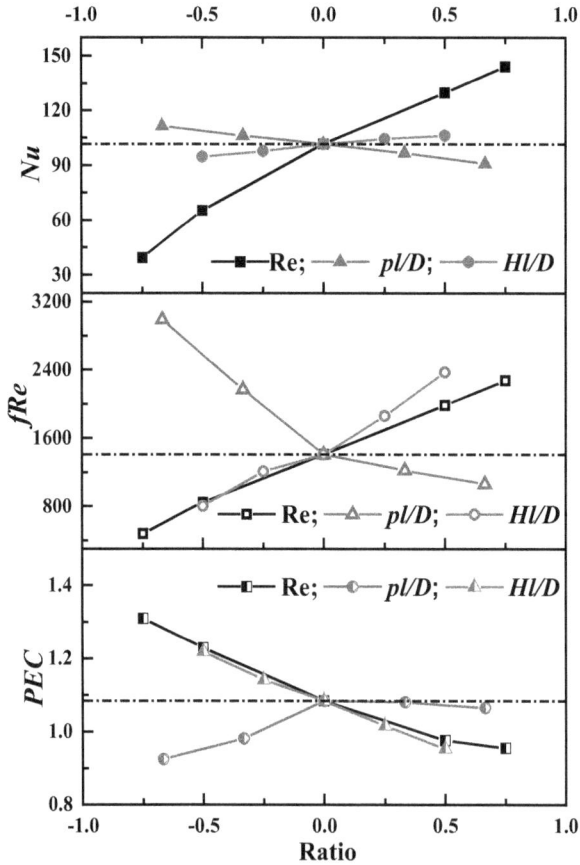

Figure 9.12 Sensitivity analysis of three indexes for HCTs.

Table 9.5 Sensitivity degrees of three index for HCTs

Factors	SC-Nu		SC-f_{Re}		SC-PEC	
	Range	Average	Range	Average	Range	Average
Re	0.552-0.816	0.660	0.796-0.881	0.829	0.159-0.277	0.225
pl/D	0.135-0.157	0.146	0.366-1.684	1.015	-0.220-0.026	-0.117
Hl/D	0.093-0.156	0.124	0.423-1.368	0.984	0.211-0.248	0.238

variation ratio of the factors ratio = $(x\text{-}x_0)/x_0$, and the reference value of the zero level is Re = 25,040, $pl/D = 1.5$, and $Hl/D = 0.1$. The sensitivity degrees of the three indexes are also listed in Table 9.5. The results present that the Re is a big influence on the Nu, that is, 4.5 and 5.3 times larger than pl/D and Hl/D. The sensitivity degrees of the three factors corresponding to f_{Re} are similar, and the pl/D is slightly larger than the other two factors. The sensitivity degrees of Re and Hl/D corresponding to PEC are similar, and pl/D is the minimum sensitive factor.

$$SC - Nu_{Re} = \frac{(Nu - Nu_0)/Nu_0}{(Re - Re_0)/Re_0} \qquad (9.25)$$

The interaction effect on Nu is, as presented in Figure 9.13, two-dimensional. It shows that the Re and the Hl/D are positive effects on Nu, while pl/D is a negative effect on Nu. The maximum Nu value can reach to 160, which is obtained at the smallest corrugation pitch ($pl/D = 0.5$–1.0), the largest corrugation height ($Hl/D = 0.125$–0.150), and a Re value of near 40,000.

After establishing the regression model for the targets, a nondominated sorting genetic algorithm II (NSGA-II) is used to deal with the three objectives optimization equations [45, 46]. The set of the GA parameters are population size = 200, maximum evolutionary algebra = 200, and optimal individual coefficient = 0.3. The Pareto front solutions [81, 82] are obtained as in Figure 9.14, which include the amount of optimal solutions and can be selected according to the particular demand of the three indexes.

Points A and D are the two extreme solutions of the Pareto front, where the A point possesses the highest PEC value and the lowest Nu and f_{Re} values, and the target performances of D point present the opposite of A point. This implies that the target relationship between PEC and f_{Re} can be attained simultaneously, and the values of Nu oppose that. Point B is decided by a mapping method [83, 84] which is considered as the equilibrium point of PEC and Nu targets, and Point C is the equilibrium point of f_{Re} and Nu targets. The target and design parameter values of the five representativeness points on Pareto front are listed in Table 9.6.

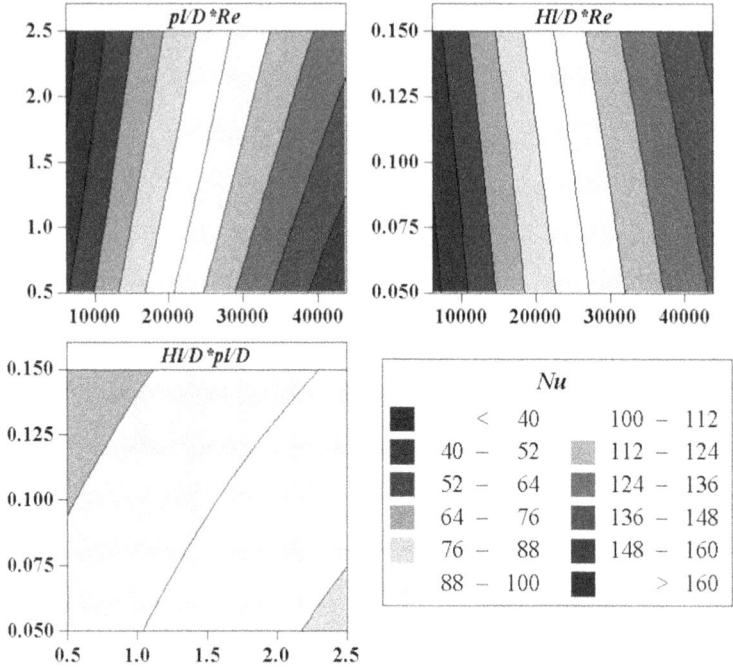

Figure 9.13 The interaction effect on Nu of the three factors.

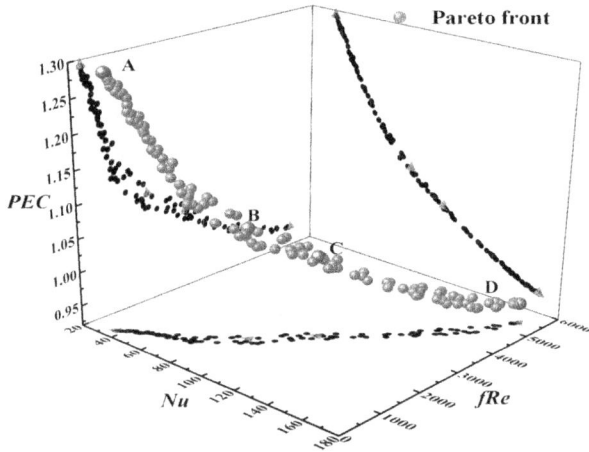

Figure 9.14 Pareto front solutions of the three-objective optimization for the HCTs.

Table 9.6 Values of target and design parameters on Pareto front

	A	B	C	D
Nu	35.77	87.55	108.81	168.12
f_{Re}	240.23	1,595.88	2,502.69	5,094.77
PEC	1.29	1.08	1.04	0.95
Re	6,352	18,375	23,858	41,420
pl/D	2.45	0.92	0.75	0.52
Hl/D	0.05	0.11	0.13	0.15

Points B, C, and D give values of Nu that are 2.45, 3.04, and 4.70 times larger than at point A, respectively. The values of f_{Re} increases with 6.64, 10.42, and 22.46 times compared with point A, respectively. The PECs of point A are declined by 16.28%, 19.38%, and 26.36%, respectively, compared with the values of point B, C, and D. The decision-making can be decided according to the specific requirement of the three targets. In a comprehensive balanced view, the design parameters between points B and C can be regularly used.

Current literature has reported multi-objective optimization on heat exchanger with two or three objectives including heat transfer, flow resistance, and overall heat transfer performance [85, 86]. For future work, more objectives, that is, weight, compactness, and cost, are needed for consideration simultaneously.

9.5 CONCLUSIONS AND OUTLOOK

Evaluation of the performance of heat transfer and flow resistance for engineering design and optimization is an essential step. The existing various evaluation techniques introduce confusion. In this chapter, the various used evaluation techniques were reviewed, and examples were presented to illustrate the utilization of some of these techniques. The evaluation techniques reviewed here on augmentation of heat transfer all have their unique physical meaning, features, and wide applications. Using different methods may result in different optimal results, so it is difficult to choose the best evaluation method presently. But the traditional evaluation technique based on the first law of thermodynamics is still most popular and used in academic studies and engineering design. The entropy production index is widely employed for vortex flow heat transfer and heat transfer of nanofluids. It is helpful for understanding the enhanced heat transfer mechanisms. The multi-objective optimization method is a new evaluation technique on heat transfer and presents a vitality as well as a great potential for engineering design. For future study, detailed distinctions between the several evaluation techniques must be considered. An integral evaluation technique from performance evaluation, mechanism analysis, and engineering optimal design is preferable.

REFERENCES

[1] Bengt Sundén, *Introduction to Heat Transfer*, WIT Press, Southampton, Boston, 2012.

[2] T. Cebeci, *Convective Heat Transfer*, Springer, Berlin, Heideleberg, 2002.

[3] S. Kakaç, R.K. Shah, W. Aung, *Handbook of Single-Phase Convective Heat Transfer*, OSTI.GOV., USA, 1987.

[4] W. Wang, Y. Zhang, J. Liu, B. Li, B. Sundén, Numerical investigation of entropy generation of turbulent flow in a novel outward corrugated tube, *Int. J. Heat Mass Transf.*, Vol. 126, 836–847, 2018.

[5] A.P. Colburn, A method of correlating forced convection heat-transfer data and a comparison with fluid friction, *Int. J. Heat Mass Transf.*, Vol. 7, 1359–1384, 1964.

[6] R.L. Webb, Performance evaluation criteria for use of enhanced heat transfer surfaces in heat exchanger design, *Int. J. Heat Mass Transf.*, Vol. 24, 715–726, 1981.

[7] J. Dong, J. Chen, W. Zhang, J. Hu, Experimental and numerical investigation of thermal -hydraulic performance in wavy fin-and-flat tube heat exchangers, *Appl. Therm. Eng.*, Vol. 30, 1377–1386, 2010.

[8] Z.S. Kareem, S. Abdullah, T.M. Lazim, M.N. Mohd Jaafar, A.F. Abdul Wahid, Heat transfer enhancement in three-start spirally corrugated tube: Experimental and numerical study, *Chem. Eng. Sci.*, Vol. 134, 746–757, 2015.

[9] S. Pethkool, S. Eiamsa-ard, S. Kwankaomeng, P. Promvonge, Turbulent heat transfer enhancement in a heat exchanger using helically corrugated tube, *Int. Commun. Heat Mass Transf.*, Vol. 38, 340–347, 2011.

[10] J.F. Fan, W.K. Ding, J.F. Zhang, Y.L. He, W.Q. Tao, A performance evaluation plot of enhanced heat transfer techniques oriented for energy-saving, *Int. J. Heat Mass Transf.*, Vol. 52, 33–44, 2009.

[11] G.I. Mahmood, P.M. Ligrani, Heat transfer in a dimpled channel: Combined influences of aspect ratio, temperature ratio, Reynolds number, and flow structure, *Int. J. Heat Mass Transf.*, Vol. 45, 2011–2020, 2002.

[12] G.I. Mahmood, M.L. Hill, D.L. Nelson, P.M. Ligrani, H.-K. Moon, B. Glezer, Local heat transfer and flow structure on and above a dimpled surface in a channel, *J. Turbomach.*, Vol. 123, 115–123, 2000.

[13] J. Chen, H. Müller-Steinhagen, G.G. Duffy, Heat transfer enhancement in dimpled tubes, *Appl. Therm. Eng.*, Vol. 21, 535–547, 2001.

[14] G.M. Belen'kiy, Lekakh B., Fokin B., Dolgushin K., Heat transfer augmentation using surfaces formed by a system of spherical cavities, *Heat Transf. Res.*, Vol. 25, 196–203, 1993.

[15] P.M. Ligrani, M.M. Oliveira, T. Blaskovich, Comparison of heat transfer augmentation techniques, *AIAA J.*, Vol. 41, 337–362, 2003.

[16] Z.Y. Guo, W.Q. Tao, R.K. Shah, The field synergy (coordination) principle and its applications in enhancing single phase convective heat transfer, *Int. J. Heat Mass Transf.*, Vol. 48, 1797–1807, 2005.

[17] J.E.X. Zhao, Z. Zhang, J. Chen, G. Liao, F. Zhang, E. Leng, D. Han, W. Hu, A review on heat enhancement in thermal energy conversion and management using field synergy principle, *Appl. Energy*, Vol. 257, 113995, 2020.

[18] Y. Zhai, Z. Li, H. Wang, J. Xu, Analysis of field synergy principle and the relationship between secondary flow and heat transfer in double-layered microchannels with cavities and ribs, *Int. J. Heat Mass Transf.*, Vol. 101, 190–197, 2016.

[19] L. Zhang, J. Li, Y. Li, J. Wu, Field synergy analysis for helical ducts with rectangular cross section, *Int. J. Heat Mass Transf.*, Vol. 75, 245–261, 2014.

[20] J. Guo, M. Xu, L. Cheng, Numerical investigations of curved square channel from the viewpoint of field synergy principle, *Int. J. Heat Mass Transf.*, Vol. 54, 4148–4151, 2011.

[21] Y.B. Tao, Y.L. He, Z.G. Wu, W.Q. Tao, Three-dimensional numerical study and field synergy principle analysis of wavy fin heat exchangers with elliptic tubes, *Int. J. Heat Fluid Flow*, Vol. 28, 1531–1544, 2007.

[22] B. Zhang, J. Lv, J. Zuo, Compressible fluid flow field synergy principle and its application to drag reduction in variable-cross-section pipeline, *Int. J. Heat Mass Transf.*, Vol. 77, 1095–1101, 2014.

[23] Y.P. Cheng, T.S. Lee, H.T. Low, Numerical simulation of conjugate heat transfer in electronic cooling and analysis based on field synergy principle, *Appl. Therm. Eng.*, Vol. 28, 1826–1833, 2008.

[24] G.M. Chen, C.P. Tso, Y.M. Hung, Field synergy principle analysis on fully developed forced convection in porous medium with uniform heat generation, *Int. Commun. Heat Mass Transf.*, Vol. 38, 1247–1252, 2011.

[25] Z.C. Liu, W. Liu, A. Nakayama, Flow and heat transfer analysis in porous wick of CPL evaporator based on field synergy principle, *Heat Mass Transf.*, Vol. 43, 1273–1281, 2007.

[26] U. Lucia, Entropy and exergy in irreversible renewable energy systems, *Renew. Sust. Energ. Rev.*, Vol. 20, 559–564, 2013.

[27] I. Dincer, Y.A. Cengel, Energy, entropy and exergy concepts and their roles in thermal, *Engineering*, Vol. 3, 116–149, 2001.

[28] L. Borel, D. Favrat, *Thermodynamics and Energy Systems Analysis: From Energy to Exergy*, EPFL Press, Lausanne, 2010.

[29] M. F.A., The design of heat exchangers for minimum irreversibility, ASME Paper, 51-A-108, 1951.

[30] O.N.S.M. Yilmaz, S. Karsli, Performance evaluation criteria for heat exchangers based on second law analysis, *Exergy, Int. J.*, Vol. 1, 278–294, 2001.

[31] A. Bejan, A study of entropy generation in fundamental convective heat transfer, *J. Heat Transf.*, Vol. 101, 718, 1979.

[32] Z.Y. Guo, D.Y. Li, B.X. Wang, A novel concept for convective heat transfer enhancement, *Int. J. Heat Mass Transf.*, Vol. 41, 2221–2225, 1998.

[33] Z.-Y. Guo, H.-Y. Zhu, X.-G. Liang, Entransy—A physical quantity describing heat transfer ability, *Int. J. Heat Mass Transf.*, Vol. 50, 2545–2556, 2007.

[34] Q. Chen, X.-G. Liang, Z.-Y. Guo, Entransy theory for the optimization of heat transfer – A review and update, *Int. J. Heat Mass Transf.*, Vol. 63, 65–81, 2013.

[35] X. Chen, T. Zhao, M.-Q. Zhang, Q. Chen, Entropy and entransy in convective heat transfer optimization: A review and perspective, *Int. J. Heat Mass Transf.*, Vol. 137, 1191–1220, 2019.

[36] X. Zhu, X. Du, Y. Ding, Q. Qiu, Analysis of entropy generation behavior of supercritical water flow in a hexagon rod bundle, *Int. J. Heat Mass Transf.*, Vol. 114, 20–30, 2017.

[37] W. Wang, Y. Zhang, J. Liu, Z. Wu, B. Li, B. Sundén, Entropy generation analysis of fully-developed turbulent heat transfer flow in inward helically corrugated tubes, *Numer. Heat Transf.; A: Appl.*, Vol. 73, 788–805, 2018.

[38] X. Cheng, X. Liang, Heat transfer entropy resistance for the analyses of two-stream heat exchangers and two-stream heat exchanger networks, *Appl. Therm. Eng.*, Vol. 59, 87–93, 2013.

[39] X. Cheng, Entropy resistance minimization: An alternative method for heat exchanger analyses, *Energy*, Vol. 58, 672–678, 2013.

[40] Y. Wang, X. Huai, Heat transfer and entropy generation analysis of an intermediate heat exchanger in ADS, *J. Therm. Sci.*, Vol. 27, 175–183, 2018.

[41] M. Rooman, M.A. Jan, Z. Shah, P. Kumam, A. Alshehri, Entropy optimization and heat transfer analysis in MHD Williamson nanofluid flow over a vertical Riga plate with nonlinear thermal radiation, *Sci. Rep.*, Vol. 11, 18386, 2021.

[42] M. Sheikholeslami, M. Jafaryar, A. Shafee, Z. Li, R.-U. Haq, Heat transfer of nanoparticles employing innovative turbulator considering entropy generation, *Int. J. Heat Mass Transf.*, Vol. 136, 1233–1240, 2019.

[43] A. Hajatzadeh Pordanjani, S. Aghakhani, A. Karimipour, M. Afrand, M. Goodarzi, Investigation of free convection heat transfer and entropy generation of nanofluid flow inside a cavity affected by magnetic field and thermal radiation, *J. Therm. Anal. Calorim.*, Vol. 137, 997–1019, 2019.

[44] E. Abu-Nada, Entropy generation due to heat and fluid flow in backward facing step flow with various expansion ratios, *Int. J. Exergy*, Vol. 3, 419–435, 2006.

[45] A. Alimoradi, F. Veysi, Optimal and critical values of geometrical parameters of shell and helically coiled tube heat exchangers, *Case Stud. Therm. Eng*, Vol. 10, 73–78, 2017.

[46] Z. Li, M. Sheikholeslami, M. Jafaryar, A. Shafee, A.J. Chamkha, Investigation of nanofluid entropy generation in a heat exchanger with helical twisted tapes, *J. Mol. Liq.*, Vol. 266, 797–805, 2018.

[47] Q. Chen, J. Ren, Generalized thermal resistance for convective heat transfer and its relation to entransy dissipation, *Chinese Sci. Bull.*, Vol. 53, 3753–3761, 2008.

[48] S. Wang, Q. Chen, B. Zhang, An equation of entransy transfer and its application, *Chinese Sci. Bull.*, Vol. 54, 3572, 2009.

[49] J. Wu, X. Liang, Application of entransy dissipation extremum principle in radiative heat transfer optimization, *Sci. China Series E: Technol. Sci.*, Vol. 51, 1306–1314, 2008.

[50] X. Cheng, X. Wang, X. Liang, Role of viscous heating in entransy analyses of convective heat transfer, *Sci. China Technol. Sci.*, Vol. 63, 2154–2162, 2020.

[51] J. Guo, L. Cheng, M. Xu, Entransy dissipation number and its application to heat exchanger performance evaluation, *Chinese Sci. Bull.*, Vol. 54, 2708–2713, 2009.

[52] Y.L. He, W.Q. Tao, Numerical studies on the inherent interrelationship between field synergy principle and entransy dissipation extreme principle for enhancing convective heat transfer, *Int. J. Heat Mass Transf.*, Vol. 74, 196–205, 2014.

[53] Q. Chen, H. Zhu, N. Pan, Z.-Y. Guo, An alternative criterion in heat transfer optimization, Vol. 467, 1012–1028, 2011.

[54] Q. Chen, M. Wang, N. Pan, Z.-Y. Guo, Optimization principles for convective heat transfer, *Energy*, Vol. 34, 1199–1206, 2009.

[55] A. Bejan, Discipline in thermodynamics, *Energies*, Vol. 13, 2487, 2020.

[56] A. Bejan, Thermodynamics of heating, *Proc. Royal Soc. A*, Vol. 475, 20180820, 2019.

[57] M.M. Kostic, Nature of heat and thermal energy: From caloric to Carnot's reflections, to entropy, *Exergy, Entransy and Beyond*, Vol. 20, 584, 2018.

[58] J. Guo, L. Cheng, M. Xu, Multi-objective optimization of heat exchanger design by entropy generation minimization, *J. Heat Transf.*, Vol. 132, 081801, 2010.

[59] S. Sanaye, H. Hajabdollahi, Multi-objective optimization of shell and tube heat exchangers, *Appl. Therm. Eng.*, Vol. 30, 1937–1945, 2010.

[60] M. Mirzaei, H. Hajabdollahi, H. Fadakar, Multi-objective optimization of shell-and-tube heat exchanger by constructal theory, *Appl. Therm. Eng.*, Vol. 125, 9–19, 2017.

[61] R.V. Rao, A. Saroj, Multi-objective design optimization of heat exchangers using elitist-Jaya algorithm, *Energy Syst.*, Vol. 9, 305–341, 2018.

[62] S. Wang, G. Jian, J. Xiao, J. Wen, Z. Zhang, Optimization investigation on configuration parameters of spiral-wound heat exchanger using genetic aggregation response surface and multi-objective genetic algorithm, *Appl. Therm. Eng.*, Vol. 119, 603–609, 2017.

[63] X. Cao, R. Zhang, D. Chen, L. Chen, T. Du, H. Yu, Performance investigation and multi-objective optimization of helical baffle heat exchangers based on thermodynamic and economic analyses, *Int. J. Heat Mass Transf.*, Vol. 176, 121489, 2021.

[64] Y. Yuan, X. Wang, X. Meng, Z. Zhang, J. Cao, A strategy for helical coils multi-objective optimization using differential evolution algorithm based on entropy generation theory, *Int. J. Therm. Sci.*, Vol. 164, 106867, 2021.

[65] H. Xu, C. Duan, H. Ding, W. Li, Y. Zhang, G. Hong, Multi-objective optimization based on economic analysis for a printed circuit heat exchanger with application to Brayton cycle, *J. Nucl. Sci. Technol.*, Vol. 58, 1038–1047, 2021.

[66] Y. Yang, H. Li, M. Yao, Y. Zhang, C. Zhang, L. Zhang, S. Wu, Optimizing the size of a printed circuit heat exchanger by multi-objective genetic algorithm, *Appl. Therm. Eng.*, Vol. 167, 114811, 2020.

[67] C. Shi, M. Wang, J. Yang, W. Liu, Z. Liu, Performance analysis and multi-objective optimization for tubes partially filled with gradient porous media, *Appl. Therm. Eng.*, Vol. 188, 116530, 2021.

[68] C. Liu, Y.D. Deng, X.Y. Wang, X. Liu, Y.P. Wang, C.Q. Su, Multi-objective optimization of heat exchanger in an automotive exhaust thermoelectric generator, *Appl. Therm. Eng.*, Vol. 108, 916–926, 2016.

[69] J. Guo, X. Huai, X. Li, J. Cai, Y. Wang, Multi-objective optimization of heat exchanger based on entransy dissipation theory in an irreversible Brayton cycle system, *Energy*, Vol. 63, 95–102, 2013.

[70] A.I. Khuri, S. Mukhopadhyay, Response surface methodology, *Wiley Interdisc. Rev. Comput. Stat.*, Vol. 2, 128–149, 2010.

[71] H. Han, R. Yu, B. Li, Y. Zhang, Multi-objective optimization of corrugated tube inserted with multi-channel twisted tape using RSM and NSGA-II, *Appl. Therm. Eng.*, Vol. 159, 113731, 2019.

[72] M.H. Alrefaei, A.H. Diabat, A simulated annealing technique for multi-objective simulation optimization, *Appl. Math. Comput.*, Vol. 215, 3029–3035, 2009.

[73] I.D.I.D. Ariyasingha, T.G.I. Fernando, Performance analysis of the multi-objective ant colony optimization algorithms for the traveling salesman problem, *Swarm Evol. Comput.*, Vol. 23, 11–26, 2015.

[74] Z. Song, B.T. Murray, B. Sammakia, L. Shuxia, Multi-objective optimization of temperature distributions using artificial neural networks, in *13th InterSociety Conference on Thermal and Thermomechanical Phenomena in Electronic Systems*, 1209–1218, 2012.

[75] A. Benyekhlef, B. Mohammedi, D. Hassani, S. Hanini, Application of artificial neural network (ANN-MLP) for the prediction of fouling resistance in heat exchanger to MgO-water and CuO-water nanofluids, *Water Sci. Technol*, Vol. 84, 538–551, 2021.

[76] H. Wu, S.A. Bagherzadeh, A. D'Orazio, N. Habibollahi, A. Karimipour, M. Goodarzi, Q.-V. Bach, Present a new multi objective optimization statistical Pareto frontier method composed of artificial neural network and multi objective genetic algorithm to improve the pipe flow hydrodynamic and thermal properties such as pressure drop and heat transfer coefficient for non-Newtonian binary fluids, *Physica A Stat. Mech. Appl.*, Vol. 535, 122409, 2019.

[77] B. Chegari, M. Tabaa, E. Simeu, F. Moutaouakkil, H. Medromi, Multi-objective optimization of building energy performance and indoor thermal comfort by combining artificial neural networks and metaheuristic algorithms, *Energy Build.*, Vol. 239, 110839, 2021.

[78] W. Wang, Y. Zhang, Y. Li, H. Han, B. Li, Multi-objective optimization of turbulent heat transfer flow in novel outward helically corrugated tubes, *Appl. Therm. Eng.*, Vol. 138, 795–806, 2018.

[79] Y. Hang, M. Qu, S.J.E. Ukkusuri, Buildings, Optimizing the design of a solar cooling system using central composite design techniques, *Energy Build.*, Vol. 43, 988–994, 2011.

[80] M. Bashiri, A.F.J.S.I. Geranmayeh, Tuning the parameters of an artificial neural network using central composite design and genetic algorithm, *IEEE Trans. Neural Netw.*, Vol. 18 (2011) 1600–1608.

[81] P. Ngatchou, A. Zarei, A. El-Sharkawi, Pareto multi objective optimization, in *Proceedings of the 13th International Conference on, Intelligent Systems Application to Power Systems*, 84–91, IEEE, 2005.

[82] A. Konak, D.W. Coit, A.E. Smith, Multi-objective optimization using genetic algorithms: A tutorial, *Reliab. Eng. Syst. Saf.*, Vol. 91, 992–1007, 2006.

[83] I. Giagkiozis, P.J. Fleming, Pareto front estimation for decision making, *Evol. Comput.*, Vol. 22, 651–678, 2014.

[84] Y. Tian, X. Xiang, X. Zhang, R. Cheng, Y. Jin, Sampling reference points on the Pareto fronts of benchmark multi-objective optimization problems, in *2018 IEEE Congress on Evolutionary Computation (CEC)*, 1–6, IEEE, 2018.

[85] S. Chamoli, R. Lu, P. Yu, Thermal characteristic of a turbulent flow through a circular tube fitted with perforated vortex generator inserts, *Appl. Therm. Eng.*, Vol. 121, 1117–1134, 2017.

[86] H. Safikhani, S. Eiamsa-ard, Pareto based multi-objective optimization of turbulent heat transfer flow in helically corrugated tubes, *Appl. Therm. Eng.*, Vol. 95, 275–280, 2016.

Chapter 10

Heat transfer measurement techniques

10.1 INTRODUCTION

Most of the important measuring methods relevant for experimental investigations of heat transfer and heat exchangers are presented. Such methods are frequently used in R&D work related to the augmentation of heat transfer in a variety of applications. Opportunities, challenges, and limitations are discussed. A few examples are displayed for illustration. A few books covering a lot of measurement methods in heat transfer are Refs. [1, 2].

10.2 INFRARED IMAGING, IR

Infrared (IR) thermography is a process where a thermal imager is used to detect radiation (heat) coming from an object, converting it to temperature and displaying an image of the temperature distribution. Such images show the detected temperature distributions, which are called thermograms. All objects above absolute zero temperature release thermal IR energy, then thermal imagers can easily detect and display IR wavelengths regardless of ambient light. IR thermography is commonly used in a variety of industries and applications. However, even though IR imagers are simple to use, interpreting the data they produce can be more challenging to break down. It is important to have knowledge of how IR imagers work, but also baseline knowledge of radiometry and heat transfer processes is needed.

A thermal-imaging camera is an advanced type of radiation thermometer used for measuring temperature at multiple points across a large area and creating two-dimensional thermographic images. Thermal-imaging cameras are considerably more software- and hardware-based than a spot thermometer. Most cameras display real-time images and can be connected to specialized software for deeper evaluation and accuracy etc. Modern thermal-imaging cameras are handheld. Figure 10.1 shows a picture of thermal-imaging camera of forward-looking infrared (FLIR) type.

DOI: 10.1201/9781003229865-10

Figure 10.1 FLIR SC640 thermal-imaging camera.

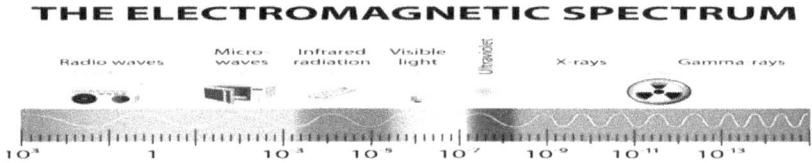

Figure 10.2 The electromagnetic spectrum.

In Figure 10.2, the electromagnetic spectrum is shown. The names of the radiation depending on wavelength are indicated and the ranges of the wavelength intervals are given.

For the thermal-imaging camera, the recordable temperature range is important. This means the minimum and maximum temperature the camera can measure. Detector resolution and thermal sensitivity are important issues. The detector resolution tells the number of pixels displayed in the images. The camera should include the most common resolutions of 160 × 120, 320 × 240 and 640 × 480. A 640 × 480 imager displays an image made up of 307,200 pixels. Thermal sensitivity refers to the smallest temperature difference the thermal-imaging camera can detect. For an IR camera showing a sensitivity of 0.05 degrees, this means that it can observe a temperature difference between two surfaces with a five-hundredth of a degree temperature difference.

10.2.1 How to use infrared thermography

IR thermography is a valuable tool for condition monitoring and preventive maintenance. In addition, IR thermography is an important measuring method in heat transfer research. Not only does it allow one to detect thermal abnormalities of machines, but it lets one do so in a nonintrusive, hands-off way while still getting results in real time. Thermographers usually employ one of three methods when performing thermal inspections, namely *comparative, baseline*, and *thermal trending*. These are shortly described here. Comparative thermography is used to measure the temperature of similar components under similar conditions. Baseline thermography is used to set a precedent or establish a reference point for an asset by taking temperature readings when the asset is in good working order. It is used in comparison with other thermal images to identify potential issues early. It is recommended to take baseline measurements on all critical assets when they are new or have just been repaired. Thermal-trending thermography shows how temperature is dispersed in a component or asset over time. It is a great method for looking at mechanical equipment with complex thermal signatures or when thermal signatures develop slowly.

10.2.2 Calibration

The calibration of uncooled thermal IR cameras with respect to absolute temperature measurement is a time-consuming and complicated process that significantly influences the cost of an IR camera. Temperature-measuring IR cameras display a temperature value for each pixel in the thermal image. Full Calibration is available for almost every FLIR thermal-imaging camera on the market today.

10.2.2.1 Thermal-imaging camera calibration

A temperature reference source is an idealized physical body that absorbs all incident electromagnetic radiation, independent of frequency or angle of incidence. At equilibrium, the device emits electromagnetic radiation which is emitted according to Planck's law meaning that it has a spectrum that is determined by the temperature alone, not by the shape of the body or its composition.

A temperature reference in thermal equilibrium (black body) has two notable properties: (a) it is an ideal emitter, which means that at every frequency it emits maximum energy and, accordingly, more than any other body at the same temperature; (b) it is a diffuse emitter, i.e., the radiated energy is independent of direction, i.e., isotropic.

An approximate black surface is created by a hole in a wall of a large enclosure. Any light entering the hole is reflected indefinitely or absorbed inside and is unlikely to escape, making the hole a nearly perfect absorber. Real materials emit energy at a fraction of a black body. The emissivity defines this fraction in the range of 0 to 1. By definition, a temperature reference in thermal equilibrium (black body) has an emissivity of $\varepsilon = 1.0$. A radiative source with lower emissivity independent of frequency is often referred to as a gray body. Establishing a temperature reference with an emissivity as close to unity as possible remains a topic of interest.

If emissivity is not considered, the surface temperature of an object may appear to be emitting more radiation than it really is, due to the addition of, e.g., reflected radiation. This can give an incorrect perception of the emissive power or temperature of the considered surface.

Figure 10.3 Use of IR camera to detect heat leakage in a family house.

10.2.2.2 Showcase of IR thermal images

In Figure 10.3, an illustration of IR thermal images on the surfaces of a family house is depicted. Such pictures are useful in detecting heat leakage. In this case, it is obvious that high leakages occur through the windows and the roof.

Further details on IR thermography are available in references [3, 4].

10.3 LIQUID CRYSTAL THERMOGRAPHY, LCT

LCT has been used extensively in convective heat transfer research. In general, the liquid crystal technique is nonintrusive, inexpensive, and capable of high spatial resolution and accuracy in temperature measurement. Typical engineering applications of LCT in duct flows are heat exchangers, internal cooling of gas turbine components, and electronics cooling.

The active bandwidths of most commercially available LCT products are limited to the temperature range from −30°C to 150°C. As thermochromic liquid crystals (TLCs) are used in advanced implementations like research on heat transfer coefficients in duct walls, one needs to heat the surface of interest, apply calibration for the temperature–color response, acquire images, and then analyze the images by suitable software.

The colors of the TLCs are observed by an RGB-data (red–green–blue) acquisition system (CCD camera, frame grabber, and computer) but commonly the RGB color space is transformed to the hue (H), saturation (S), and intensity (I) color space. The hue value is a monotonic function of the dominant wavelength reflected by the TLC and is best suited for a unique color versus temperature. When TLCs are used in experiments for determining convective heat transfer coefficients, two types of experiments prevail. The first one implies a steady-state experiment, while the second one is transient in terms of thermal boundary conditions. The steady-state heat transfer experiment with a constant heat flux density is conceptually simple and relatively easy to analyze. It has found frequent usage.

TLCs are designated by giving two colors or temperatures. The designation R40C5W implies that the activation temperature, i.e., red color appearance, is 40°C, and 5 W means that the blue color appears 5°C above the start temperature of the red color. 5 W is a crude estimation of the bandwidth of the liquid crystal, i.e., in this case the operating temperature regime is 40–45°C. Beyond this range the material will not exhibit any colors. If the bandwidth is less than 2°C, the LC is said to be narrow banded.

10.3.1 Calibration of LCT

Calibration of the liquid crystals is crucial for the LCT with high resolution. In the calibration process, the overall influences of the liquid crystal material, illuminating lights, camera system, etc. are considered. The goal of the calibration is to obtain an accurate and computationally efficient color–temperature curve, which can be used as the heat transfer measurements are evaluated.

The calibration experiment can be conducted in a separate system (on which controlled thermal conditions are imposed) or directly on the test surface (in situ), using the same optical arrangement as in the thermal measurement section. The calibration conducted in a separate system may result in some problems as it might be difficult to enable precisely the same imposed conditions in the calibration procedure. The in situ calibration can be more beneficial if properly carried out. However, the arrangement of in situ calibration is more complicated because, in addition, several thermocouples must be fitted into the test section. Therefore, it is more practical to perform the calibration in a separate unit where several thermocouples are fitted and the thermal boundary conditions can be more easily and accurately controlled.

LCs reflect different colors (red, green, blue) selectively when subject to temperature changes, for example, 35–36°C. The color response allows the surface temperature to be measured accurately. Figure 10.4 displays the

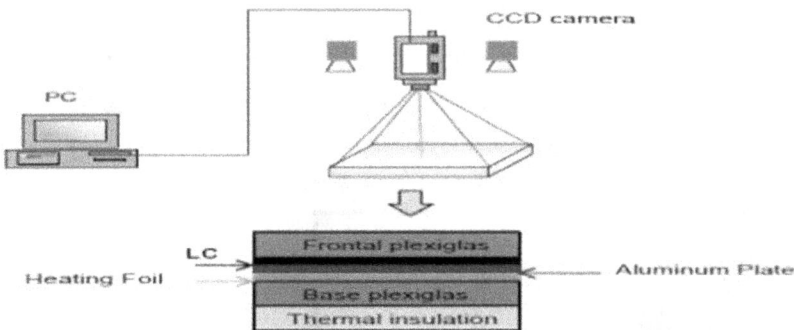

Figure 10.4 Illustration of a separate calibration unit. (Based on Refs. [5–8].)

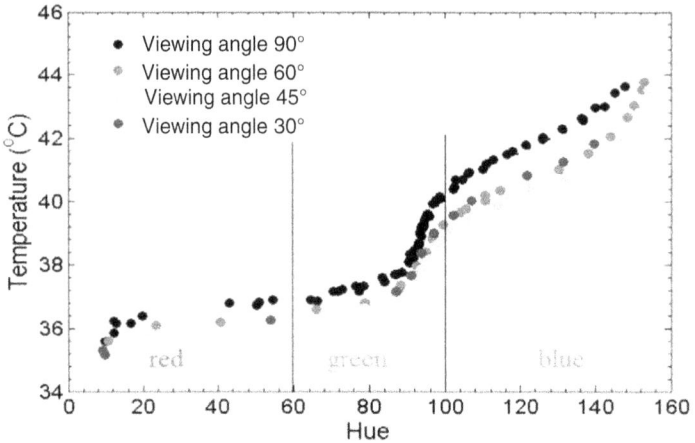

Figure 10.5 Calibrated temperature-hue profiles for various viewing angles. (From Ref. [9].)

layout of a separate calibration unit. The composite structure of the calibration plate is highlighted. The complementary equipment like the CCD camera, lightening lamps, computer, and frame grabber as well as the location of the heating foil are shown.

Figure 10.5 shows calibration curves, i.e., temperature versus hue for a specific liquid crystal. This Figure shows the influence of the viewing angle.

Figure 10.6 shows the color display, temperature, and hue values, and how it changes from the start at about 35°C and the end at 36°C.

Figure 10.7 shows a set up for investigation of a so-called pocket geometry appearing at the rear part of an aircraft engine. The base channel is a

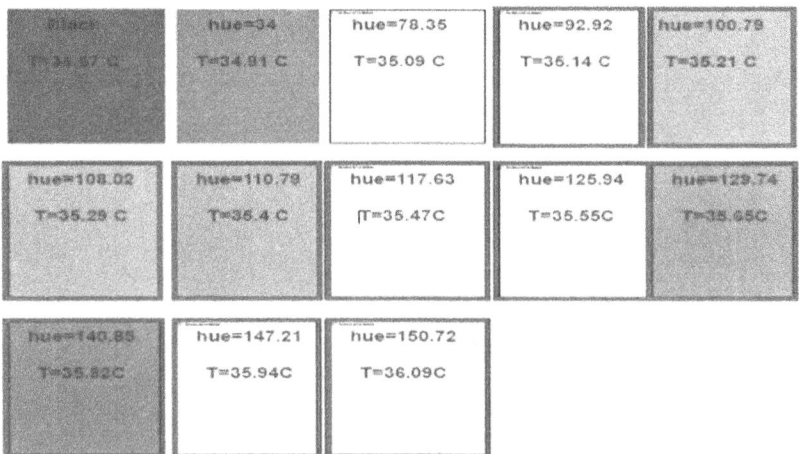

Figure 10.6 Color display for a liquid crustal R35C1W, temperatures and corresponding hue values. (From Ref. [5–8].)

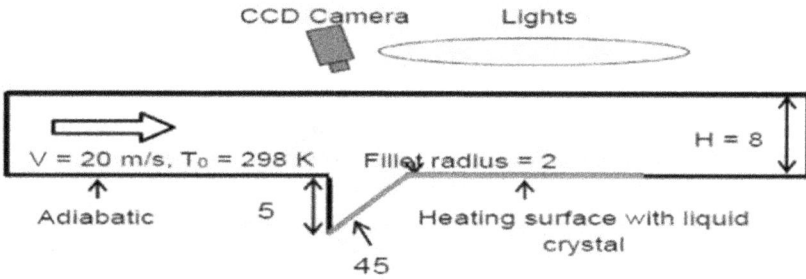

Figure 10.7 LCT measurements in pocket geometry related to an aircraft engine. From Ref. [10].

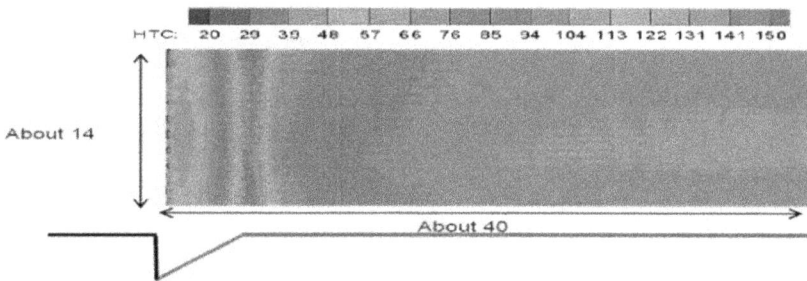

Figure 10.8 Distribution of the heat transfer coefficient downstream of the pocket. (From Ref. [10].)

rectangular one with an aspect ratio of 4. The LCT is used to determine the local heat transfer coefficients in the pocket and on the wall downstream of the pocket.

Figure 10.8 displays the local heat transfer coefficient distribution in the pocket and on the downstream flat wall. It is obvious that high heat transfer coefficients are achieved just downstream of the pocket exit as the flow reattaches on the flat wall. Further downstream, boundary layer develops and gradually the heat transfer coefficient diminishes as the boundary later thickness grows.

More information on LCT in heat transfer research can be found in refs. [11–13].

10.4 THERMOCOUPLES, TCs

A thermocouple is an electrical device consisting of two dissimilar electrical conductors forming an electrical junction. It is used for measuring the temperature at a particular point/location. It is a type of sensor used for measuring the temperature in the form of an electric current or the electromotive force (EMF).

10.4.1 Working principle of thermocouples

The working principle of thermocouples depends on three phenomena, namely the Seebeck, Peltier, and Thompson phenomena. These are briefly presented here. The Seebeck phenomenon occurs between two different metals. As heat is supplied to any of the two metals, electrons start to flow from the hot metal to the cold one. Accordingly, a direct current is induced in the circuit. The Peltier phenomenon is the inverse of the Seebeck phenomenon. The Peltier phenomenon means that a temperature difference can be created between two different conductors by imposing a potential difference between them. The Thompson phenomenon tells that when two dissimilar metals are joined, and if they create two junctions, then a voltage is induced over the entire conductor length because of the temperature gradient.

The thermocouple consists of two dissimilar metals. These metals are welded together at the junction point. This junction is considered as the measuring point. The materials being used in a thermocouple depend on the temperature-measuring range. Figure 10.9 shows the principle for a copper–iron thermocouple. Figure 10.10 shows a more detailed sketch of an

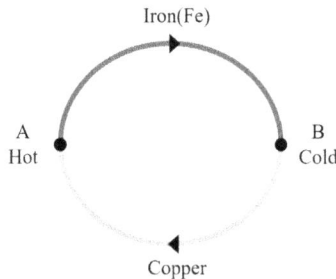

Figure 10.9 Principle of thermocouple working principle – Cu-Fe metals.

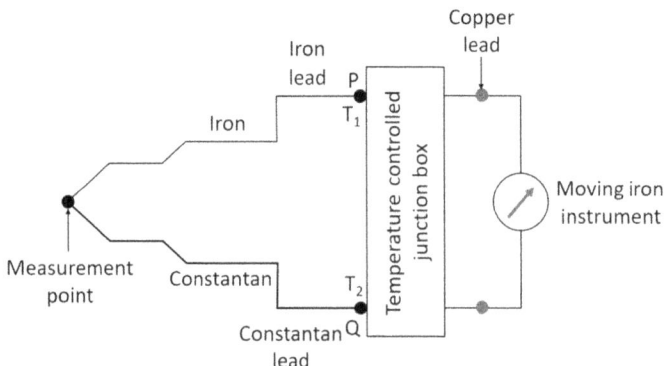

Figure 10.10 Sketch of an iron–constantan thermocouple.

iron–constantan thermocouple. This figure also indicates complementary components.

In Figure 10.10, P and Q are the two junctions of the thermocouples. T_1 and T_2 denote the temperatures at the junctions. An EMF is induced in the thermocouple circuit as the temperature of the junctions is different from each other.

The EMF (E) induced in the thermocouple circuit is given by the equation.

$$E = a(\Delta\theta) + b(\Delta\theta)^2 \tag{10.1}$$

where $\Delta\theta$ is the temperature difference between the hot thermocouple junction, and the reference thermocouple junction a and b are constants.

The output EMF obtained from the thermocouples can be measured by the following instruments: (a) a multimeter, which is a simple method to measure the output EMF of the thermocouple, and is connected to the cold junctions of the thermocouple; (b) a potentiometer – the DC output of the thermocouple can be recorded by a DC potentiometer; (c) an amplifier with output devices – the output obtained from the thermocouples is amplified through an amplifier and then fed to the recording instrument.

10.4.1.1 Advantages and disadvantages of thermocouples

Generally, there are some pros and cons concerning thermocouples. Thermocouples are regarded to have advantages like (1) being cheaper than other temperature-measuring devices, (2) having fast response times, and (3) having a wide temperature range. Thermocouples are also considered to have disadvantages like (1) having low or limited accuracy but can be calibrated and (2) recalibrating of thermocouples might be difficult or cumbersome.

10.4.1.2 Types of thermocouples

Thermocouples are commonly given a letter to define the type. In addition, one distinguishes between thermocouples using base metals and those using noble metals.

10.4.1.2.1 Base metal TCs

For base metal thermocouples, the base metal is made from common and inexpensive materials such as nickel, iron, and copper.

Type K Thermocouple is based on nickel–chromium or nickel–alumel. It is very common as it is inexpensive and has a wide temperature range as well as providing accuracy and reliability. The temperature range is from −330 to 2,300°F (−250 to 1,250°C). The wire color code is yellow and red.

Type J Thermocouple is based on iron/constantan. It permits a smaller temperature range and shows a shorter lifespan at high temperatures. Due to oxidation problems associated with iron, some precautions must be taken as this type of thermocouple is used in an oxidizing environment. The temperature range is between **32 and 1,400°F** (−210 to +760°C), and the wire color code is white and red.

Type T Thermocouple is based on copper/constantan. This thermocouple type is very stable and is commonly used in very low-temperature applications such as in cryogenics. The temperature range is between **−330 and 700°F** (−270 to +370°C), and the wire color code is blue and red.

10.4.1.2.2 Noble metal TCs

This type of thermocouple can withstand high temperatures and is manufactured of expensive materials.

Type S Thermocouple is composed of platinum rhodium – 10%/ platinum. It is mostly used in high-temperature applications. However, it can be used for lower-temperature applications due to high stability. It is frequently used in biotech industries.

Type R Thermocouple is composed of platinum rhodium – 13%/ platinum. It is more expensive as it contains a higher percentage of rhodium as compared to Type S. Its performance is like that of Type S.

Type B Thermocouple is composed of platinum rhodium – 30%/ platinum rhodium – 6%. It is known to be applicable to the highest temperature. It provides the highest stability and accuracy at high temperatures.

Table 10.1 summarizes the common thermocouple types and their operating temperature ranges.

Table 10.1 Common thermocouple types and their operating temperature ranges

Type notation	Metal combination	Temperature range
K	Nickel–chromium, Nickel–alumel	−270°C to +1,260°C
J	Iron–constantan	−210°C to +760°C
T	Copper–constantan	−270°C to +370°C
E	Nickel–chromium/constantan	−270°C to +870°C
N	Nicrosil/nisil	−270°C to +390°C
S	Platinum rhodium – 10 %/platinum	−50°C to +1,480°C
R	Platinum rhodium – 13%/platinum	−50°C to +1,480°C
B	Platinum rhodium – 30%/platinum rhodium – 6%	0°C to +1,700°C

Nicrosil/nisil is based on NiCrSi alloy and has high stability, and aimed to replace the K-type TC.

10.4.1.3 Factors affecting the choice of the thermocouple

10.4.1.3.1 Operating temperature

The operating temperature is the full ambient temperature range over which a thermocouple has a linear output. This range decides which metal combination of thermocouple being most suitable in a certain application. As the temperature range is higher, the cost becomes higher.

10.4.1.3.2 Response time

Three types of thermocouples exist, namely *Exposed, Grounded,* and *Ungrounded.* An exposed thermocouple provides the fastest response time, but if the thermocouple is to be used in a corrosive gas or at high pressure, then an exposed thermocouple should be avoided.

An ungrounded thermocouple gives a slow response time but, still, it is regarded as the best choice in applications where one desires to have a thermocouple electronically isolated and shielded by the sheath.

10.4.1.3.3 Accuracy

Accuracy is a very important issue for any instrument. Accuracy is an indication of how close the measured value is compared to the real temperature value. This is also referred to as error or tolerance.

10.4.1.3.4 Chemical, abrasion, or vibration resistance

As a thermocouple is used in a corrosive environment, the sheath material should be chemically resistant.

Reference [14] gives more information on thermocouples.

10.5 NAPHTHALENE SUBLIMATION TECHNIQUE

The naphthalene sublimation technique has been demonstrated by several investigators to be an excellent method to obtain heat transfer results experimentally. It has been around for many years but started to become common in experiments already in the 1950s.

The basis is to use mass transfer tests relying on the analogy between heat and mass transfer. Dimensional analysis of the governing equations for heat and mass transfer demonstrates that for any geometrically similar configuration and for similar boundary conditions the equations have the same form.

Then a simple relation exists between the Sherwood number (describing mass transfer) and the Nusselt number (describing heat transfer).

The differential equations of heat and mass transfer are presented below for a two-dimensional turbulent boundary layer. If the boundary conditions are similar, the solutions of the equations have the same form.

Heat transfer

$$u\frac{\partial T}{\partial x} + v\frac{\partial T}{\partial y} = \left(a + \varepsilon_T\right)\frac{\partial^2 T}{\partial^2 y} \tag{10.2}$$

Mass transfer

$$u\frac{\partial c}{\partial x} + v\frac{\partial c}{\partial y} = \left(D + \varepsilon_M\right)\frac{\partial^2 c}{\partial^2 y} \tag{10.3}$$

In Equations (10.2) and (10.3), u and v represent the velocities in the boundary layer, while a is the thermal diffusivity and D the diffusion coefficient. The is the turbulent eddy viscosity for heat transport and is the turbulent eddy viscosity for mass transport. T is the local temperature and c is the concentration of a species. The Nusselt and Sherwood numbers are defined as,

$$\mathrm{Nu} = \frac{hL}{k_f} \tag{10.4}$$

$$S_h = \frac{h_M}{D_f} \tag{10.5}$$

The Lewis number Le and the Prandtl number Pr are introduced and defined as,

$$\mathrm{Le} = \frac{\rho c_p D_f}{k_f} \tag{10.6}$$

$$\mathrm{Pr} = \frac{\mu c_p}{k_f} \tag{10.7}$$

If Le = 1 and Nu = S_h, then the Lewis relation is obtained as,

$$h_M = \frac{h}{\rho c_p} \tag{10.8}$$

The analogy requires ideally that the Prandtl and Schmidt numbers are equal. Nevertheless, the naphthalene mass transfer method is applicable to many heat transfer problems under certain conditions. The Sc number of naphthalene vapor is 2.28 at 25°C and is decently close to the Pr of many gases and liquids. For air, Pr = 0.72, while for water it is in the range of Pr = 1–14. It is then stated that the analogy can be applied with reasonable confidence.

The analogy method relies on accurate values of the naphthalene properties. The basic properties of significance are the density of solid naphthalene, the saturated vapor density or saturated vapor pressure, mass diffusion coefficient in air, or the Schmidt number. The saturation vapor pressure of naphthalene is very sensitive to temperature. Then special care is needed to obtain an accurate value of the surface temperature during the measurements.

The procedure in using the naphthalene sublimation technique is as follows: (a) provide a specimen with naphthalene, (b) measure the initial naphthalene surface shape or the initial mass of the naphthalene specimen, (c) conduct the experiment with the naphthalene specimen, (d) measure the naphthalene surface shape or mass of specimen after the experiment. This gives the mass transfer rates from the sublimation depths or from the difference in mass together with the time the specimen was exposed to the flow, (e) reduce the data to obtain the mass/heat transfer coefficient and Sherwood number.

Note that the vapor pressure of naphthalene is sensitive to temperature so precautions must be taken to ensure that thermal equilibrium prevails with the air temperature in the laboratory and temperatures must be measured accurately. Figure 10.11 illustrates the principle of the naphthalene sublimation technique.

10.5.1 Limitations in application of the naphthalene sublimation technique

There are some limitations to be aware of in the application of this measuring technique: (a) in low-velocity applications, long run time is needed; (b) in high-velocity applications, aerodynamic or viscous heating may affect the measurement; (c) in high shear flow situations, mechanical erosion might be a concern; (d) change of shape due to sublimation – this is noticed as

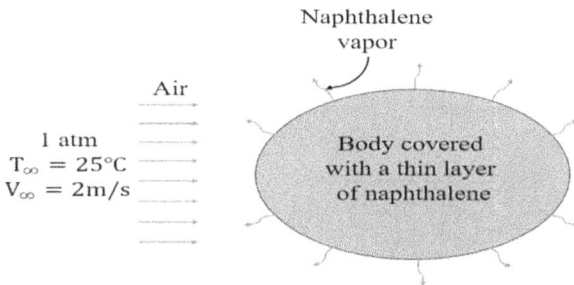

Figure 10.11 Principle of the naphthalene technique.

a gradual shape change due to the sublimation; (e) mainly time-averaged values can be obtained accurately, while transient values might be hard to achieve; (f) for complicated geometries, casting and machining might be difficult; (g) it is generally recommended to use a wind tunnel of the suction type as the test section is placed upstream of the blower, and it is then more easy to maintain a constant flow temperature; (h) due to the latent heat of sublimation, the vapor temperature near the wall is different from that in the freestream – this is stated as a concern mostly for natural convection tests; (i) the uncertainty is generally reported as 7% at the 95% confidence level.

An early investigation using the naphthalene sublimation technique to study evaporation from falling drops was carried out by Frössling [15]. The technique has been further developed and applied extensively in gas turbine heat transfer research, not the least at University of Minnesota, Minneapolis, Minnesota, USA. Review articles are available, [16, 17].

10.6 PRESSURE-SENSITIVE PAINT TECHNIQUE

The pressure-sensitive paint (PSP) technique is an optical method providing measurement of pressure on surfaces without using disturbing sensors, and there are no effects on the considered surface. To obtain the surface pressure, the model is coated with a specific paint containing luminophore molecules, which is illuminated by, for example, a UV light source emitting light of sufficient short wavelength to excite the luminophore. The intensity of the resulting fluorescence light is related to the local oxygen concentration and allows for determination of a two-dimensional pressure field. This technique has often been used in investigations of film cooling effectiveness for gas turbine heat transfer applications. The theoretical basis of the PSP technique relies on the heat and mass transfer analogy already presented in Section 10.5.

Figure 10.12 shows the boundary layer analogy between heat and mass transfer as film cooling is considered. T_{aw} denotes the adiabatic wall temperature and T_{∞} denotes the freestream temperature.

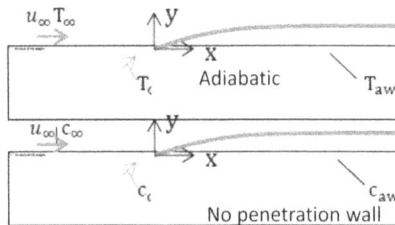

Figure 10.12 Film cooling using heat transfer and mass transfer procedures.

In the heat transfer measurements, the film cooling effectiveness is given by,

$$\eta = \frac{T_\infty - T_{aw}}{T_\infty - T_c} \qquad (10.9)$$

In the corresponding mass transfer experiments, the film cooling effectiveness is given by,

$$\eta = \frac{c_\infty - c_{aw}}{c_\infty - c_c} \qquad (10.10)$$

When the mainstream is air and the secondary flow is nitrogen (N_2), the film cooling effectiveness can be derived as:

$$\eta = \frac{T_\infty - T_{aw}}{T_\infty - T_c} = \frac{c_\infty - c_{aw}}{c_\infty - c_c} = \frac{c_{O2,Air} - c_{O2\,N2+Air}}{c_{O2,Air} - c_{O2,N2}} = 1 - \frac{c_{O2\,N2+Air}}{c_{O2,Air}} \qquad (10.11)$$

It is also possible to use C_{O2} or Ar, etc. as the secondary flow.

The PSP characteristics go like this. After being irradiated by excitation light, the paint emits fluorescence, and the intensity of the fluorescence light is related to the partial pressure of oxygen on the paint surface. On the test surface, as the partial pressure of the oxygen goes up, the luminous intensity I goes down. The modified Stern–Volmer equation can be used to describe the relationship between the oxygen partial pressure and the luminous intensity. This equation reads.

$$\frac{I_R - I_B}{I - I_B} = \alpha + \beta \frac{p_{O2}}{p_{O_2R}} \qquad (10.12)$$

where α and β are obtained by calibration. Finally, one has

$$\eta = 1 - \frac{p_{O2,N2+Air}}{p_{O2,Air}} = 1 - \frac{p_{O2,N2+Air}/p_{O2,R}}{p_{O2,Air}/p_{O2R}} = 1 - f\left(\frac{I_R - I_B}{I_{N2+Air} - I_B}\right) \Big/ f\left(\frac{I_R - I_B}{I_{Air} - I_B}\right) \qquad (10.13)$$

10.6.1 Experimental procedure of the PSP to measure the film cooling effectiveness

The procedure is illustrated by the four steps in Figure 10.13. First, the excitation light, mainstream, and secondary flows are turned off to obtain background luminous intensity (I_B). In the second step, the mainstream and secondary flows are turned off, while the excitation light is turned on to obtain the reference luminous intensity (I_R). In the third step, the excitation light, mainstream, and secondary flow (air) are turned on to obtain

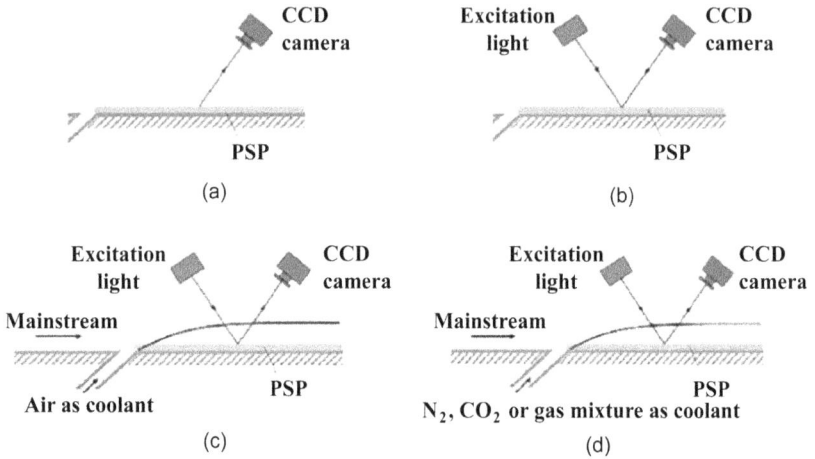

Figure 10.13 The four steps in the PSP experimental procedure. (a) Image of Background; (b) Image of Reference; (c) Image of Air as Coolant; and (d) Image of N_2, CO_2 or Gas mixture as Coolant.

the luminous intensity $I_{O2,\,air}$. Finally in the fourth step, the excitation light, mainstream, and secondary flow (nitrogen) are turned on to obtain the luminous intensity $I_{O2,\,fg}$.

After completing the above four steps, the film cooling effectiveness can be calculated by Eqn. (10.13). Recent investigations using the PSP technique are exemplified by Refs. [18, 19].

10.7 PARTICLE IMAGE VELOCIMETRY, PIV

PIV techniques have been developed significantly and are now widely used in studies of fluid mechanics and convective heat transfer. PIV is often used to measure the instantaneous velocity field in a two-dimensional plane with rather high spatial resolution. The velocity is computed by finding the displacement of tracer particles in a certain time interval. A principal sketch is shown in Figure 10.14.

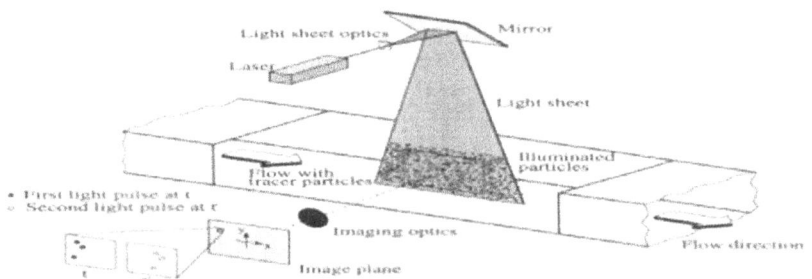

Figure 10.14 Principles of PIV.

Seeding of small tracer particles in the fluid flow takes place. A pulsed light sheet is applied to illuminate the interesting area of the flow field twice to catch the fluid movement. The light scattered by the tracer particles is recorded by a CCD sensor. The digital PIV recording is then divided into small interrogation windows. The particle displacement is calculated by cross correlating the two corresponding interrogation windows. By fast Fourier transformation (FFT), the position of the highest peak in the correlation domain indicates the mean displacement of the particles in a particular interrogation window. The process is repeated and then the displacement vectors of all interrogation windows are transformed into a complete instantaneous velocity map.

10.7.1 Measuring procedure

The measuring procedure is illustrated based on a specific experimental setup for flow measurements in duct flows [10]. A commercially available PIV system developed by Dantec Dynamics was used. The system is arranged to measure the velocity fields in the vertical symmetry plane and the horizontal plane of a duct with ribs aimed for heat transfer augmentation. Oil-based aerosol with mean particle diameter of 1 μm is generated by a TSI 9306 six-jet atomizer. The concentration of the seeding particles is regulated by compressed air pressure and number of Laskin nozzles. To obtain homogeneous trace particles, the duct inlet is preceded by a plenum with two screens. A Quantel Q-switched Nd:YAG laser provides the pulsed illumination with a wavelength of 532 nm. The duration of each pulse is 10 ns, and the maximum output energy is 120 mJ. The light-sheet thickness in the test section is kept at 0.8 mm. A digital camera containing a CCD chip with 1,280 × 1,024 pixels and a Nikon AF Micro 60f/2.8D lens with an optical filter were used to record the particle images. Typically, the magnification factor ranges from 50 μm/pixel to 70 μm/pixel.

The time interval between the laser pulses is set such that the particles move at most 8 pixels between a pair of images. In the cited experiments, the time interval varies from 10 to 20 μs depending on the velocity magnitude and the desired resolution. Within each window, the number of seeding particles is greater than 5, and the particle image size projected onto the CCD sensor is approximately 3 pixels. FFT-based cross-correlation coupled with a two-dimensional Gaussian fit was applied to find the position of the correlation peak.

10.7.2 Estimation of uncertainties in PIV

Assuming that the measured particle displacement is accurate to the extent between 0.01 to 0.1 pixels, the experimental uncertainty in the instantaneous velocity measurement is estimated to be less than 2%. The sample size is another error source for the mean and fluctuating velocity statistics.

Commonly 500 generated images have been used for each measurement plane.

The uncertainties of the velocity gradient and shear stress (-u'v') are estimated to be within 5% and 10%, respectively.

Figure 10.15 shows recorded streamline distributions for a duct with a solid rib and a perforated rib (enabling penetration through the rib). The measurements clearly indicate the flow physics.

A general description of the PIV technique can be found in Ref. [20].

10.8 HOT-WIRE ANEMOMETRY

Hot-wire anemometry is a technique to measure the velocity in flow fields. Figure 10.16 shows the principle for a single wire as well as the prongs and probe body. Nevertheless, wires can appear in many different configurations.

Hot wires can be made very small to minimize their disturbances on the measured flow. They are sensitive to rapid changes in flow velocity as the wire has a small time constant. An energy balance equation is used to describe the heating and cooling of the wire. Such an equation can then be evaluated to determine the velocity of the fluid flowing across the wire.

Hot-wire anemometers can be operated in either constant current or constant temperature configurations. For the constant current operation, there is a risk of burning out the wire if the cooling flow is too low. If the flow is too high, the wire will not be heated up sufficiently to provide data of sufficient quality. The constant temperature configuration is accordingly the most common.

Figure 10.17 displays a Wheatstone bridge used for the constant temperature operation of a hot wire.

10.8.1 Heat balance equation

The heat generation by Joule heating is given by:

$$H_g = I^2 R_w \tag{10.14}$$

where I stand for the current through the circuit and R_w is the wire resistance at temperature T_w.

The wire resistance is assumed to vary linearly with temperature as

$$R_w = R_0 \left[1 + C \left(T_w - T_o \right) \right] \tag{10.15}$$

where C is the temperature coefficient of the resistivity.

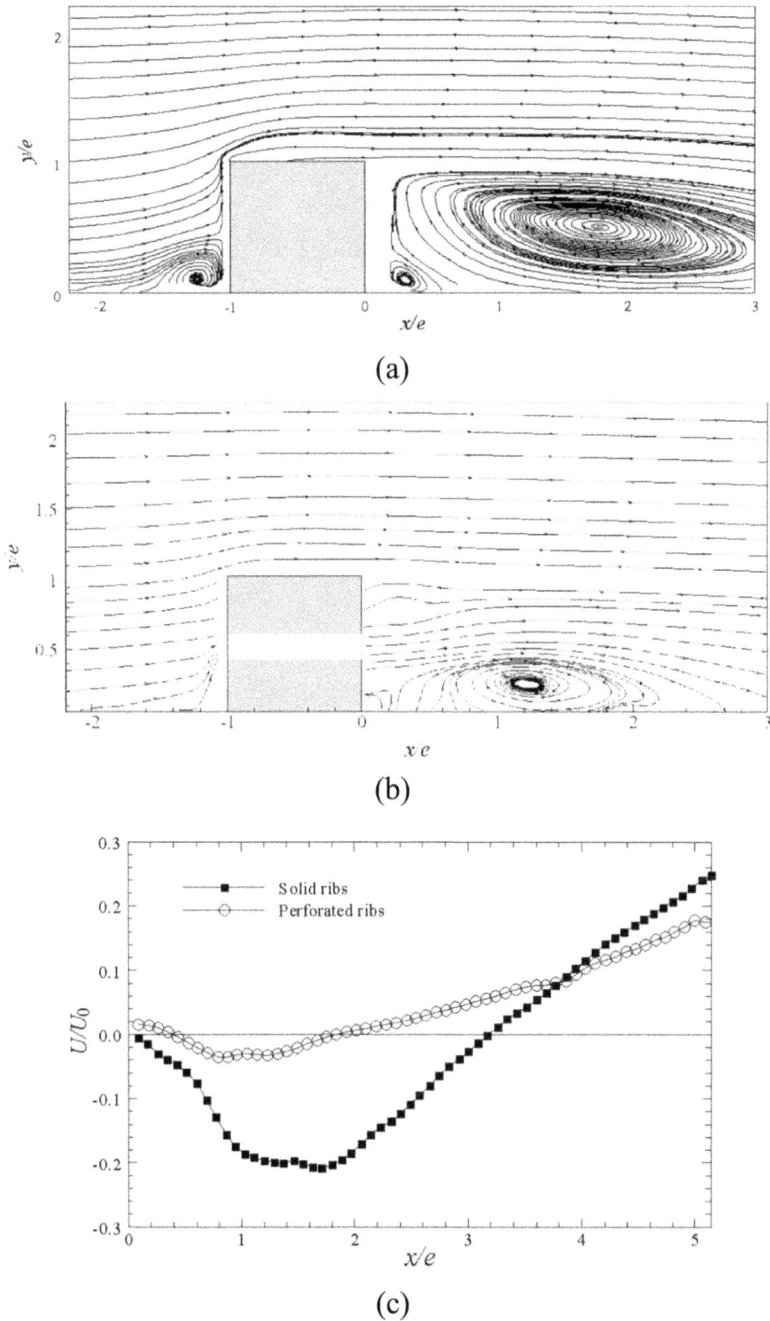

(a)

(b)

(c)

Figure 10.15 Illustration of application of the PIV technique: (a) the flow field around a square solid rib, (b) corresponding flow field for a perforated rib, (c) velocities profiles along the centerline downstream the ribs in (a) and (b). (From Refs. [9, 10].)

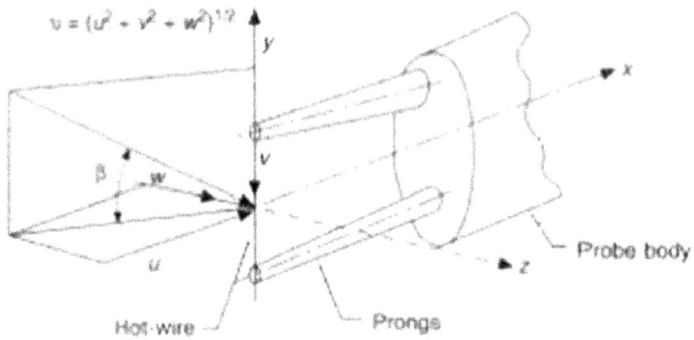

Figure 10.16 Sketch of hot-wire anemometry.

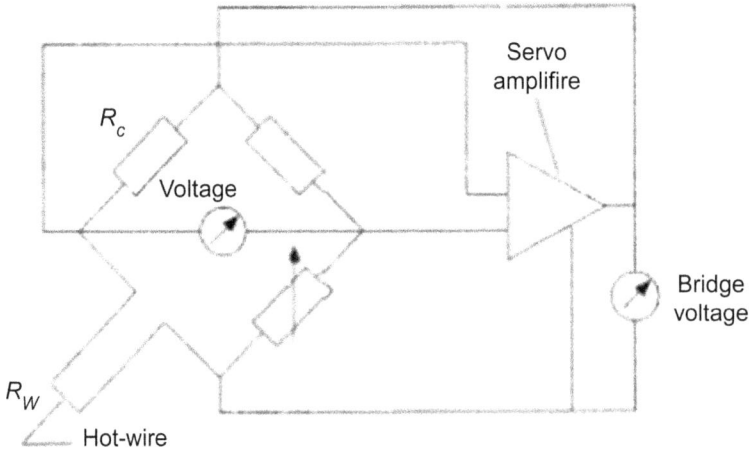

Figure 10.17 Constant temperature circuit diagram Wheatstone bridge.

The convective heat transfer coefficient as a fluid is passing across an infinite rod and is given by the Nusselt number correlation:

$$\text{Nu} = \frac{hd}{k_f} = \text{Re}^{0.5} + 0.42\,\text{Pr}^{0.2} + 0.57\,\text{Pr}^{0.33} \tag{10.16}$$

where

$$\text{Re} = \frac{ud}{\mu/\rho} \tag{10.17}$$

The heat balance yields,

$$H_g = hA_s\left(T_w - T_{gas}\right)$$ (10.18)

Accordingly, a relation between the velocity and current can be found as,

$$I^2 = A + B\sqrt{U}$$ (10.19)

This equation is referred to as King's law. The coefficients A and B are found by calibration. Conveniently fluctuating components can also be evaluated to find, e.g., the turbulence intensity.

In Ref. [21], more details about hot-wire anemometry can be found.

10.9 UNCERTAINTY ANALYSIS IN MEASUREMENTS

The uncertainty analysis heat transfer measurements can be performed by applying the estimation method described by Moffat [22]. For LCT measurements, the uncertainty (at the 95% confidence level) in local Nusselt number values considers the effects of errors in the measurement of voltage and current, fluid temperature and wall temperature. In addition, the calculation of heat losses by conduction and radiation should be included in estimation of uncertainty. The uncertainty (at the 95% confidence level) in temperature by LCT can be estimated as ±0.2 K. This value also considers the uncertainty of the thermocouples being used to calibrate the liquid crystal. In addition, the imperfection of the experiment, i.e., conduction along the calibration plate, effect of illumination, lack of perfect repeatability, etc. can be considered. The uncertainty in the local Nusselt number is commonly estimated to be within ± 6.5%. The Reynolds number may have a calculated uncertainty of ± 6.2%.

Similar uncertainties of the other measurement techniques can be found in the cited references.

REFERENCES

[1] E.R.G. Eckert, R.J. Goldstein, *Measurements in heat transfer*, Hemisphere Publishing Corporation, USA, 1976.
[2] J.C. Han, L.M. Wright, *Experimental methods in heat transfer and fluid mechanics*, CRC Press, Taylor & Francis Group, Boca Raton, USA, 2020.
[3] C. Falsetti, M. Sisti, P.F. Beard, Infrared thermography and calibration techniques for gas turbine applications: A review, *Infrared Phys. Technol.*, Vol. 113, 103574, 2021.
[4] A. Schulz, IR thermography as applied to film cooling in gas turbine components, *Meas. Sci. Technol.*, Vol. 11(7), 946–956, 2000.

[5] Z. Ghorbani-Tari, B. Sunden, G. Tanda, On liquid crystal thermography for determination of the heat transfer coefficients in rectangular ducts, in *Computational Methods and Experimental Measurements XV* (eds. G.M. Carlomagno, C.A. Brebbia), 255–266, WIT Press, 2011.

[6] S. Li, Z. Ghorbani-Tari, G. Xie, B. Sunden, An experimental and numerical study of flow and heat transfer in ribbed channels with large rib pitch-to-height ratios, *J. Enhanc. Heat Transf.*, Vol. 20(4), 305–319, 2013.

[7] Z. Ghorbani-Tari, B. Sunden, G. Tanda, Experimental study of convection heat transfer in the entrance region of a rectangular duct with transverse ribs, ASME HT2012-58178, 2012.

[8] Z. Ghorbani-Tari, B. Sunden, Experimental study of heat transfer control around an obstacle by using a rib, *Heat Transf. Res.*, Vol. 47(8), 781–795, 2016.

[9] C. Wang, Experimental study of outlet guide vane heat transfer and gas turbine internal cooling, PhD thesis, Lund University, Sweden, 2016.

[10] J. Liu, S. Hussain, W. Wang, G. Xie, B. Sundén, Experimental and numerical investigations of heat transfer and fluid flow in a rectangular channel with perforated ribs, *Int. Comm. Heat Mass Transf.*, Vol. 121, 105083, 2021.

[11] D. Kumar, R. Choudhury, A. Layek, Application of liquid crystal thermography for temperature measurements of the absorber plate of solar air heater, *Mater. Today: Proc.*, Vol. 59, 605–611, 2022.

[12] J. Schmid, M. Gaffuri, A. Terzis, P. Ott, J. von Wolfersdorf, Transient LCT using a time varying heat flux, *Int. J. Heat Mass Transf.*, Vol. 179, 121718, 2021.

[13] S.V. Ekkad, P. Singh, LCT in gas turbine heat transfer: A review on measurement technology and recent investigations, *Crystals*, Vol. 11, 2021.

[14] T.W. Kerlin, Practical thermocouple thermometry, ISA, 1999.

[15] N. Frössling, Über die Verdünstung fallender Tropfen, *Gerlands Beiträge Zür Geophysik*, Vol. 52, 170–216, 1938.

[16] P.R. Sousa Mendes, The naphthalene sublimation technique, *Exp. Therm. Fluid Sci.*, Vol. 4(5), 510–523, 1991.

[17] R.J. Goldstein, H.H. Cho, A review of mass transfer measurements using naphthalene sublimation technique, *Exp. Therm. Fluid Sci.*, Vol. 10(4), 416–434, 1995.

[18] L. Gao, G. Yang, T. Gao, R. Li, X. Hu, Experimental investigation of a linear cascade with large solidity using PSP and dual camera system, *Therm. Sci.*, Vol. 30, 682–695, 2021.

[19] B.L. Zhang, H.R. Zhu, C.Y. Yao, B. Sunden, Experimental study on film cooling performance of a turbine blade tip with a trapezoidal slot cooling scheme in transonic flow using PSP technique, *Exp. Therm. Fluid Sci.*, Vol. 130, 110513, 2021.

[20] M. Paffel, C.L. Willert, F. Scarano, C.J. Kähler, S.T. Wereley, J. Kompenhaus, *Particle image velocimetry- A practical guide*, 3rd Ed., Springer, Cham, 2018.

[21] C.G. Lomas, *Fundamentals of hot wire anemometry*, Cambridge University Press, Cambridge, UK, 2021.

[22] R.J. Moffat, Describing the uncertainties in experimental results, *Exp. Therm. Fluid Sci.*, Vol. 1,1, 3–17, 1988.

Chapter 11

Computational methods used in heat transfer

11.1 INTRODUCTION

In the late 1960s, computational methods began to have a significant impact on the analysis of fluid flow and heat transfer as well as on thermal design. Initially, so-called panel methods were introduced in aerospace applications. These methods were based on the distribution of surface singularities on given configurations. Nonviscous flows around immersed bodies could be solved. Later, more capabilities were added to surface panel methods and accordingly it was possible to include more accurate higher-order formulations, lifting capability, unsteady flows and coupling with boundary layer formulations. Nevertheless, the panel methods could not offer accurate solutions for high-speed nonlinear flows, and thus more advanced models of the flow field equations were needed. Gradually this development led to what is nowadays called computational fluid dynamics (CFD). In [1], further details can be found.

CFD is interdisciplinary with a wide spectrum of applications in science and engineering. Momentum transfer in fluid flow, heat and mass transfer, combustion and chemical reactions appear frequently, and these processes are of strong significance in automotive, space and aviation, chemical process industries as well as in atmospheric science, energy, medicine, and micro- and nanotechnology. Rapid development and extended applications of CFD have emerged, and it is nowadays used as a modeling or simulation tool as well as for R&D in many industries. Besides, developments for new challenges continue, and they are used in fundamental research at universities.

In some applications, for example aerospace, the fluid flow is compressible, and the density of the fluid varies with pressure. Commonly the flow velocity is high, and the Mach number is greater than 0.3. If the Mach number is between 0.3 and 0.8, the subsonic regime prevails. The relation between pressure and density is then weak, and no shocks will be found within the flow. On the other hand, a highly compressible flow has a Mach number greater than 0.8. The density is then strongly sensitive to pressure, and shocks may be formed. A transonic flow regime is defined for Mach numbers in the range of $0.8 < Ma < 1.2$, while a supersonic regime is stated

DOI: 10.1201/9781003229865-11

in the range of 1.2 < Ma< 3.0. In a supersonic flow, the pressure effects are only transported downstream, and accordingly the upstream flow is not affected by conditions and disturbances downstream.

The total temperature, T_t, is the key parameter, and it is the sum of the static and dynamic temperatures. The total temperature can be calculated in two ways, namely

$$T_t = T + \frac{V_i^2}{2c_p} \text{ or } T_t = T\left(1 + \frac{\gamma - 1}{2}\text{Ma}^2\right) \tag{11.1}$$

where V is the flow velocity and c_p the specific heat of the gas.

Another useful quantity in analysis of compressible flows is the total pressure, P_t. It is defined as the sum of the static and the dynamic pressures.

Compressible flows are generally much more sensitive to the boundary conditions and material properties than incompressible flows. If the applied settings are not defining a physically real flow situation, then the analysis might be very unstable and even fail to reach a converged numerical solution. Formulation of the boundary conditions and material properties in an appropriate manner will improve the opportunities to achieve a successful analysis.

In analysis of heat transfer in compressible cases, it is recommended to apply total (stagnation) temperature boundary conditions at the inlets instead of static temperatures. Total temperature should also be applied to any solids or walls with known temperature boundary conditions. Note that when heat transfer is present in compressible convective flow analysis, viscous dissipation, pressure work, and kinetic energy terms are included.

This chapter summarizes CFD methods, including modeling of turbulent convective flows, and associated problems and limitations, as well as providing presentations of CFD applications in heat transfer engineering. Available commercial computer software and in-house codes are briefly mentioned.

11.2 GOVERNING EQUATIONS

The well-established governing partial differential equations of mass conservation, transport of momentum, energy, and mass fraction of species can all be formulated by a general partial differential equation as [2, 3],

$$\frac{\partial \rho \varphi}{\partial t} + \frac{\partial}{\partial x_j}\rho \varphi u_j = \frac{\partial}{\partial x_j}\left(\Gamma \frac{\partial \varphi}{\partial x_j}\right) + S \tag{11.2}$$

where φ is the arbitrary dependent variable, for example, the components of the velocity field, temperature, and others. Γ is a generalized diffusion coefficient, while S represents the source term for φ. This general differential equation has four terms. These are referred to as, from left to right in Eqn. (11.2), unsteady term, convection term, diffusion term, and source term.

11.3 ON NUMERICAL METHODS TO SOLVE PARTIAL DIFFERENTIAL EQUATIONS

There are methods well established to numerically solve the governing equations of fluid flow, heat transfer, and mass transport phenomena. These are referred to as the finite difference method (FDM) [4], the finite volume method (FVM) [2, 3], the finite element method (FEM) [5, 6], the control volume finite element method (CVFEM) [7], and the boundary element method (BEM) [8]. In this chapter, only details of the FVM are highlighted.

11.3.1 The finite volume method

As the FVM is applied, the computational domain is divided into several so-called control volumes. The integral forms of the conservation equations are used for all control volumes. At the control volume center, a node point is located. At this node, the variables are stored. Then, the values of the variables at the control volume faces are found by interpolation. The surface and volume integrals are evaluated by quadrature formulas. Algebraic equations are obtained for each control volume. These equations include values of the variables for neighboring control volumes.

For complex geometries, the FVM has been found quite suitable as it is conservative if the surface integrals are the same for control volumes sharing boundary.

The FVM appears frequently in particular for analysis and simulations of convective fluid flow and heat transfer. It appears also in many of the existing commercial CFD codes. Further details are available in [2, 3]. Below a brief illustration is presented, and an arbitrary control volume is depicted in Figure 11.1.

A formal integration of the general equation across the control volume can be written as,

$$\iiint_V \frac{\partial \rho U_j \phi}{\partial x_j} dV = \iiint_V \frac{\partial}{\partial x_j}\left(\Gamma_\phi \frac{\partial \phi}{\partial x_j}\right) dV + \iiint_V S_\phi dV \qquad (11.3)$$

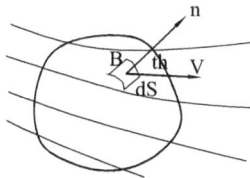

Figure 11.1 Sketch of an arbitrary control volume.

By applying the Gaussian theorem or the divergence theorem, one finds,

$$\iint_S \rho\phi\vec{U} \cdot d\vec{S} = \iint_S \Gamma_\phi\Delta\phi \cdot d\vec{S} + \iiint_V S_\phi dV \qquad (11.4)$$

Summing up over all the faces of the control volume, this equation can be transferred to

$$\sum_{f=1}^{nf} \phi_f C_f = \sum_{f=1}^{nf} D_f + S_\phi \Delta V \qquad (11.5)$$

where the convection flux is denoted as C_f and the diffusion flux as D_f, and the scalar value of the arbitrary variable ϕ at a face, Φ_f, has to be found.

11.3.1.1 Convection-diffusion schemes

A convection-diffusion numerical scheme must have the properties of conservativeness, boundedness, and transportiveness to ensure that physically realistic results will be reached and that the iterative solutions are stable. The schemes called upstream, hybrid, and power-law possess these properties and are commonly found to be stable, but, unfortunately, they suffer from numerical or false diffusion in multidimensional flows when the velocity vector is not parallel with one of the coordinate directions. On the other hand, the central difference scheme lacks transportiveness and is well known to generate unrealistic solutions at large Peclet numbers. Higher-order schemes, for example, QUICK, van Leer, and others, may minimize the false diffusion but are less numerically stable. In addition, implementation of boundary conditions can be problematic with the higher-order schemes, and the computational demand can become extensive because additional grid points are needed and the expressions for the coefficients in Eqn. (11.5) will be more complex.

11.3.1.2 Source term

The source term S depends in general on the considered variable φ. In the discretized equation, it is desirable to account for such a dependence. Commonly, the source term is assumed to follow a linear relationship to φ.

At a grid point P, S is then written as,

$$S = S_C + S_P\varphi \qquad (11.6)$$

It is required that S_P is negative to avoid divergence.

11.3.1.3 Solution of discretized equations

The discretized equations have the form of Eqn. (11.5) where the ϕ-values at the grid points are the unknowns. At boundaries without fixed ϕ-values, the boundary values can be eliminated by using prescribed conditions of the fluxes at such boundaries. Gauss elimination is a so-called direct method to solve algebraic equations. For one-dimensional cases, the coefficients create a tridiagonal matrix, and an efficient algorithm has been developed. This algorithm is called the Thomas algorithm, or the tridiagonal matrix algorithm (TDMA). For two-dimensional and three-dimensional cases, direct methods require large computer memory and long computational time. Accordingly, iterative methods are commonly applied to solve the algebraic equations. A popular method is the line-by-line technique combined with a block correction procedure. The TDMA is applied to solve the equations along the chosen line. Because the equations are nonlinear and sometimes interlinked, the iterative methods are needed

In turbulent forced convection, the change in the value of the variable φ from one iteration to another is very high so that convergence of the iterative process may not be achieved at all. To circumvent this and to limit the magnitude of the changes, so-called under-relaxation factors (between 0 and 1) are introduced.

11.3.1.4 Pressure in momentum equations

A pressure gradient term appears in every coordinate direction (i.e., a source term S). If these gradients would be known a priori, solution of the discretized equations for the fluid flow velocities could follow the same procedure as for any scalar. However, commonly the pressure gradients are not known but must be determined as part of the solution. Thus, the pressure and velocity fields are interlinked, and the continuity equation (mass conservation equation) must be used to develop a solution strategy.

Also, other related difficulties exist in solving the momentum and continuity equations. It is known that if the velocity components and the pressure are calculated at the same grid points in a straightforward manner, some physically unrealistic fields, like checker-board solutions, may appear in the numerical solution. A remedy to avoid such problems is the application of so-called staggered grids. Then, the velocity components are placed at staggered or displaced locations. These locations are at the control volume faces that are perpendicular to them. All the other variables are determined at the ordinary grid points. Another remedy has been developed and then a non-staggered or collocated grid is used. All the variables are stored at the ordinary grid points. However, a special interpolation scheme is applied to find the velocity components at the faces of the control volumes. The Rhie–Chow interpolation method is commonly applied; see Ref. [9].

11.3.1.5 Procedures for solution of the momentum equations

In the preceding section, it was stated that the velocity and pressure fields are coupled. A strategy must be developed in the solution procedure of the momentum equations. The semi-implicit-method-pressure-linked-equations (SIMPLE) algorithm is the oldest one developed. First, a pressure field is guessed, and then the momentum equations are solved to obtain a corresponding velocity field. Then pressure correction and velocity corrections are introduced. An algebraic equation for the pressure correction can be derived from the continuity equation. The velocity corrections are linked to the pressure corrections, and coefficients linking the velocity corrections to the pressure correction depend on the considered algorithm; see below.

The momentum equations are then solved again but now with the corrected pressure as the guessed pressure. One obtains updated velocities and updated pressure, and velocity corrections are determined. This process is repeated until convergence is reached.

There are some other but similar algorithms available today. So, for instance, SIMPLEC (SIMPLE-consistent) and SIMPLEX (SIMPLE-extended) are common. The difference from SIMPLE is mainly in the coefficients linking velocity corrections to the pressure correction, see, e.g., Jang et al. [10].

The algorithm named pressure implicit splitting operators (PISO) (see, e.g., Issa [11]), has received popularity more recently. Originally it was developed as a pressure–velocity coupling strategy for unsteady compressible flow. In contrast to the SIMPLE algorithm, it has one predictor step and two corrector steps.

Another algorithm is the SIMPLER one (SIMPLE-revised). In this algorithm, the mass conservation equation is used to derive a specific discretized equation for the pressure. To update the velocities through the velocity corrections, still the pressure correction is used.

In the standard pressure discretization procedure, interpolation of the pressure on the faces using cell center values is carried out. In contrast, the presto! discretization for pressure calculates pressure on the faces. As staggered grids are used, this is enabled, and then velocities and pressure are not co-located. The presto! is stated to provide more accurate results because errors of interpolation and assumptions of pressure gradient on boundaries are eliminated. This algorithm has been found to operate well for flows with strong body forces (i.e., swirl) and high Rayleigh number flows (i.e., buoyant convection). Presto! however, has been found to be computationally demanding, because one needs to allocate additional memory for alternate grids. When one solves buoyant convection flows using the standard discretization, one assumes that a zero-pressure gradient exists at the wall which results in erroneous velocities at the boundary. On the other hand, if one uses the presto! algorithm, one accounts for the pressure gradient at the boundary and accordingly obtains a correct solution. More details about presto! can be found in [12].

11.3.1.6 Convergence

The numerical solution procedure is commonly iterative, and a criterion must be met to claim that a converged solution has been achieved. One way is to calculate the residual R, which is defined as,

$$R = \sum_{NB} a_{NB}\phi_{NB} + b - a_P\phi_P \tag{11.7}$$

for all the variables. NB here stands for neighboring grid points, for example, E, W, N, S (east, west, north, south). If a solution has reached convergence, $R = 0$ everywhere. In practice, it is common to state that the largest value of the residuals $[R]$ should be less than a certain value. Then, the solution is claimed to be converged.

11.3.1.7 Number of grid points and control volumes

The sizes of the control volumes do not need to be uniform, and the successive grid points can be unequally spaced. However, often it is desirable to keep the grid spacing uniform. A fine grid is employed where steep gradients appear, while a coarser grid spacing may suffice where small variations are expected. Various turbulence models require certain conditions on the grid structure closest to solid walls. High and low Reynolds number versions of such models demand alternate conditions.

The solution procedure is recommended to be carried out on many setups of the grids with different fineness and varying degrees of nonuniformity. Then opportunities to estimate the accuracy of the numerical solution procedure are established.

Adoption of the adaptive grid techniques can provide advantages to increase the resolution in critical areas where, e.g., resolution of the pressure jump downstream of a shockwave is needed. In hypersonic flow modeling, meshing of external flows needs special care, especially near shock conditions as capturing shock effects needs a fine mesh.

11.3.1.8 Complex geometries

In complex or irregular geometries, the CFD methods based on Cartesian, cylindrical, or spherical coordinate systems have obvious limitations. Using Cartesian, cylindrical, and/or spherical coordinates forces the boundary surfaces to be treated in a stepwise manner. Approaches based on body-fitted or curvilinear orthogonal, nonorthogonal grid systems are needed to overcome this problem. These grid systems can be unstructured, structured, or block-structured or of a composite/hybrid type. Because the grid lines follow the boundaries, it becomes simpler to implement boundary conditions.

Nonorthogonal grids also present disadvantages. The transformed equations contain more terms, and the grid nonorthogonality may cause unphysical solutions. Vectors and tensors may be defined as Cartesian, covariant, contravariant, and physical or nonphysical coordinate oriented. The arrangement of the variables on the grid affects the efficiency and accuracy of the solution algorithm.

Dividing the domain of interest in small parts or small subdomains is called grid generation or meshing. It is an important issue as it affects the simulation time as well as the accuracy of the computed results. Presently many commercial CFD packages have their own grid generators but also several grid generation packages are available. These are compatible with some of the CFD codes. Interaction with various computer-aided-design (CAD) packages is also a requested feature today. In the first paragraph above, the basics of various algorithms were mentioned, and such basics have enabled creation of high-quality grids. Additional information on how to handle complex geometries is available in Refs. [13, 14].

11.4 THE CFD APPROACH

The FVM approach described in the previous sections is a popular particularly for convective fluid flow and heat transfer. Several of the commercial CFD codes apply this approach. In heat exchangers and other heat transfer equipment, both laminar and turbulent flows occur. Turbulent flow and heat transfer normally require modeling approaches, while laminar convective flow and heat transfer can be simulated. By applying turbulence modeling, the goal is to consider all the relevant physics by using as simple mathematical model as possible. The instantaneous mass conservation, momentum, and energy equations provide a closed set of the five unknown variables, i.e., u, v, w, p, and T. However, the computing demands concerning resolution in space and time for direct solution of the time-dependent fully turbulent flows at high Reynolds numbers (so-called DNS calculations) are huge and major developments in computer hardware are still needed. Accordingly, DNS is viewed as a research tool for rather simple flows at moderate Reynolds number, and calculations on supercomputer are required. In practice, thermal engineers request computational procedures supplying information about the turbulent processes, but there is no need to know the effects of every single eddy in the flow. This means that information about the time-averaged properties of the flow and temperature fields (e.g., mean velocities, mean stresses, mean temperature etc.) are sufficient. Commonly, then the so-called Reynolds decomposition is executed. Any variable is then written as the sum of its time-averaged value and a superimposed fluctuating value. Additional unknowns appear in the governing equations, six for the momentum equations and three for the temperature field (energy) equation. The appearing terms are called turbulent stresses and turbulent heat fluxes, respectively. The aim of turbulence modeling is to enable prediction of these additional unknowns, i.e., the turbulent stresses and turbulent heat fluxes, with sufficient generality and accuracy. The methods established by the Reynolds averaged equations are commonly referred to as RANS (Reynolds Averaged Navier–Stokes equations) approaches. The large eddy simulation (LES) falls between the DNS and RANS approaches concerning the computational demand. Like DNS, 3D simulations are carried out over quite many time steps but only the larger eddies are resolved. An LES grid is in general coarser in space, and the time steps can be bigger than for DNS as the small-scale fluid motions are treated by a sub-grid-scale (SGS) model.

11.4.1 Turbulence models

Turbulence models commonly appearing in industrial and aerospace applications can be classified as (a) zero-equation models, (b) one-equation models, (c) two-equation models, (d) Reynolds stress models, (e) algebraic stress models (ASMs), and (f) LES. Besides those mentioned, direct numerical solution (DNS) and detached eddy simulations (DES) have also been developed. These will also be described briefly below.

The models (a), (b), and (c) account for the turbulent stresses and heat fluxes by introducing a turbulent viscosity (or eddy viscosity) and a turbulent diffusivity (or eddy diffusivity). Linear and nonlinear versions of the models have been presented; see [15–17]. A popular one-equation model is the Spalart–Allmaras model [18] in which a specific transport equation is solved for the eddy viscosity. This model has been used extensively for aerospace and turbomachinery applications but is not very common for heat exchangers and heat transfer applications. In two-equation models, two parameters are determined by solving two additional differential equations. Nevertheless, one should remember that these equations are not exact but only approximate and involve constants which should be determined by justification with experiments. In addition, models using the eddy viscosity and eddy diffusivity approach are isotropic in nature and are not able to evaluate effects of anisotropy. Various modifications and alternate modeling concepts have been proposed accordingly. Such models are exemplified by the k-ε, and k-ω models at high or low Reynolds number as well as in linear and nonlinear versions. A model, which has gained certain interest, is the so-called V2F model introduced by Durbin [19]. It extends the applicability of the k-ε model by incorporating near-wall turbulence anisotropy and nonlocal pressure-strain effects, but, still, it retains a linear eddy viscosity assumption. Two more transport equations are solved. These represent the velocity fluctuation normal to walls and a global relaxation factor, respectively. The shear stress transport k-ω model (SST k-ω) by Menter [20] has become popular recently as it introduces a blending function of gradual transition from the standard k-ω model near solid surfaces to a high Reynolds number version of the k-ε model valid far away from a solid surface. Accordingly, the prediction of the onset and the size of separation under adverse pressure gradients becomes more accurate.

Reynolds stress equation models (RSM) solve differential equations for the turbulent stresses (Reynolds stresses), and directional effects are accounted for. Six modeled differential equations (i.e., not exact equations) for the turbulent stress transport are solved in conjunction with an equation for the turbulent scalar dissipation rate ε. Generally, RSM models are complex and require huge computing efforts, and they are not widely used for industrial fluid flow and heat transfer applications like heat exchangers.

An economic way to account for the anisotropy of the turbulent stresses is to apply ASMs and explicit algebraic stress models (EASMs). These do not solve the Reynolds stress transport equations. The basic idea is that the convective and diffusive terms are modeled or even neglected. Then the Reynolds stress equations are reduced to a set of algebraic equations.

The simple eddy diffusivity concept (SED) is often applied to calculate the turbulent heat fluxes. The turbulent diffusivity for the transport of heat is then obtained by the turbulent viscosity divided by a turbulent Prandtl number. However, this model is not able to account for nonisotropic effects in the thermal field but nevertheless this model is frequently applied in engineering computations. However, there are models in the literature accounting for nonisotropic heat transport, for example, the generalized gradient diffusion hypothesis (GGDH) and the WET method. Such higher-order models require that the Reynolds stresses are calculated accurately by considering nonisotropic effects. If not, improvement of the performance is not achieved. In addition, partial differential equations can be formulated for the three turbulent heat fluxes, but numerical solutions of the corresponding modeled equations are rarely found. Further details can be found in, e.g., [21].

In the LES model, the time-dependent flow equations are solved for the mean flow and the largest eddies, while the effects of the smaller eddies are modeled. It was expected that the LES model would emerge as the future model for industrial applications but still it is limited to low Reynolds number and simple geometries. To handle wall-bounded flows with focus on the near-wall phenomena, like heat and mass transfer and shear, at high Reynolds number present an issue because of the near-wall resolution requirements. Complex wall topologies create problems for the LES approach. More recently, attempts to combine LES and RANS based methods have been presented. The so-called hybrid models then arise, for example the detached eddy simulation (DES) [22].

Recently ([23]), it has been stated that good prospects exist for steady problems with RANS turbulence modeling to be solved accurately even for very complex geometries if the technologies for solution adaptation develop to be mature for three-dimensional problems. It has been conjectured that the future growth rate for supercomputers will be lower than the rate during recent years. This will probably slow down the pure LES reliability for high Reynolds number applications, but the hybrid RANS-LES approaches have been judged to create a great potential. The breakthrough in turbulence modeling to accurately predict separation and transition from laminar to turbulent flow has not been foreseen to occur soon.

11.4.2 Wall effects

Accounting for wall effects in numerical calculations of turbulent flow and heat transfer follows two standard procedures. In one of the procedures, low Reynolds number modeling is employed, while in the other the wall function method is adopted. In the wall functions approach, empirical formulas and functions link the dependent variables at the near-wall cells to the corresponding parameters on the wall. These functions are based on wall laws for the mean velocity and temperature, and expressions for the near-wall turbulence quantities. With increasing Reynolds number, the accuracy of the wall function approach is increasing. The wall function approach is commonly efficient and requires limited CPU time and memory size but on the other hand it becomes inaccurate at low Reynolds numbers. When low Reynolds number effects are important, so-called low Reynolds number versions of the turbulence models are introduced and then the molecular viscosity appears in the diffusion processes. Damping functions are also introduced. So-called two-layer models have been suggested to enhance the wall treatment. The turbulent kinetic energy transport equation is solved while an algebraic equation is used, e.g., for the turbulent dissipation rate.

11.4.2.1 Enhanced wall treatment

The enhanced wall treatment is a blended wall model of wall functions. It blends the separate models in the two-layer approach by using a damping function to enable a smooth transition between the models.

For the velocity/momentum, the wall models are given below.

In the two-layer approach, if $y^+ < 10$, the linear law of the wall is applied, i.e., $u^+ = y^{+°}$, while if $y^+ > 10$ for the wall adjacent cell, then the log law of the wall is used, i.e.,

$$u^+ = \frac{1}{\kappa} \ln\left(y^+\right) + C \tag{11.8}$$

For the enhanced wall treatment approach, there is no control function to determine if y^+ is greater or lower than a certain value. The u^+ value is calculated from a single wall model. The enhanced wall function is written as,

$$u^+ = e^{\Gamma} u^+_{\text{lam}} + e^{1/\Gamma} u^+_{\text{turb}} \tag{11.9}$$

where Γ is a blending function allowing the two different models to be smoothly blended. For example, in the ANSYS FLUENT code, the blending function is

$$\Gamma = -\frac{0.01\left(y^+\right)^4}{1 + 5y^+} \tag{11.10}$$

It should be noted that there is an additional two-layer approach for the k equation (and ε). For the ω equation, there is no two-layer approach. Additional blending is carried out for the turbulence quantities (k, ε), and that is where the most significant differences appear.

11.4.3 CFD codes

Several engineering and consulting companies as well as industries worldwide are using commercially available general purpose so-called CFD codes for simulation of fluid flow, heat and mass transfer, and combustion in various applications. Such codes, being around for a long time, are ANSYS FLUENT, ANSYS-CFX, CFD++, STAR CCM+, and others. Also, many research institutes and universities worldwide apply commercial codes besides using their in-house developed codes. Even open-source codes like OPEN-FOAM are available. Specialized codes exist, e.g., for aerospace applications.

Successful application of such codes and interpretation of the computed results require understanding of the fundamental concepts of computational methods. Important issues are also how to handle complex geometries and the generation of suitable computational grids. Commercial codes have often their own grid generation tool, for example, ANSYS-ICEM, but also stand-alone software like Pointwise are popular. The codes are generally compatible with various CAD tools.

NUMECA is a newer code and has been recognized for CFD software, optimization, and heat transfer technologies. NUMECA is now included in the portfolio of the Cadence company (www.cadence.com). Cadence customers are found in innovative companies, automotive, electronics, mobile, aerospace, industrial, and healthcare.

CONVERGE is another, relatively new, CFD software package. CONVERGE is designed for research and design processes for many applications, for example, gas turbine engines. Since 3D CFD simulations normally require long runtimes, CONVERGE (www.convergecfd.com) enables highly parallel simulations on many processors.

11.5 ADVANCED TOPICS NOT TREATED

There are several more topics important for successful CFD modeling and simulations of engineering heat transfer problems. Among these not discussed in this chapter are (a) boundary condition implementation, (b) adaptive grid methods, (c) local grid refinements, (d) how to solve the algebraic equations, (e) convergence and accuracy criteria, (f) parallel computing opportunities, and (g) animation of results.

11.6 EXAMPLES

In this section, a few examples of applications of the CFD methodology are described.

11.6.1 A plate heat exchanger: Compabloc

The opportunities to use computational methods for plate heat exchangers (PHE) are illustrated by considering a PHE named Compabloc. The geometry is shown in Figure 11.2.

Different channels are formed based on the profiles of the plates. The frame of Compabloc has four columns, four side panels, a top plate, and a base plate. In this section, a method to improve the mesh quality for PHE simulations is considered. A suitable clearance between two adjacent plates is outlined to improve the mesh quality around the contact points. In Figure 11.3a, a 3D model of the geometry is shown, and Figure 11.3b depicts the complicated passage structure.

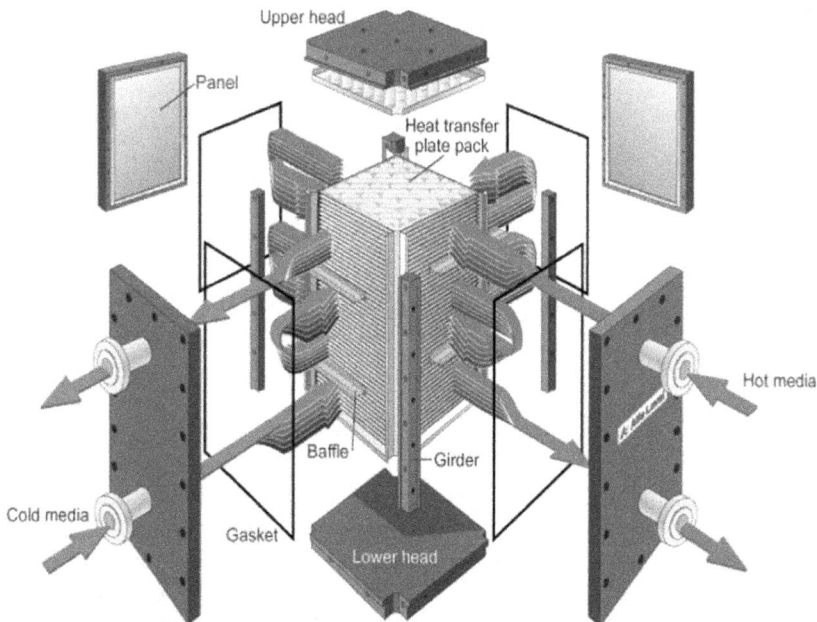

Figure 11.2 Plate heat exchanger: Compabloc (Courtesy Alfa Laval).

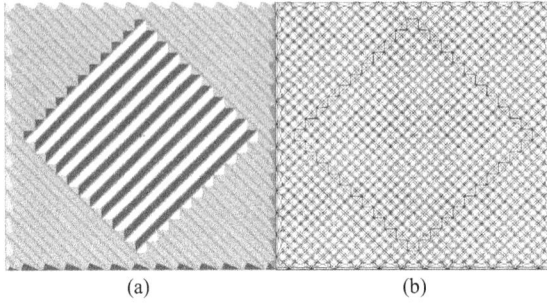

Figure 11.3 (a) Corrugated plate and (b) two stacked plates.

Figure 11.4 A unitary computational cell.

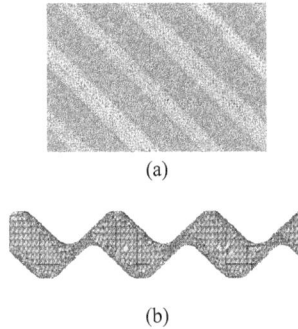

Figure 11.5 (a) Shell mesh display and (b) volume mesh display.

Figure 11.4 highlights a so-called unitary cell of the PHE flow passages and the surface contact points at the corners are clearly visible. Meshes were generated by ANSYS-ICEM. A tiny clearance c between two adjacent plates is used to improve the mesh performance around the contact points. Figure 11.5 presents a shell mesh and a volume mesh by tetrahedrons.

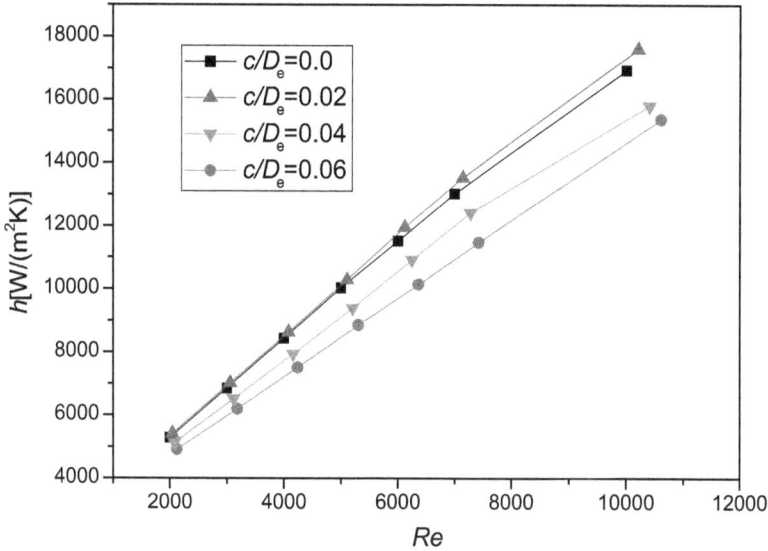

Figure 11.6 Influence of various relative clearance values on convective heat transfer coefficient versus Reynolds number.

Figure 11.6 shows heat transfer coefficients for various clearance ratios (c/D_e, D_e equivalent diameter). The ratio c/D_e = 0.02 is found to agree well with c/D_e = 0. Further details can be found in [24].

11.6.2 Film cooling at the trailing edge of a gas turbine blade

To decrease the trailing edge temperature and extend operating life of gas turbines, the cutback film cooling is an important issue. In this section, results from DES are demonstrated for calculation of the film cooling efficiency η close to the cutback region for various incident angles. In this study investigation, impacts of five different incident angles (0°, 5°, 10°, 15°, 20°) and three blowing ratios (0.2, 0.8, 1.25) were studied. It was found that the incident angle changes the flow structure significantly. By increasing the incident angle, the separation region becomes suppressed, but the η changed in a nonlinear manner. At a low blowing ratio, changing the incident angle mainly affected the film cooling efficiency η near the slot. Nevertheless, at high blowing ratio, the incident angle alters η downstream the slot exit. An increase of the blowing ratio makes the η near the centerline to become larger. Figures 11.7–11.10 show the geometries, sketch of cutback film cooling, the computational mesh, and the average film cooling efficiency. Further details are provided in [25].

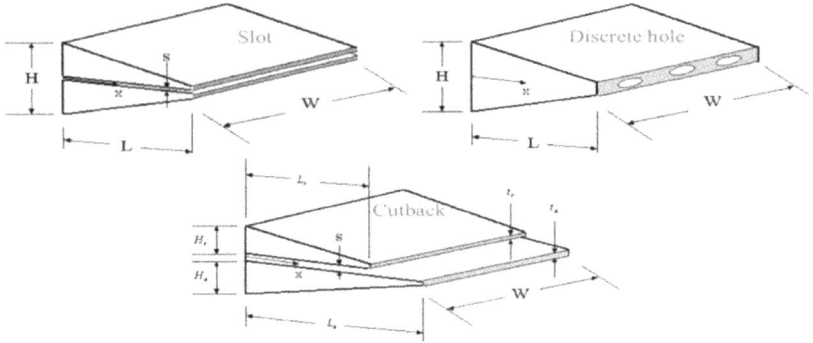

Figure 11.7 Different film cooling types at the trailing edge of a gas turbine blade.

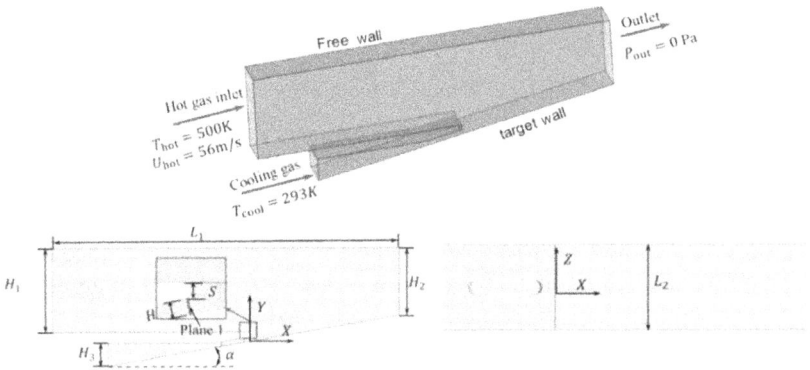

Figure 11.8 Geometry of the cutback film cooling.

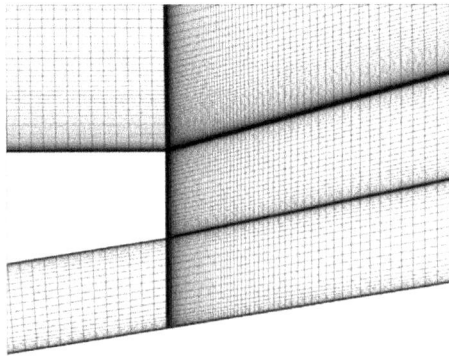

Figure 11.9 Mesh schematic.

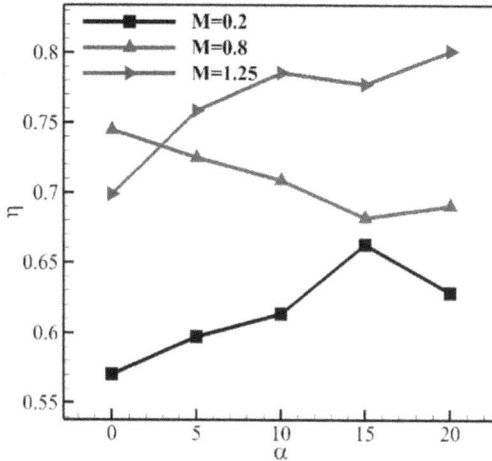

Figure 11.10 Averaged η for different incident angles and blowing ratios.

11.7 CONCLUSIONS

In this chapter, analysis of transport phenomena based on CFD computational approaches was summarized and recent findings were considered. It is concluded that many challenges must be overcome before CFD emerges as the primary tool in research and engineering work on augmented heat transfer and in thermal engineering design. Improvements to the computer codes and their performances need to be addressed. Similarly, multiprocessor technology needs to be massively introduced, improvements in grid generation technique including unstructured grids, adaptive grids, and deformable boundaries must be gained. Other issues concern validation and improvement of turbulence modeling, handling of boundary layer transition, and improvement of multiphase flow modeling. Besides, well-documented validation data bases (from experiments) are needed.

The results of various investigations have revealed that the CFD approach is able to demonstrate important physical effects as well as provide satisfactory engineering and scientific results in decent agreement with corresponding experiments.

REFERENCES

[1] T.K. Sengupta, *Theoretical and Computational Aerodynamics*, Wiley, UK, 2015.
[2] S.V. Patankar, *Numerical Heat Transfer and Fluid Flow*, McGraw-Hill, New York, 1980.
[3] D.A. Andersson, J.C. Tannehill, R.H. Pletcher, *Computational Fluid Mechanics and Heat Transfer*, 2nd Ed., Taylor and Francis, USA, 1997.
[4] G.D. Smith, *Numerical Solution of Partial Differential Equations*, Oxford University Press, London, 1978.

[5] J.N. Reddy, D.K. Gartling, *The Finite Element Method in Heat Transfer and Fluid Dynamics*, CRC Press, Boca Raton, Fla, 2010.

[6] R.W. Lewis, K. Morgan, H.R. Thomas, K.N. Seetharamu, *The Finite Element Method in Heat Transfer Analysis*, J. Wiley, and Sons, UK, 1996.

[7] M.S. Kandelousi, D.D. Ganji, *Hydrothermal Analysis in Engineering Using Control Volume Finite Element Method*, Academic Press, Oxford, 2015.

[8] L.C. Wrobel, *Boundary Element Method – Volume 1 Applications Thermo-Fluids and Acoustics*, J. Wiley and Sons, UK, 2002.

[9] C.M. Rhie and W.L. Chow, Numerical study of the turbulent flow past an air-foil with trailing edge separation, *AIAA J.*, Vol. 21, 1525–1532, 1983.

[10] D.S. Jang, R. Jetli and S. Acharya, Comparison of the PISO, SIMPLER and SIMPLEC algorithms for treatment of the pressure velocity coupling in steady flow problems, *Numer. Heat Transf.*, Vol. 10(3), 209–228, 1986.

[11] R.I. Issa, Solution of the implicity discretized fluid flow equations by operator-splitting, *J. Comput. Phys.*, Vol. 62, 40–65, 1986.

[12] R. Peyret, *Handbook of Computational Fluid Mechanics*, Academic Press Limited, USA, 1996.

[13] D. McBride, N. Croftand M. Cross, Combined vertex-based-cell-centred finite volume method for flow in complex geometries, in *Third International Conference on CFD in the Minerals and Process Industries*, 351–1356, CSIRO. Melbourne, Australia, 2003.

[14] B. Farhanieh, L. Davidson, B. Sunden, Employment of the second-moment closure for calculation of recirculating flows in complex geometries with collocated variable arrangement, *Int. J. Numer. Meth. Fluids*, Vol. 16, 525–554, 1993.

[15] S. Pope, *Turbulent Flows*, Cambridge University Press, Cambridge, UK, 2000.

[16] D.C. Wilcox, *Turbulence Modeling for CFD*, 2nd Ed., DCW Industries, Inc., La Canada, CA, 2002.

[17] P.A. Durbin, T. I.-P. Shih, An overview of turbulence modeling, in *Modeling and Simulation of Turbulent Heat Transfer* (eds. B. Sunden, M. Faghri), 3–31, WIT Press, Southampton, UK, 2005.

[18] P.R. Spalart, S.R. Allmaras, One-equation turbulence model for aerodynamic flows, AIAA Paper -92-0439, 1992.

[19] P.A. Durbin, Separated flow components with k-ε-v2 model, *AIAA J.*, Vol. 33(4), 659–664, 1995.

[20] F.R. Menter, Zonal two-equation k-ω models for aerodynamic flows, AIAA Paper 93-2906, 1993.

[21] B.E. Launder, On the computation of convective heat transfer in complex turbulent flows, *ASME J. Heat Transf.*, Vol. 110, 1112–1128, 1988.

[22] P.R. Spalart, W.-H.M. Stretlets, S.R. Allmaras, Comments on the feasibility of LES for wings and the hybrid RANS/LES approach, advances in DNS/LES, in *Proceedings of the First AFOSR International Conference on DNS/LES*, 1997.

[23] P.R. Spalart, V. Venkatakrishnan, On the role of and challenges of CFD in aerospace industry, *The Aeronautical J.*, Vol. 120(1223), 209–232, 2016.

[24] V. Sekhar-Gullapalli, B. Sunden, CFD simulation of heat transfer and pressure drop in compact brazed plate heat exchangers, *Heat Transf. Eng.*, Vol. 35(4), 358–366, 2014.

[25] W. Du, L. Luo, S. Wang, B. Sunden, Film cooling in the trailing edge cutback with different land shapes and blowing ratios, *Int. Comm. Heat Mass Transf.*, Vol. 125, 105311, 2021.

Index

Pages in *italics* refer to figures and pages in **bold** refer to tables.

For Product Safety Concerns and Information please contact our EU
representative GPSR@taylorandfrancis.com
Taylor & Francis Verlag GmbH, Kaufingerstraße 24, 80331 München, Germany

www.ingramcontent.com/pod-product-compliance
Lightning Source LLC
Chambersburg PA
CBHW060336220326
41598CB00023B/2725

* 9 7 8 1 0 3 2 1 3 5 6 2 5 *